U0211863

影印版说明

本书介绍了镁合金冶金的基本理论和技术，主要包括：镁的物理冶金、镁合金的热力学性能、镁合金的沉淀析出过程、镁的合金化及镁合金设计、镁合金成形、镁合金腐蚀及表面处理等，重点介绍了镁合金及镁基复合材料在航空、汽车、医疗器械等领域的应用。

本书对镁合金熔炼及成形加工技术的研究和生产具有参考价值，适合冶金、材料加工等行业的工程技术人员使用，也可供高等院校相关专业的师生参考。

Mihriban O. Pekguleryuz　加拿大麦吉尔大学采矿和材料工程系金属研究实验室教授。

Karl U. Kainer　德国亥姆霍兹联合会（GKSS）材料研究所镁研究中心主任、教授。

A. Arslan Kaya　土耳其穆拉大学冶金和材料工程专业的创始人、教授。

材料科学与工程图书工作室

联系电话　0451-86412421

　　　　　0451-86414559

邮　　箱　yh_bj@aliyun.com

　　　　　xuyaying81823@gmail.com

　　　　　zhxh6414559@aliyun.com

WOODHEAD PUBLISHING IN MATERIALS

影印版

镁合金冶金基本原理

Fundamentals of magnesium alloy metallurgy

Edited by Mihriban O. Pekguleryuz, Karl U. Kainer
and A. Arslan Kaya

哈爾濱工業大學出版社
HARBIN INSTITUTE OF TECHNOLOGY PRESS

黑版贸审字08-2017-080号

Fundamentals of magnesium alloy metallurgy
Mihriban O. Pekguleryuz, Karl U. Kainer, A. Arslan Kaya
ISBN: 978-0-85709-088-1

Elsevier (Singapore) Pte Ltd.
3 Killiney Road
#08-01 Winsland House I
Singapore 239519
Tel: (65) 6349-0200
Fax: (65) 6733-1817

First Published 2017

图书在版编目（CIP）数据

镁合金冶金基本原理=Fundamentals of magnesium alloy metallurgy：英文 /（加）米罕班·O.P.，（德）凯尔·U.肯纳，（土）A.阿斯兰·卡亚主编.—影印本.—哈尔滨：哈尔滨工业大学出版社，2017.10

ISBN 978-7-5603-6391-2

Ⅰ.①镁… Ⅱ.①米… ②凯… ③A… Ⅲ.①镁合金-冶金-研究-英文 Ⅳ.①TG146.22

中国版本图书馆CIP数据核字（2017）第002891号

责任编辑 杨 桦 张秀华 许雅莹
出版发行 哈尔滨工业大学出版社
社　　址 哈尔滨市南岗区复华四道街10号 邮编 150006
传　　真 0451-86414749
网　　址 http://hitpress.hit.edu.cn
印　　刷 哈尔滨市石桥印务有限公司
开　　本 660mm×980mm 1/16 印张 23.75
版　　次 2017年10月第1版 2017年10月第1次印刷
书　　号 ISBN 978-7-5603-6391-2
定　　价 260.00元

Fundamentals of magnesium alloy metallurgy

Edited by
Mihriban O. Pekguleryuz,
Karl U. Kainer and A. Arslan Kaya

Oxford Cambridge Philadelphia New Delhi

Contents

Contributor contact details

(* = main contact)

Editors

Prof. Mihriban O. Pekguleryuz
McGill University
3610 University Street
Montreal
Quebec H3A 2B2
Canada

Email: Mihriban.pekguleryuz@mcgill.ca

Karl U. Kainer
Helmholtz-Zentrum Geesthacht
Zentrum für Material und Küstenforschung GmbH
MagIC – Magnesium Innovation Centre
Max-Planck-Str. 1
21502 Geesthacht
Germany

Email: karl.kainer@hzg.de

and

Materials Technology
Hamburg University of Technology
Germany

A. Arslan Kaya
Metallurgy and Materials Engineering Department
Faculty of Engineering
Mugla University
48000 Mugla
Turkey

Email: aakaya@mu.edu.tr

Chapter 1

Neale R. Neelameggham
Ind LLC
9859 Dream Circle
South Jordan
UT 84095
USA

Email: rneelameggham@gmail.com

Chapter 2

A. Arslan Kaya
Metallurgy and Materials Engineering Department
Faculty of Engineering
Mugla University
48000 Mugla
Turkey

Email: aakaya@mu.edu.tr

Chapter 3

Dr ShunLi Shang and Professor Zi-Kui Liu*
Department of Materials Science and Engineering

The Pennsylvania State University
University Park
Pennsylvania 16802
USA

Email: sus26@psu.edu; dr.liu@psu.edu; liu@matse.psu.edu

Chapter 4

C. L. Mendis*
Research Scientist
Helmholtz Zentrum Geesthacht
MagIC Magnesium Innovation Centre
Institute for Materials Research
1 Max Planck Straße
Geesthacht 21502
Germany

Email: chamini.mendis@hzg.de

and

Formerly of National Institute for Materials Science
1–2-1 Sengen
Tsukuba 305–0047
Japan

K. Hono
National Institute for Materials Science
1–2-1 Sengen
Tsukuba 305–0047
Japan

Email: kazuhiro.hono@nims.go.jp

Chapter 5

M. Pekguleryuz
McGill University
3610 University Street
Montreal
Quebec H3A 2B2
Canada

Email: Mihriban.pekguleryuz@mcgill.ca

Chapter 6

M. R. Barnett
ARC Centre of Excellence for Design in Light Metals
Centre for Material and Fibre Innovation
Institute for Technology Research and Innovation
Deakin University
Geelong
VIC 3217
Australia

Email: matthew.barnett@deakin.edu.au

Chapter 7

S. Bender
iLF Institut für Lacke und Farben e.V.
Fichtestr. 29
39112 Magdeburg
Germany

Email: susanne.bender@lackinstitut.de

J. Göllner
Otto von Guericke University Magdeburg
Institute of Materials and Joining Technology (IWF)
P.O. Box 4120
39016 Magdeburg
Germany

Email: joachim.goellner@ovgu.de

A. Heyn
Federal Institute for Materials Research and Testing (BAM)
Division 6.1 – Corrosion in Civil Engineering
Unter den Eichen 87
12205 Berlin
Germany

Email: andreas.heyn@bam.de

C. Blawert* and P. Bala Srinivasan
Helmholtz-Zentrum Geesthacht
Zentrum für Material und Küstenforschung GmbH
Institut für Werkstoffforschung
Max-Planck-Str. 1
21502 Geesthacht
Germany

Email: carsten.blawert@hzg.de; bala.srinivasan@hzg.de

Chapter 8

Alan A. Luo
GM Technical Fellow
Chemical and Materials Systems Laboratory
General Motors Global Research & Development
30500 Mound Road
Warren, Michigan
USA

Email: alan.luo@gm.com

Chapter 9

Hajo Dieringa
Helmholtz-Zentrum Geesthacht
MagIC – Magnesium Innovation Centre
Max-Planck-Str. 1
21502 Geesthacht
Germany

Email: hajo.dieringa@hzg.de

Chapter 10

Frank Witte
Laboratory for Biomechanics and Biomaterials
Hannover Medical School
Anna-von-Borries- Str. 1–7
30625 Hannover
Germany

Email: witte.frank@mh-hannover.de

1 Primary production of magnesium

R. NEELAMEGGHAM, IND LLC, USA

DOI: 10.1533/9780857097293.1

Abstract: This chapter reviews the production technology for a variety of magnesium processes developed over the past 150 years on a commercial scale. It discusses why processes vary considerably in the case of magnesium, unlike the case of aluminum production.

Key words: magnesium, light-weight structural metal, molten chloride electrolytic process, thermal reduction, Pidgeon process, electro-thermal.

1.1 Introduction

Magnesium ion is the most abundant structural metal ion in the ocean; it is the fifth most abundant element in the hydrosphere (3.1×10^{15} tons). In the earth's crust (lithosphere) magnesium is considered to be the eighth most abundant element. If we consider the topmost 3.8 km, magnesium is the third most abundant 'structural metallic element'. It should be noted that the average depth of the ocean is 3.8 km – this is the hydrosphere, where magnesium is the only extractable structural metal. This makes magnesium a unique structural element, which can be extracted from either the hydrosphere or the lithosphere. Aluminum is sparse in the ocean, and is extracted from the lithosphere only.

As we all know, manmade materials are made by processes using the raw materials available, or which can be acquired at a low cost while converting them to a value-added material. Irrespective of the source of the raw material, additional energy matter is required to effect the conversion of the mineral into metal. The nature and cost of the energy and energy materials have been important factors in the choice of the process development of magnesium.

Since magnesium is available from the lithosphere and the hydrosphere, various routes are available for extraction into metal. This chapter is written so as to take us through the history of commercial processes in the nineteenth and twentieth centuries, before discussing the chemistry of process evolution, steps involved in production methods, and major equipment needed for different processes. Following this, future possibilities are discussed.

It took over 18 years of laboratory and pilot research, with personal attention given by Herbert H. Dow between 1896 and 1915, before a commercial line for producing magnesium came on line. It took another 16 years before reducing the cost of a pound of magnesium from 5 dollars to about 30 cents by the early 1930s. The process was further refined over the years in reducing operating costs (Campbell and Hatton, 1951). Dow Magnesium had a production of over 100 000 tons per year in its peak years during its 80 plus years of operation before being shut down in 1997.

The same period saw the development of magnesium for structural applications, both in the USA as well as in Germany. Dow Chemical is credited with introducing Dow-Metal pistons for the automotive sector in the 1920s, while the Germans developed a magnesium alloy engine for Volkswagen in the 1930s, helped by I.G. Farbenindustrie's magnesium process. Herbert H. Dow also pioneered the introduction of magnesium into the construction of aircraft in the early 1920s (Campbell and Hatton, 1951), even though this pioneering effort was not able to compete with aluminum – which is 1.5 times heavier than magnesium. We still continue to revisit this subject time and again, even to the present day, in educating the public about the benefits of magnesium alloys as a structural metal and the fact that magnesium can be safely used (Gwynne, 2010). With the advent of higher strength magnesium alloys, magnesium composites can compete with fiber reinforced composites in alternative energy generation such as wind power, etc.

Unlike for other metals, the processes used in the production of magnesium have gone through several historic changes – almost following the changes in the economic dominance history on a global scale – whether it be the world wars, or the cold war through the 1980s, or the emergence of the global economy in the 1990s, and through the recent commodity rise and fall during 2006–9. In 1935, John A. Gann, Chief Metallurgist of The Dow Chemical Co., noted the following '... our light metals occur only in the form of compounds so stable that their discovery, isolation, commercial production, and use were forced to await some of the modern advances in chemistry and engineering. Under such conditions, the evolution of a new industry is often a romance in which scientific and industrial difficulties and near failures add to the thrill of success' (Gann, 1935). The truth of this statement has been proved time and again in the production processes, even in recent times, and in the further development and uses of magnesium.

All magnesium metal production processes go through the following unit process steps (see Fig. 1.1):

i. Raw material upgrading
ii. Removal of unwanted and undesirable impurities

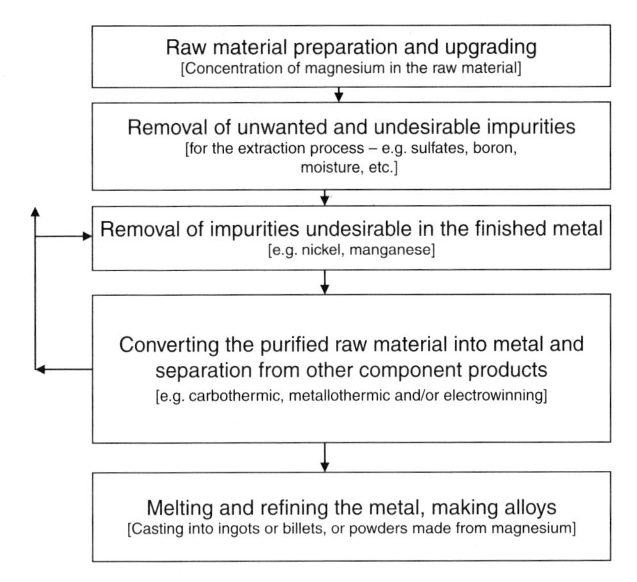

1.1 General flow sheet for magnesium production.

iii. Removal of impurities undesirable in the finished metal
iv. Converting the purified raw material into metal and separation from other component products – along with processing and or reuse of other raw material components
v. Melting, refining and casting metal and/or alloys
vi. Granular magnesium and alloys.

1.2 Raw materials and production methods

Most of the metallic elements are usually extracted or reduced from their respective oxides, or oxide compounds. The lithospheric compounds from which magnesium is extracted are: dolomite ($CaCO_3 \cdot MgCO_3$), magnesite ($MgCO_3$), periclase (magnesium oxide) (MgO), hydro-magnesite ($3MgCO_3 \cdot Mg(OH)_2 \cdot 3H_2O$), brucite ($MgO \cdot H_2O$), and silicates of magnesium (olivine($(Mg,Fe)_2SiO_4$), serpentine $3MgO \cdot 2SiO_2 \cdot 2H_2O$ with partial iron substitution of magnesium, fosterite, *biotite micas,* etc.). The lithospheric minerals magnesium sulfate (epsomite- $MgSO_4 \cdot 7H_2O$), kieserite ($MgSO_4 \cdot H_2O$), langbeinite ($K_2SO_4 \cdot 2MgSO_4$), and kainite ($KCl \cdot MgSO_4 \cdot 3H_2O$), carnallite ($KCl \cdot MgCl_2 \cdot 6H_2O$) are of hydrospheric origin found in evaporites.

The hydrosphere – oceans, and terminal lakes – has magnesium as the second most abundant metallic cation in the salinity. Sodium, present in a larger quantity, usually provides the ionic balance for the chloride ion in saline waters; sulfate is needed to provide ionic balance of magnesium along with chloride ions. Magnesium minerals found from the evaporites in the chloride form include carnallite ($KCl{\cdot}MgCl_2{\cdot}6H_2O$) and bischofite ($MgCl_2{\cdot}6H_2O$). Many of these were identified initially in Stassfurt, Germany in the mid-nineteenth century. Most of the process variations have been caused by the choice of raw material, whether it is oxide or a chloride type material, as we will see in the forthcoming discussions.

1.2.1 Nineteenth century magnesium production processes

In 1808, Humphry Davy took moistened magnesium sulfate and electrolyzed it onto a mercury cathode. He also converted red hot magnesium oxide with potassium vapor, collecting the magnesium into mercury. Both processes produced magnesium amalgam, from which he made the metal by distilling out the mercury. In 1828, Bussy reduced magnesium chloride with potassium metal in a glass tube; when the potassium chloride was washed out, small globules of magnesium were present.

Faraday in 1833 electrolyzed impure magnesium chloride in a molten state to get magnesium metal; but it took two more decades before Robert Bunsen made a commercial quantity in a small laboratory cell using molten anhydrous magnesium chloride. He noted the need to dehydrate the magnesium chloride for improving the electrolysis by avoiding sludge formation. Bunsen demonstrated in 1852 that it is easier to dehydrate magnesium chloride in a potassium chloride bath – this later led to the use of naturally occurring carnallite as a source for making magnesium. Commercial production of magnesium on a larger scale was initiated in 1886 – about the same time as the beginnings of the Hall–Heroult cell for aluminum.

Since oxide magnesium ores, such as MgO, are found in high grade (90% plus purity), attempts were made to use this as feed material using a molten fluoride melt – similar to the Hall–Heroult cell during the late nineteenth century. But the high melting point of magnesium fluoride above 950°C, along with the low solubility of magnesium oxide even in these fluorides, and the high vapor pressures of magnesium at these temperatures, made the growth of these processes uneconomical and difficult.

Molten dehydrated carnallite ($KCl{\cdot}MgCl_2$) was electrolyzed to magnesium metal in 1886 by the Aluminium und Magnesium Fabrik, Germany. This was further developed by Chemische Fabrik Griesheim-Elektron starting in 1896 – this became I.G. Farbenindustrie in the twentieth century. Molten carnallite electrolysis still continues in the twenty-first century, with various improvements made in the twentieth century.

1.2.2 Commercial magnesium production processes of the twentieth century

Several in-depth articles, as well as books, are available on the commercial production technologies of magnesium. These references are highlighted, avoiding duplication of detail on the processes currently used.

In 1938, Haughton and Prytherch noted that the extraction of magnesium followed three processes – electrolysis of fused chlorides, electrolysis of the oxide in solution of molten fluorides, and direct reduction of the oxide by carbon in an arc furnace with a hydrogen atmosphere followed by re-distillation in inert atmosphere. World War II brought the silico-thermic reduction of oxides to the fore. At this time, the production of magnesium was one third that of aluminum worldwide (Haughton and Prytherch, 1938).

Unlike aluminum, the demand for magnesium took a precipitous drop following World War II, causing the variation of production technology processes. Carbo-thermic and fluoride-melt electrolytic processes exited commercial production. Silico-thermic processes took a backseat, until the mid 1990s, to the electrolytic conversion of magnesium chlorides.

China entered the global economy in the early 1990s, interested in developing uses for its *apparently* very low-cost ferro-silicon. China, a non-market economy, thus revived magnesium oxide conversion by the silico-thermal method into a dominant process of the first decade of the twenty-first century.

Some of the early history of processes in the first half of the twentieth century is given in articles or chapters in books (Emeley, 1967; Ball, 1956; Beck, 1939; Schambra, 1945). At the time of writing (2010), the production of magnesium has narrowed to two main processes – one from lithospheric magnesium mineral dolomite by the thermal process, and the other from hydrospheric magnesium chloride. It is now felt that the abundance of magnesium resources, the evolution of non-fossil alternative energy, the realization of a global market economy (where costs of raw materials are a significant issue), along with the development of new uses for magnesium and its alloys, can alter this dominance by two main processes during the next 70 years.

Evans gives a concise summary of the evolution of commercial processes in the light metals aluminum, magnesium and lithium over the last five decades (Evans, 2007). Habashi presented a history of magnesium in a 2006 symposium (Magnesium technology in the Global Age) held in Montreal (Habashi, 2006). Production technologies of magnesium, a chapter by, Eli Aghion and Gilad Golub, discusses present day electrolytic as well as thermal reduction processes (Aghion and Golub, 2006). An in-depth history of magnesium by Robert E. Brown discusses the production of magnesium through the year 2003 (Brown, available from http://www.magnesium.com, and Brown, 2000; Brown, 2003).

The articles by Wallevik *et al.* (2000) and by Hans Eklund *et al.* (2002), presented a detailed summary of the 415 Kamp magnesium electrolytic cells operated by Norsk Hydro Canada from 1991 through early 2007, using purchased magnesium carbonate converted to magnesium chloride onsite. R.L. Thayer discussed the improvements on the electrolytic process for magnesium production at US Magnesium (Thayer and Neelameggham, 2001). The raw material for the US Magnesium process is the magnesium ion in the Great Salt Lake brine.

The Lloyd M. Pidgeon memorial session (2001) presentations of magnesium technology 2001 discussed advances in the Pidgeon silico-thermic process in China. R.E. Brown describes the 240 retort silico-thermic magnesium plant of the 1960s in Selma, Alabama. (Brown, 1997). X. Mei's presentation in the TMS annual meeting describes the Pidgeon process as practiced in China in 2001, along with some of the developments being made to increase the unit throughput of retorts (Mei *et al.*, 2001). The Pidgeon process pilot plant studies were published in 1944 by the inventor in the Transactions of AIME (Pidgeon and Alexander, 1944).

In the Magnola process, the Noranda magnesium plant of 1999–2001 used serpentine, the asbestos plant waste containing high magnesium content as the raw material. The magnesium is converted from a silicate form into molten anhydrous magnesium chloride and then electrolyzed in a multipolar cell. (Avedesian, 1999; Watson *et al.*, 2000). Dow Chemical had proposed a bipolar magnesium electrolysis cell in 1944 (Blue *et al.*, 1949) to reduce power consumption in the cells. Ishizuka (1981, 1982) presented other concepts of bipolar cells for making magnesium. This concept was further modified into multipolar cells in the early 1980s by a cooperative effort between Osaka Titanium and Alcan's development of monopolar cells, leading to multipolar cells. O. Sivilotti discussed these multipolar cells in comparison with monopolar cells in his paper. At this time, the yet-to-come Noranda magnesium plant, and the never-started Australian magnesium plant, Queensland, Australia of the 1990s were both planning to use the multipolar technology (Sivilotti, 1997). Japanese titanium producer Toho uses its own multipolar cell design. Alcoa produced magnesium by Magnetherm technology (Faure and Marchal, 1964; Jarrett, 1981). This process was practiced during 1977–2002 period at Addy, Washington.

The book by Strelets, 'Electrolytic production of magnesium', is a compendium of electrolytic magnesium technology developed by the Germans, later improved by the Soviet Union by VAMI (the Soviet aluminum and Magnesium Institute) and the Soviet magnesium plants in Solikamsk, Russia, Zaporozhye, Ukraine, and Kazakhstan until 1970 (Muzhzhavlev *et al.*, 1965, 1971, 1977; Strelets, 1977). Variation of these processes with further improvements has been practiced through the present time in the Dead Sea magnesium plant, Israel from 1995 as well as at Solikamsk, Russia. Both use carnallite ($KCl \cdot MgCl_2 \cdot 6H_2O$) as the starting material – mined evaporite mineral in Russia or the solar evaporated Dead Sea carnallite. The Russian technology for use with molten recycled magnesium chloride from titanium production is practiced by Avisma magnesium, Berzniki, Perm, Russia, as well as in Kazakhstan Ust-Kamenogorsk titanium–magnesium combine.

The electrolytic process for magnesium production is well advanced for both magnesium chloride from raw magnesium feed stocks, as well as the Kroll process recycled molten anhydrous magnesium chloride. The latter continues in commercial practice through the present day. The multipolar cell developed by the Japanese producer Sumitomo Sitix – or Osaka Titanium in conjunction with Alcan, is described by Sivilotti in his paper in the Light Metal Symposium (Sivilotti, 1988), and by Christensen (Christensen *et al.*, 1997). O.G. Sivilotti, M. Vandermulen, J. Iseki and T. Izumi, provided (the mid 1970s) the developments in their monopolar Alcan-type magnesium electrolytic cells (Sivilotti *et al.*, 1976). A variation of this concept cell is still practiced at Timet, Henderson, Nevada.

The economic comparison among carbo-thermal, silico-thermal and electrolytic processes as of 1967 was provided by engineers at the US Bureau of Mines (Elkins *et al.*, 1968). Some of these comparisons are still true 40 years later. Information on these three processes which prevailed in the 1930–1960 period in the United States is provided in detail in the US Bureau of Mines Reports of Investigations (Elkins *et al.*, 1965, 1967). The carbothermic process developed by Hansgirg was in commercial production only during the 1930–50 period, and efforts to revive it in an economic fashion still continue to date on a laboratory scale and at times pilot scale (Hansgirg, 1932; Brooks *et al.*, 2006). In the 1930s, magnesium oxide was converted to magnesium using calcium carbide as a reductant, by the Murex process (Beck, 1939; Emeley, 1966).

1.3 Chemistry of extraction of magnesium from raw material

The scientific basis of extraction defined by thermo-chemistry has been evolving side by side with changes in the commercial production of magnesium. Some of these are shown. There are two main chemical aspects in the extraction of minerals into metals. First, the minerals are usually mixed with other compounds and these have to be reacted or modified before the single compound can be energetically split into metal, and secondly removing the other elements forming the compound, as well as the new compounds formed if any. It should be noted that commercial processes consume 3 to 4 times the thermodynamic energy required for splitting the raw material into

Table 1.1 Thermodynamic properties of several magnesium compounds

Compound	Heat of Formation, ΔH, kcal/g.mole	Entropy S, cal/K/mole	Melting Pt. °C	Boiling Pt. °C
Mg	0	7.77	650	1105
MgF_2	–266.0	13.7	1263	2320
$MgCl_2$	–153.4	21.2	714	1418
MgO	–143.7	6.55	sub.	2770
$MgSO_4$	–305.5	21.9	1130	
$MgCO_3$	–262.0	15.7	dec.	
$MgSiO_3$	–8.7	16.2	1560	
Mg_2SiO_4	–15.1	22.75	1100	
MgC_2	21.0	14.0	dec.	
C	0	1.36	sub.	
CO	–26.4	47.3	–205	–192
CO_2	–94.05	51.1	sub.	–79
CaO	–151.6	9.5	2615	3500
$CaSiO_3$	–21.5	19.6	1540	
Ca_2SiO_4	–30.1	30.5	2130	
CaC_2	–14.1	–16.8	2300	
$CaCO_3$	–288.4	21.2		
H_2	0	31.21	–259	–253
$H_2O(g)$	–57.8	45.1	0	100
HCl	–22.0	44.65	–114	–85
Si	0	4.5	1410	3280
SiO_2	–217.0	10.0	1723	
Al	0	6.77	659	2450
Al_2O_3	–400.0	12.2	2050	
Fe	0	6.49	1536	3070
FeO	–63.0	14.05	1378	

Source: Kubaschewki (1967).

the finished metal. Table 1.1 gives the temperatures and entropies for the formation of some important magnesium compounds from which magnesium metal is formed, along with some of the common reductants.

1.3.1 Chemical basis in magnesium oxide as raw material

Magnesium oxide raw material can be obtained from (a) magnesium hydroxide,(b) from magnesium carbonate, or (c) from dolomite $CaCO_3 \cdot MgCO_3$. It is possible to make precipitated magnesium hydroxide from sea water – which would require calcined limestone or calcined dolomite. Calcination of limestone or dolomite is an endothermic process which takes place around 1000°C, consuming considerable energy. The fuel or the energy matter-carbonaceous material such as coal – releases additional carbon dioxide in the flue gases besides the carbon dioxide released from the base mineral.

$$Mg(OH)_2 = MgO + H_2O \quad [1.1]$$

$$MgCO_3 = Mg + CO_2 \quad [1.2]$$

$$MgCO_3 \cdot CaCO_3 = MgO \cdot CaO + 2CO_2 \quad [1.3]$$
(dolomite = calcined dolomite + carbon dioxide)

$$C \text{ (fuel/energy matter)} + O_2 = CO_2 \quad [1.4]$$

Unlike many non-ferrous metals it is difficult to reduce magnesium oxide with carbon. Magnesium oxide carbo-thermic reduction takes place above 1900°C.

$$MgO + C = Mg \text{ (v)} + CO \quad [1.5]$$

The product magnesium is in the gaseous state, as is carbon monoxide. This requires special separation techniques. Hansgirg developed the rapid quenching of the gaseous mixtures, along with other gases such as hydrogen or methane, to effect a condensation of magnesium. This process, even after 20 years of commercial operations and developments by Hansgirg, still continued to be difficult as there were considerable side reactions and back reactions resulting in further energy-intensive purification schemes.

The use of calcium carbide as a reductant is also useful in making quality magnesium, as the oxygen from magnesium oxide is taken up by calcium rather than by carbon as shown in Equation [1.5], without reverse reactions as in the carbo-thermic approach.

$$MgO + CaC_2 = Mg(v) + CaO + 2C \quad [1.6]$$

It should be mentioned that the reaction is carried out in vacuum retorts.

It has also been surmised by Bleecker and Morrison that magnesium oxide can be reduced at high temperatures using metals like silicon or aluminum per Equations [1.7] and [1.8] (Bleecker and Morrison, 1919)

$$3MgO + 2Al = 3Mg(v) + Al_2O_3 \quad [1.7]$$

$$2MgO + Si = Mg(v) + SiO_2 \quad [1.8]$$

These reactions require temperatures higher than 1400–1500°C to be of value as an atmospheric pressure reaction. The reactions are complicated, due to the formation of intermediate compounds of magnesium ortho-silicate or magnesium aluminates – depending on the reductant. This was further refined to be a vacuum-assisted reaction around 1200°C in the form of the well-known Pidgeon process. The Pidgeon process further established the usefulness of lower-cost dolomite which would yield $CaO \cdot MgO$ when calcined, and form the reactant along with lower-cost ferro-silicon in the place of silicon metal. Equation [1.9] shows the reaction which is conducted in evacuated retorts.

$$2MgO \cdot CaO + FeSi = 2Mg(v) + Ca_2SiO_4 + Fe \qquad [1.9]$$

Several detailed studies have been made on the thermodynamics of the Pidgeon process as to why the reaction is possible on a fundamental basis. (Hopkins, 1954; Kubaschewski, 1967; Ray *et al.*, 1985; Thompson, 1997). The batch silico-thermic process was upgraded to be a semi-continuous process by the Magnetherm process in the early 1960s, by converting the solid product silicate of Pidgeon process into a molten slag by adding alumina and/or aluminum. This facilitates electro-slag melting using a water-cooled electrode, allowing continual feeding of calcined dolomite and ferro-silicon into the molten pool, and continuous removal of magnesium vapor into large condensers.

It should be noted that these metallo-thermic reactions are indirect carbo-thermic reactions, in that carbon is used to produce the reductant. Aluminum is made from aluminum oxide (electrolytic method) using carbon anodes. Silica and carbon make silicon reductant and carbon oxides via the electro-thermal approach. Prior removal of carbon oxide in the gaseous form provides the metal reductant which would make a solid or liquid product when taking the oxygen away from magnesium compound, allowing clean magnesium vapor to condense.

1.3.2 Chemical basis in magnesium chloride as raw material

The magnesium source from the hydrosphere typically is present in the ionic form in brines. From this one can either precipitate it as hydroxide or upgrade it as concentrated magnesium chloride. Earlier magnesium chloride based processes, such as the Dow process or the Norsk Hydro – Norway – both used an alkali – such as milk of lime ($Ca(OH)_2$) or waste alkali sodium hydroxide. The hydroxide is then converted to magnesium chloride using hydrochloric acid (Dow process) (Mantell, 1950). Oxide minerals such as magnesite – $MgCO_3$, or magnesium silicates can also be converted into magnesium chloride solutions and processed further.

$$Mg^{++} + 2OH^- = Mg(OH)_2 \qquad [1.10]$$

$$Mg(OH)_2 + 2HCl = MgCl_2 + H_2O \qquad [1.11]$$

These reactions show that hydrospheric magnesium can also result in magnesium oxide making them applicable to the oxide raw material reactions.

The hydrospheric magnesium ion present in brine can be concentrated to magnesium chloride rich liquor by evaporative crystallization. This involves removal of large quantities of sodium chloride, potassium chloride/sulfate salts by the initial crystallization. This process is economical wherever vast solar evaporation fields are available. This has been practiced by US Magnesium since 1972. The solar evaporation upgrading of the Great Salt Lake (GSL) brine – which is 3 to 4 times the concentration of magnesium in sea water, is described by Barlow (Barlow *et al.*, 1980). Sources such as the GSL brines have considerable amounts of sulfates, which continue to be present in the solar concentrated magnesium chloride, require further sulfate removal, and therefore added reagents.

The hydrospheric magnesium ions can be extracted as a double chloride salt carnallite, $KCl{\cdot}MgCl_2{\cdot}6H_2O$ from marine evaporites, underground potash minerals which were formed from the marine evaporites, or from terminal lakes containing lower sulfate ions. We have already noted that the processes originally developed in Germany in the 1880s, and those in Solikamsk, Russia (operating since 1936) and that in Dead Sea, Israel (operating since 1995) utilize carnallite as a source of magnesium.

Magnesium oxide from rocks – magnesite, silicates, or magnesium oxide from hydroxides converted by carbo-chlorination into magnesium chloride – was also practiced in Germany before World War II, at Basic Magnesium plant in Henderson, Nevada during the World War II, and later in Norsk Hydro, Norway (during 1950–1987) (Mantell, 1950; Streletz, 1977).

$$MgO + Cl_2 + 0.5C = MgCl_2 + 0.5CO_2 \qquad [1.12]$$

1.4 Fused salt electrolysis

From the basic thermo-chemistry information, it is known that several metallic elements cannot be made from their ions in aqueous solution. The free energy of reaction of water when converted to electromotive force gives a decomposition voltage of 1.23 volts for water into hydrogen and oxygen. Many of the reactive metal ions in solution have a greater decomposition voltage than the water splitting reaction – examples are aluminum, titanium, magnesium, alkali metals, etc.

When solutions containing these reactive metal ions are electrolyzed in aqueous solution, one ends up forming the hydroxides of these metals at the cathode. Only when the cathode is mercury will these metal ions form an amalgam with mercury, requiring further distillation steps separating the metal. This was realized from the time of Faraday in the early 1800s, who then started developing an alternate avoiding water altogether in the electrolysis. This is fused salt electrolysis. Here the compounds of metal with a higher decomposition potential than the metal ion to be electrolyzed act as diluents (similar to water in aqueous electrolysis). The diluents help to reduce melting points, and to increase conductivity of the melt, allowing electrolysis to take place in the molten state.

In the case of the Hall–Heroult system for aluminum electrolysis, aluminum oxide is dissolved in the diluent sodium aluminum fluoride (cryolite), allowing molten aluminum to form on the carbon cathode. In the case of magnesium, the choice was made to use magnesium chloride, as it was difficult to find molten salt solvents which could dissolve magnesium oxide at a lower temperature than the fluoride melts. The chloride melts need to be free of oxides for efficient electrolysis. The melting point of mixed magnesium chloride, along with alkali and at times calcium chloride, is less than 500°C compared to fluoride melts requiring over 1000°C. However, molten magnesium chloride does not dissolve magnesium oxide, thus the need to feed the electrolysis using purified magnesium chloride.

1.5 Impurity removal chemistry in thermal processing

In the case of magnesium oxide minerals processed by the silico-thermal route, the most undesirable material is the alkali metal silicate that stays in the calcined dolomite. This vaporizes along with magnesium (in the silico-thermal reduction) and the alkali condensate causes unwanted fires when the crown is removed from the retort, leading to higher metal losses. This is addressed partly by selecting high grade dolomite low in alkali impurities, and partly by retort condenser design. Chemical removal of these impurities is not practiced.

In the silico-thermal process, it is easy to make several of the commercial grade magnesium products based on charge quality (Froats, 1980), except in some cases, under inadequately controlled process conditions, the silicon content tends to be high in the finished magnesium product in spite of the distillation nature of the process. It is necessary to avoid manganese and nickel impurities in the dolomite as well as ferro-silicon, to keep the purity of magnesium under control.

1.5.1 Impurity removal chemistry in electrolytic process

Even though Dow Magnesium and Norsk Hydro Magnesium are no longer in operation, these plants operated for several decades during the twentieth century, and this chapter would be remiss if the fundamentals used in these were not discussed. In the case of making magnesium chloride for the electrolytic reduction, it is necessary to remove sulfates from the mix, as sulfates have a tendency to decompose into oxides in the molten salt mix, causing problems with other purifying reactions, as well as in the molten salt electrolysis. Sulfates are typically removed from magnesium chloride solutions using calcium chloride, which forms gypsum precipitate and which is filtered out. In the case of the Dow Chemical process, excess calcium ion was removed by using sulfuric acid followed by removal of excess sulfate using barium compounds.

Other impurities which are removed from magnesium chloride solutions are boron compounds which typically come in sea water or marine evaporites. Parts-per-million quantities of boron can cause magnesium coalescence problems in the electrolysis step. The borates deposit high melting boride on magnesium metal formed on the cathodic surface which rises from the cathode in the form of droplets of molten metal instead of coalesced metal going towards the magnesium compartment. The Dow process adjusted the alkalinity of magnesium hydroxide precipitation to remove most of the boron. Final traces of boron were then removed by ion exchange from magnesium chloride solutions. Other processes utilize solvent extraction schemes to remove boron as very dilute boron compounds in solution in aqueous stripping solutions.

Magnesium chloride originating from marine evaporites and other aqueous sources do not have many heavy metal impurities. But the magnesium chloride generated from magnesite or silicate sources using acid digestion contains heavy metals. These heavy metals, such as iron and nickel, can be removed by precipitation as hydroxides or as sulfides from magnesium chloride solutions, as required by the further downstream process of magnesium chloride purification. The limit of nickel in pure magnesium metal product is specified not to exceed 14 ppm. Higher nickel content makes the magnesium more corrosive. Nickel is mutually soluble in magnesium, unlike iron which has a limited solubility in molten magnesium. In most cases the upstream process reagent chemicals are controlled to give nickel-levels of less than 3 ppm, preferably less than 2 ppm in the magnesium chloride feed to electrolysis.

1.5.2 Water removal from magnesium chloride

There are two types of magnesium chloride cells. Dow Magnesium cells used $MgCl_2 \cdot 1.5H_2O$ (between mono and dihydrate of magnesium chloride) prills as cell feed. The second type uses anhydrous magnesium chloride – this type has several variations in the design.

Most other types of electrolytic cells use anhydrous magnesium chloride as cell feed. Norsk Hydro cells used magnesium chloride prills, which had less than 0.5% magnesium oxide and water each as cell feed. US Magnesium and most of the recycled magnesium chloride fed cells in titanium plants use molten anhydrous magnesium chloride as the cell feed. The Russian cell technology, practiced in Russia, Ukraine and Israel, utilize anhydrous carnallite in the molten state as feed to electrolytic cells. A synopsis of magnesium chloride dehydration fundamentals is given below.

When magnesium chloride is evaporated and dehydrated, it forms several crystalline hydrates with increasing melting points for lower hydrates.

$$MgCl_2 \text{ solution} \rightarrow MgCl_2 \cdot 6H_2O \rightarrow MgCl_2 \cdot 4H_2O \rightarrow MgCl_2 \cdot 2H_2O \rightarrow MgCl_2 \cdot H_2O$$

Control of temperature of drying, as practiced by Dow, results in a mixture of $MgCl_2.2H_2O$ and $MgCl_2.H_2O$. The magnesium chloride mono-hydrate does not become anhydrous; it tends to hydrolyse forming magnesium hydroxyl chlorides MgOHCl, and MgO plus HCl. The oxide and the hydroxyl-chloride tend to form undesirable sludge during electrolysis, and make the preparation of molten magnesium chloride difficult.

$$MgCl_2 \cdot H_2O \text{ or } MgCl_2 + H_2O \rightarrow MgOHCl + HCl \qquad [1.13]$$

$$MgCl_2 + H_2O \rightarrow MgO + 2\,HCl \qquad [1.14]$$

These reactions indicate that keeping higher partial pressure of anhydrous HCl will suppress the hydrolysis, as practiced by Norsk Hydro in their essentially anhydrous magnesium chloride prill process. This technique, of using hydrogen chloride gas to make anhydrous magnesium chloride, was first applied by Herbert H. Dow in 1920, who later found it economical to electrolyze $MgCl_2 \cdot 1.5\ H_2O$ which had very little hydroxychloride. Particle size in the drying step also controls the amount of hydroxyl-chloride and oxide coming from hydrolysis.

Use of other hydrolysis-inhibiting chlorides, such as potassium chloride or ammonium chloride, has been practiced in making the anhydrous magnesium chloride during the past 150 years by several intermediate process step variations. During the 1970s Braithwaite proposed using organic adducts of magnesium chloride glycolates, which are then dehydrated in a distillation tower, from which they are precipitated as a magnesium chloride ammoniate which has to be deammoniated in additional steps effecting the production of anhydrous magnesium chloride – this was later called Nalco process (Allain, 1980). The economics of this process depended on effective recycling of expensive organics. This was later modified and proposed to for use by the never-built Australian magnesium plant in the late 1990s.

Anhydrous magnesium chloride can also produced by carbo-chlorination in the molten state. The following reactions apply whether pure magnesium oxide is chlorinated or magnesium oxide in molten magnesium chloride is carbo-chlorinated.

$$MgO + 0.5C + Cl_2 \rightarrow MgCl_2 + 0.5CO_2 \quad [1.15]$$

$$H_2O + 0.5C + Cl_2 \rightarrow 2HCl + 0.5CO_2 \quad [1.16]$$

Details of the energetics and thermodynamics involved in this dehydration were developed by K. Kelly, USBM in the 1940s, long after commercial production had been in practice based on an empirical approach. But this study helped develop improvements in the dehydration process (Kelley, 1945).

One should note that making anhydrous magnesium chloride from hydrated magnesium chloride is an energy-intensive process step, even though less than that required in the electrolytic step. This is somewhat analogous to calcination energy prior to applying the reducing energy in heating the retorts to reaction temperatures in the silico-thermic process.

1.6 Process equipment

Here only highlights of the process equipment are given, as the information on details of the equipment required and energies involved in this process can be found in several of the references mentioned earlier.

1.6.1 Silico-thermal process major equipment

Pidgeon process

The batch retort Pidgeon process is labor intensive, besides requiring condensed energy in the form of ferro-silicon (which is formed by reduction of iron oxide, carbon, silica and electricity), energy for calcining dolomite double carbonate to double oxide, followed by thermal energy required for the silico-thermal reaction of heating the retorts to 1150–1200°C. This heating can be done by coal, coal gas or natural gas firing. The Pidgeon retort made of high nickel steel is about 10 inch ID × 12 inch OD × 10 feet long with a cold end about two feet outside the furnace. This is shown in Fig. 1.2. Vertical retorts with higher heat transfer have been tried, but not many are used in practice.

The range of factors are shown in Table 1.2 for the process as practiced in China in producing 1 MT of magnesium, as derived from Mei's article (Mei, 2000) and several news reports seen in the 2008–10 period. The calcination

Table 1.2 Factors of Pidgeon process in China

Dolomite	10.5–13 MT
Ferro-silicon	1.15–1.35 MT*
Fluorite	0.15–0.19 MT
Flux	0.18 MT
Retort	0.2
Electricity	1800–3000 kwh
Coal**	9 to 12T
Or Natural Gas	66.3 DTh (only a few plants practice this)
Vacuum oil	10 kg
Maintenance	60 USD
Labor	145 USD (in 2001) or 50 Man-Hours/MT

*The preparation of ferro-silicon as a reducing agent is in itself an energy-intensive process – typically consuming about 11–12 kwh per kg FeSi – ferro-silicon is made using iron oxide – silica with sufficient carbon made into electrodes for an electro-thermal reactor.

**Several of the magnesium facilities in China have been installing additional air-pre-heating equipment to use the retort furnace off gases at 1250°C. This is helping to reduce the coal consumption to 7–8 tons per ton range. Shukun recently noted that the coal consumption has come down to about 50% of what was used 10 years ago (Shukun *et al.*, 2010).

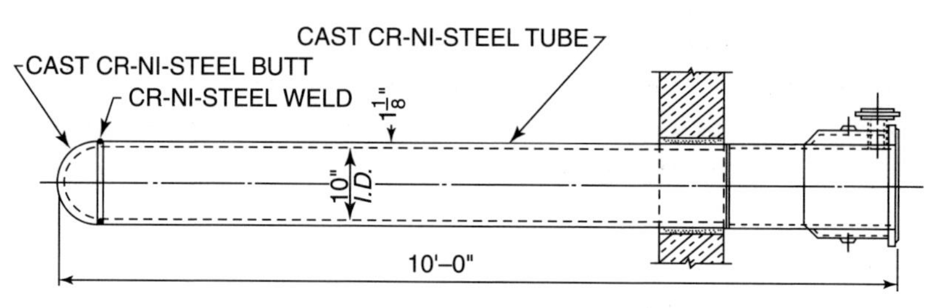

1.2 Magnesium retort for ferro-silicon reduction Pidgeon process (Mayer, 1944).

of dolomite is carried out in conventional lime kilns. Most magnesium plants in China use purchased ferro-silicon. The generation of ferro-silicon consumes considerable quantities of added reagents and electrical energy. The climate-control carbon dioxide emission reduction requirements in recent times have required the plants to reduce coal consumption by using coal gas or natural gas as well as requiring air pre-heating for improved efficiency.

The retorts, being small in size, generate only a small quantity of material for each batch, requiring multiples of retorts in a plant sized for several thousand tons per year production. In addition, these high temperature and high nickel alloy retorts have only about a four month life, impacting the cost of production. Several Chinese producers have been attempting to develop vertical retorts of larger capacity, and envision going into the Magnetherm process, etc., to alleviate the labor intensive multiple small retort production (Shukun *et al.*, 2010); but there are no announced or known commercial conversions into this approach. The reduction furnaces each accommodate about 24 retorts, each of which requires manual filling, to be made evacuation ready, with manual removal of crowns in an 8 to 12 hour cycle. For example, a 15 000 tpy magnesium process will have about 1000 retorts which have to be manually operated day in and day out.

Silico-thermic process in Brazil

Rima Corporation carried out the silico-thermic process in vertical reactor of multiple times the throughput of Pidgeon retorts. This modified Ravelli furnace interior has a steel resistive element surrounding the silico-thermal briquette charge which provides the heat to the reaction zone (Ravelli *et al.*, 1981).

1.6.2 Magnetherm process

At present there are no Magnetherm processes in operation. The Magnetherm process equipment, being an electro-slag melting technology in vacuum, is a high throughput furnace, with vacuum connections to exhaust magnesium vapor continuously into the crucible condenser. Descriptions are given in references.

1.6.3 Electrolytic process major equipment

Conventional aqueous reaction tanks and solid–liquid separation equipment are used in the purification of magnesium chloride solutions. The front end of the processes have evaporating vessels followed by either flash driers (Dow), spray driers (US Magnesium), or prilling towers (Norsk Hydro Canada) depending on the process chosen for dehydration of purified brines. It should be noted that considerable expense is involved in the feed preparation step where magnesium chloride is made anhydrous.

Carbo-chlorination of magnesium oxide pellets bound with magnesium chloride brine is carried out on tall refractory lined chlorinator vessels with varying temperature zones, practiced by Germans prior to World War II, and then from 1951 until the mid 1980s by Norsk Hydro, Norway, each capable of servicing 1500 tpy production requiring multiple chlorinators in plants designed for 30 000–50 000 tpy magnesium. In-cell chlorination of much higher throughput was developed by Toomey and Davis at N.L. Magnesium in the 1970s (Toomey *et al.*, 1976). Figure 1.3 shows the flow sheet of the US Magnesium process.

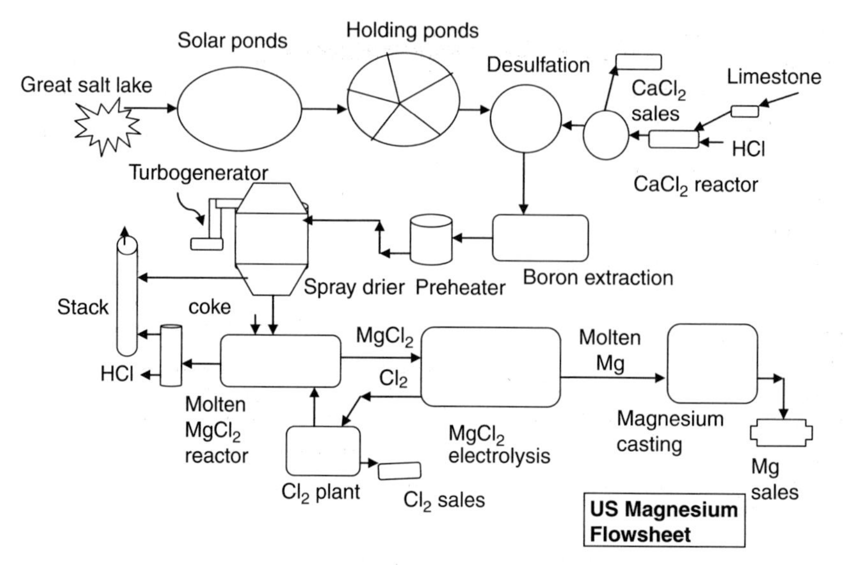

1.3 A flow sheet for an electrolytic magnesium process (US Magnesium LLC).

On a simpler note, in the fused salt electrolyzer molten magnesium chloride in a mix of other alkali and alkaline earth chlorides is electrolyzed between an anode and a cathode, with the design of the electrolyzer allowing the light magnesium metal deposited in liquid form to rise to the surface and away from the rising chlorine gas at the anodes to avoid recombination of magnesium and chlorine. Most industrial electrolyte melts are heavier than molten magnesium. Bunsen's 1852 concept of using inverted troughs or pockets cut into the cathode (he utilized a carbon cathode and a carbon anode in a crucible with a partition between cathode and anode) to help collect the ascending molten magnesium (Hock, 1953).

The cathode is made of mild steel, while graphite is used as the anode material. The molten salt is usually self-heated by the resistance-heating component of the electrolyzer, while kept inside a refractory lined vessel. The variations in cell design have followed all possible geometric arrangements of cathode and anode (top entry, side entry or bottom entry), as well as effective utilization of the gas lift action of the rising chlorine bubbles in causing the magnesium movement upward and away from the chlorine.

The 1953 treatise, 'magnesium extraction from fused salts', Hock describes the various cells used in the 1860–1950 period – dividing the chapter into electrolysis in fluoride baths, and in molten chloride baths, followed by descriptions of cells based on natural and artificial carnallite (the Graetzel cell, the early Griesham cell, the Wintershall cell, etc.), and cells based on magnesium chloride cell feed (the I.G. – M.E.L. cell, the Dow cell and other cells) (Hock, 1953).

Dow Magnesium used gas-heated steel cells to allow interruptible power supply operations. The thick-walled steel container also acted as the main cathode connector to cathodic electrodes surrounding graphite anodes. The upper portion of the cell was refractory lined allowing chlorine gas release without corroding the steel. The cathodes surrounding the graphite anodes were totally immersed in the melt with magnesium collectors going to a separate magnesium chamber. The top-entry anodes were cylindrical graphite electrodes to facilitate periodic insertions of additional length making up for the consumption of carbon by the reaction of water in the magnesium chloride hydrate fed to the cell (Mantell, 1950). The effect of heavy metal and other impurities in the fused salt electrolysis of magnesium chloride can be seen in some review articles (Kipouros and Sadoway, 1987; Strelets, 1977).

Norsk Hydro started with the multiple semi-walled (diaphragms) top-entry steel cathodes and graphite anodes hung inside refractory lined anode box compartment of the I.G. Farbenindustrie design. It took over 35 years to develop its 415 Kamp monopolar magnesium electrolytic cell with a single refractory partition wall between chlorine and magnesium compartments (diaphragmless). This was operated for about 16 years in their Canadian operation before being shut down in early 2007. These cells are self-heated – thermally balanced – and the shell is refractory lined, allowing side-entry cathodes (Wallevik *et al.*, 2000; Eklund *et al.*, 2002).

US Magnesium and Russian producers each locally developed thermally balanced diaphragmless monopolar cells. Multipolar cells developed at the Japanese titanium producers each have a steel cathode–graphite anode pair with added bipolar graphite plates in the inter-electrode space. Figure 1.4 shows a cell room at US Magnesium facilities.

The electrolyte in most cases is a combination of alkali and alkali earth chlorides – NaCl, KCl, LiCl, $CaCl_2$ containing $MgCl_2$. The cells are fed in batch mode 3 to 4 times a day if molten salt feed is used; otherwise, dehydrated powder is fed continuously. The magnesium chloride content is kept between 12% and 20%, depending on the cell type and the cell feed. Molten metal from the cells is periodically removed by using vacuum vessels.

In the early days of cell development molten magnesium was removed manually. In the early I.G. Farbenindustrie cells with top-entry cathodes, air was allowed to flow through the cathode compartment – which incidentally removed the chlorine containing fumes coming into the metal compartment. The molten metal was typically protected by the fluxing action of circulating molten salt over it in the cell. Still, a portion of it can become oxidized forming magnesium oxide/hydroxychloride resulting in sludge dropping to

1.4 An electrolytic magnesium process cell room (Courtesy of US Magnesium).

the cell floor. This sludge was removed manually. Even in the closed cells of the diaphragmless cells, some sludge gets formed from the periodic charging and metaling operations during which the lids are opened. These are removed on a very infrequent basis (once in several months) using mechanical devices. Excess electrolyte is typically removed along with the metal which is allowed to settle as part of refining molten magnesium.

The chlorine gas is removed continually from the anode compartment. Pure chlorine is produced wherever anhydrous magnesium chloride or anhydrous carnallite act as cell feed. The anode gas has to be purified to remove condensable chloride salt which entrains with chlorine, followed by compression prior to reuse. It may undergo liquefaction making liquid chlorine if required. A part of the chlorine is reused in the dehydration step in plants which use carbo-chlorination. In the case of Norsk Hydro Canada, it chose to react anode gas chlorine with hydrogen gas to convert it to hydrochloric acid needed for the digestion of the magnesium carbonate feed stock it used to make the magnesium chloride. In the case of Dow cells, the anode gas is a mixture of hydrogen chloride and chlorine – all of which is converted to hydrochloric acid needed for digesting magnesium hydroxide. The process parameters, such as power utilization, cell life, etc., are given in several of the references mentioned earlier. Comparative tables excerpted from Thayer and Neelameggham (2001) are shown in Tables 1.3 and 1.4.

Table 1.3 Electrolyzer comparison (Thayer and Neelameggham, 2001)

Item	Diaphragm cell	Diaphragmless cell	USM cell
Operating temperature	704–760°C	704–760°C	676–704°C
Chlorine strength	70%	80%	>96%
Chlorine recovery	85%	98%	99.9%
Power, kwh/kg	18.8 to 19.9	16.6 to 18.8	12.1 to 14.3
Voltage	6.5–7.0	6.0–6.3	4.5–5.0
Production (Max)-T/day	0.8–1.0	1.2–1.6	2.8–3.0
Melt capacity – tons	20	42	90
Cell life, days	180–240	300–600	1000–1500

Table 1.4 Comparison of industrial DC cells (Thayer and Neelameggham, 2001)

Company	Cell	Cell type	Voltage, V power	kwh/kg Mg	Tons/day
Norsk Hydro	DLE	Monopolar	5.3 V	13.0	>4.0 t/d
Alcan Int.	MP3	Multipolar	NA	10.0	NA
MagCorp	M-cell	Monopolar	5.0 V	12.	2.8 t/d
AVISMA	Bottom entry	Monopolar	4.7 V		0.7 t/d
UKTMP (Kaz)	Top-entry anode	Monopolar	4.8 V	13.2	10.8 t/d

Use of continuous molten salt flow line feeding into a head cell and taking out the magnesium formed and excess electrolyte at an end cell was envisioned in the 1960s in the United States, but was not developed further commercially. This approach was modified and updated in Ukraine; a version practised in Israel. The operating parameters of such a commercial process are described by Shekhovtsov in his presentation of the magnesium electrolytic production process (Shekhovtsov *et al.*, 2000).

1.7 Melting, refining and casting magnesium

In the case of silico-thermal magnesium, the metal comes in a solidified form called crown from the retorts. This is added to a steel crucible, where it is melted along with mixed chloride salt flux. The molten salt acts as a heat transfer medium as well as a flux minimizing the oxidation of metal. This step is excluded from electrolytic plants which take out molten magnesium along with some cell salt.

From this melt, molten magnesium is transferred to another crucible, allowing the salt to settle, pumping the clean magnesium into ingot molds for casting. Sometimes an inspissated flux which is essentially a mix of magnesium chloride – potassium chloride along with fluorspar (fluorite) and magnesium oxide as thickening agent to help in the settling of oxide type impurities in the molten metal. This can be mechanized in a conveyor casting setup, where it is necessary to use a cover gas, which can be a sulfur dioxide or a gaseous non-toxic fluoride, in the molten zone until a crust is formed leading to further solidification before falling off the endless casting belt for stacking.

There are various designs of the refining cell, including special refractory lined (molten magnesium attacks most refractory oxides) compartmented vessels such as that adopted by Norsk Hydro. The Russians and Ukrainians developed an inverted steel bell-jar type continuous refining cell which carried out the receiving of molten magnesium, refining and pumping to casting conveyors. A variety of furnace designs with salt – which is electrically heated, onto which molten magnesium floats in a confined fashion used for magnesium refining – is described by Barannik (Barannik and Sikorskaya, 1997). Externally heated (by gas heating or electrical heating) steel crucibles are also used industrially to hold molten metal which is refined and pumped into casting machine.

Molten magnesium requires an inhibitive gas film above it during casting to prevent it from oxidizing. Sulfur flour and sulfur oxides have shown ability to provide this inhibitive gas film, which is as tenacious as the oxide film formed on molten aluminum providing the protection. Use of fluoride containing gases as inhibitive gases was initially proposed in 1934. It was not till 1980 that sulfur hexafluoride, a non-toxic gas, was proposed to give improved protection of molten magnesium with air/CO_2/SF6 gas mixtures (Couling and Leontis, 1980), with techniques for minimal use of expensive cover gas while controlling associated corrosion on magnesium containing crucibles. But due to the very high global warming potential of sulfur hexafluoride (22 000 times that of carbon dioxide) its use is being phased out since the late 1990s (Bartos 2001).

The ingots cast by these methods are typically 99.8–99.9% pure magnesium with about 0.03% iron, 0.01% aluminum as major soluble impurities. Thermal magnesium tends to have more silicon impurity in it, giving a bluish tinge to the ingots. In cases where lower iron is required in the metal by customers, it can be accomplished by carrying out a precipitation step using titanium or zirconium chlorides, followed by settling before casting. This would produce 99.95–99.97% pure metal. Higher purities can be achieved by distillation if needed.

Instead of conveyor casting, some magnesium producers provide semi-continuous direct chill cast T-bars or billets, as in aluminum plants. These direct chill cast bars are further saw cut to desired weight as required by the customers. As about 20–30% of the magnesium is sold to the die-casting industry, primary producers providing alloy ingots of the common type –

AZ91, AZ81, AM60, AM50, etc. Sometimes special compositional ingots can also be provided depending on the simplicity and demand.

1.8 Magnesium alloy powder

A chapter on magnesium production would be incomplete without a section on magnesium or magnesium alloy powder. Magnesium is not only a structural metal, but it is also a reducing agent, and a chemical reagent. Magnesium is used as a reducing agent making several reactive metals from their compounds – primarily from the halides. Examples are making beryllium, uranium, titanium and zirconium metals. It has been a chemical reagent in the production of organic compounds through the organo-magnesium Grignard reagent. During the past 35 years it has become a widely used desulfurizing agent for iron and steel production, consuming over 10% of the magnesium produced worldwide. Magnesium or magnesium aluminum compounds are used in pyrotechnics.

In addition to the use of granular magnesium as a chemical reactant, magnesium alloy powders have been formed into structural near net shapes by unique powder metallurgical processing such as thixo-molding or rheocasting. Recently it has been found that these powder metallurgical processes can produce magnesium articles which have strengths about 80–120% greater than conventional casting techniques.

Magnesium being a soft metal is easy to machine and rasp like 'wood'. This approach has been widely developed where the melted and cast magnesium ingot is ground to powder in large tonnage. In the late 1970s and early 1980s processes were developed to make salt-coated magnesium which can be made from molten metal (Legge *et al.*, 1983; Skach and Cobel, 1983) for use in desulfurizing molten iron and steel. The boiling point of magnesium is 1100°C – and the molten iron is usually around 1200°C and molten steel around 1550°C – where the desulfurization takes place. Any solid magnesium introduced into the molten hot metal becomes a vapor instantly, which makes the process violent. To subdue the violence, the salt-coating approach was adopted for several years; this was later replaced by the use of mixed magnesium powder with a large quantity of calcium oxide [quick lime] which also acted as a flux in the desulfurizing process (Bieniosek and Zebrowski, 2002). During the past 10 years, grinding Pidgeon process crown magnesium directly, instead of from an ingot formed from the crowns, was also adopted as an alternate where feasible (Jackman, 2004).

Molten magnesium can be atomized to making fine spherical shape powder which was adopted for pyrotechnics and military applications. Such atomization requires protective gas use during the atomization to minimize burning. Ust-Kamenogorsk, Kazakhstan utilized the atomizing technique to make salt-coated magnesium.

Mechanical machining of metals causes high strain deformations which result in nanocrystalline structures. Such nanocrystalline material can be used in making finished components having higher strength than castings which have a coarser grain structure. Conventional granular magnesium, made by mechanical machining for several decades for chemical applications, can now be easily applied to making high strength components.

1.9 Future trends

1.9.1 Thermal process variations

There have been multiple variations of processes suggested as well as engineered over the years in all aspects of magnesium production – variations in silico-thermal or alumino-thermal processes, near atmospheric pressure processes, etc. There have been many processes envisioned and tested, befitting Gann's 1935 definition of romance with magnesium. Only a few reach near commercial production, some have gone through pilot plant stages.

Wynnyckyj studied a potentially low-cost continuous packed-bed reactor process where calcined dolomite and silicon are reacted using preheated hydrogen gas to carry magnesium vapor, which can be condensed into a liquid form outside the vessel while recycling hydrogen back to the reactor (Wynnyckyj and Tackie, 1988).

The Zuliani process, an atmospheric pressure variation of the Pidgeon process being developed in Canada is being piloted (Gossan, 2010), available through http://gossan.ca/). The Mintek plasma process (Mintek thermal magnesium process – MTMP), being developed in South Africa, is another atmospheric pressure variation of the ferro-silicon reduction process. The MTMP is based on silico-thermic reduction of calcined dolomite (dolime) in a DC open-arc furnace at atmospheric pressure. This project was tested in the mid 1980s at the 100 kW scale; and proceeded to the 750–850 kW scale (80–100 kg Magnesium/h) pilot in 2004 (Abdellatif, 2008).

While the Magnetherm process is a silico-thermic process carried out in a molten slag, Austherm, Australia was developing an alumino-thermic reduction of magnesium oxide in a thermal plasma-arc furnace using scrap aluminum as a reductant during the late 1990s. This process, called the Heggie–Iolaire process, is one of several alumino-thermic processes being tested on a laboratory pilot scale (Wadsley, 2000). Alcoa evaluated the possibility of making an atmospheric pressure process for the Magnetherm furnace in order to overcome the costs associated with high temperature vacuum systems (Christini and Ballain, 1988).

Cameron discusses the Magram process, which is an atmospheric pressure variation of Magnetherm, studied on a pilot scale in Europe during the 1990s, with operating temperatures in the range 1650–1750°C (Cameron *et al.*, 1996; Cameron, 1997).

While the Bolzano process, initiated in the 1980s using Ravelli furnaces, uses internal steel resistor elements, other methods of providing heat to the calcined dolomite–ferro-silicon charge, such as induction heating, have been tested on a laboratory pilot scale (Jaber, 2006). Robert Odle's nozzle based carbo-thermal approach flow sheet can be seen at www.metallurgicalviability.com/NBCMg.htm.

1.9.2 Electrolytic process variations

In the electrolytic production of magnesium, the important aspects of improvement relate to increased energy efficiency, cell component life, and higher throughput in terms of production per man hour. We have noted the multipolar electrode approaches taken by Sumitomo Sitix and Alcan in the 1980s earlier, to minimize electrical resistance from electrodes – thus improving the voltage efficiency of the cells. The paper by Neelameggham and Priscu discussed the energy reduction approaches in magnesium on a fundamental basis in 1985 (Neelameggham and Priscu, 1985), mainly for electrolytic processes.

A solid oxide membrane process, suggested by Uday Pal, has been developed over the past 10 years, whereby a dissolved magnesium oxide in a molten salt is electrolyzed using an oxygen ion conducting membrane tube, such as yttria-stabilized zirconia, between an inert anode inside the tube and an argon-sparged high-temperature-resistant metal tube cathode to separate magnesium forming at the surface of cathode and vaporizing above melt and then onto a condenser and oxygen forming out of the anode inside the solid oxide membrane (SOM). This is still on a bench scale, with the molten electrolyte being at as high a temperature as in the Pidgeon process, around 1150–1300°C (for the 10% MgO – 90% MgF_2 melt), requiring an inert anode for oxygen evolution (Woolley *et al.*, 2000; Krishnan *et al.*, 2005; Powell, 2010). Reported laboratory work has reached about 40 gm per day.

Other processes have been envisioned to use magnesium oxide as a raw material in the electrolytic approach – most of them see the lack of solubility of magnesium oxide as a stumbling block. Studies done on magnesium oxide solubility in molten salts indicate the difficulties in finding a suitable solvent at a temperature comparable to that of the mixed chloride electrolyte used with magnesium chloride feed stock.

Fundamental analytical techniques used in finding the form in which magnesium oxide and hydroxy chloride impurities were shown by Harris' students while trying to develop methods to improve anhydrous magnesium chloride impurities (Kashani-Nejad *et al.*, 2004). Protsenko suggested that, among several methods which could work in the electrolysis from oxygen containing magnesium raw materials, electrolysis of magnesium hydroxyl-chloride with a porous diaphragm between consumable carbon anode and a steel cathode would be a viable approach (Protsenko, 2000).

It was noted in the 1970s that rare earth chlorides tend to dissolve magnesium oxide by a chemical reaction making a molten rare earth–oxy-chloride and magnesium chloride mix. Kuchera reported on the solubility of magnesium oxide in calcium oxide–calcium chloride melt (Kuchera and Saboungi, 1976). Oxide solubilities in the system were studied by Kipouros *et al.* (1996). Ram Sharma proposed electrolytic cell concepts using a neodymium chloride, magnesium chloride bath to which magnesium oxide could be fed making the conventional electrolysis possible (Sharma, 1996). Huimin Lu carried out tests using a 5000 ampere magnesium reduction cell, which operated using a lanthanum chloride magnesium chloride electrolyte and magnesium oxide powder (Lu *et al.*, 2005). Fine magnesium oxide powder had a 10 weight percent solubility in a 50:50 (mole basis) lanthanum chloride – magnesium chloride melt at 700°C. The pilot cell showed 85% current efficiency and a 12 kwh/kg energy use.

Molten magnesium floats on top of the electrolyte in conventional magnesium electrolysis, as molten magnesium is lighter than the electrolyte. This particular aspect causes sludge formation and oxidation losses of 1–5% of the magnesium. Dean *et al.* disclosed some variations of electrolytes which are lighter than molten magnesium, so that molten magnesium is formed at the cell bottom similar to aluminum electrolysis cells, at around 700°C – these have a high content of lithium chloride and potassium chloride, with less than 1.5% calcium fluoride. (Dean and McCutchen, 1960). Their tests conducted with just high potassium chloride–magnesium chloride baths needed 850–900°C to assure that the electrolyte was lighter than molten magnesium, and the latter could sink to the cell bottom minimizing and avoiding sludge formation, showed it is possible to achieve 85–90% current efficiency (Dean *et al.*, 1959). Even in these cells it is necessary to avoid air flow on top of the electrolyte which is prone to hydrolysis from moisture in the air.

Several processes are discussed in the literature for using magnesium containing silicates whether natural (olivine) or man made (fly-ash containing economic quantities of magnesium), and carbonates by converting them to chlorides followed by electrolysis. There have also been several attempts made at low temperature electro-deposition of reactive metals and this quest will continue, probably leading to new methods.

1.10 Conclusion

The continued use, and obtaining the benefits, of light-weight structural alloys depends on a supply of good quality magnesium from reliable production methods. It is essential to tailor the production process to make best use of the raw materials and energy input at hand. It should be noted that, whenever a process uses an oxidic magnesium raw material, or through an oxide or hydroxide converted to chloride, use of purchased chlorine as a make-up quantity becomes an essential cost item. So it is essential to recover the costing, making them into value-added products. In cases where natural chloride based raw materials are available, outlets to utilize chlorine as is, or in the form of chloride chemicals, are essential. Movsesov *et al.* discusses some methods of chlorine utilization in magnesium production (Movsesov *et al.*, 1997). The silico-thermal process tries to utilize the calcium silicate–iron–fluorite containing retort residue as a material for road base and/or aggregate for concrete in finding an outlet to reduce waste side products.

The twenty-first century may see changes in the magnesium production process, not because of lack of raw materials or lack of process know-how, but from continued pressures in balancing the true market costs of process components, as it did in the twentieth and nineteenth centuries. The climate-control aspects of carbon dioxide release into the atmosphere in metal production have already made the producers move away from non-toxic sulfur hexafluoride inhibitive cover gas back to the originally practiced sulfur dioxide in a controlled fashion. There has been several life cycle analyses of magnesium production, along with downstream fabrication of parts for use in automobiles. Ramakrishnan and Koltun have provided an academic study on the greenhouse impacts of present day magnesium production processes (Ramakrishnan and Koltun, 2004). The pressures from the climate-control aspects might impact the status of differing processes used for magnesium production. This is yet to be seen.

1.11 References

Abdellatif M (2008), *Mintek Thermal Magnesium Process: Status and Prospective, Advanced Metals Initiative*, Department of Science and Technology, Johannesburg, South Africa, 1–14. Available through http://www.mintek.co.za/Pyromet/Files/2008Abdellatif.pdf

Aghion E and Golub G (2006), 'Production methods', in Friedrich H E and Mordike B L, *Magnesium Technology*, Berlin, Springer Verlag, 30–61.

Allain R J (1980), 'A new economical process for making anhydrous magnesium chloride', in McMinn C J, 'Light Metals 1980', 109th Annual AIME Meeting, Las Vegas, NV, 929–936.

Available from http:/www.madsci.org/posts/archives/

Available from http://www.seafreiends.org.nz/oceano/seawater.htm 1 (Accessed 15 July 2010).

Avedesian M (1999), 'Magnesium and magnesium alloys', in Avedesian M and Baker H, *ASM Specialty Handbook*, Materials Park, OH, ASM International, 1–4.

Ball C J P (1956), 'The history of magnesium', *J. Inst. Metals*, **84**, 399.

Barannik I and Sikorskaya I (1997), 'Furnaces with salt heating for magnesium refining: advantages and drawbacks, specific features of operation', in Aghion E and Eliezer D, 'Magnesium 97', *Proceedings of the first Israeli Conference on magnesium science and technology*, Magnesium Research Institute (MRI) Ltd., Israel, 15–20.

Barlow E W, Johnson S C and Sadan A (1980), 'Solar pond as a source of magnesium for electrolytic cells' in McMinn C J, 'Light Metals 1980', Las Vegas NV, *Proceedings of the TMS Light Metals Committee Symposium*, 913–927.

Bartos S C (2001), 'U.S. EPA's SF6 emission reduction partnership for the magnesium industry: an update on early success', in Hryn J, 'Magnesium Technology 2001', TMS Annual Meeting, 43–48.

Beck A (1939), *The Technology of Magnesium and Its Alloys*, (translation in English), London, F.A. Hughes & Co. Ltd, 1–19.

Bieniosek T H and Zebrowski G R (2002), *Magnesium desulfurization agent*, U.S.Patent 6352570, 2002 March.

Bleecker W F and Morrison W L (1919), *Production of alkali earth metals*, U.S. Patent, 1311378, 1919.

Blue R D, Hunter R M and Neipert M P (1949), *Electrolytic apparatus for producing magnesium*, U.S. Patent, 2468022, 1949, April 26.

Brooks G, Trang S, Witt P, Khan M H N and Nagle M (2006), 'The carbo thermic route to magnesium', *J. Metals*, **58**, 51–55.

Brown R E (1997), 'Silico-thermic magnesium in Alabama', in Bickert C M and Guthrie R L, 'Light Metals 1997', Metaux Legers, Montreal, *Proceedings of the International Symposium*, The Metallurgical Society of C.I.M, 603–613.

Brown R E (2000), 'Magnesium Industry Growth in the 1990 Period', in Kaplan H I, Hryn J and Clow B, 'Magnesium Technology 2000', Nashville, TN, TMS Annual Meeting, 3–12.

Brown R E (2003), *A History of Magnesium Production*, available from http://www.magnesium.com/w3/data-bank/index.php?mgw=196.

Cameron A M, Lewis L A and Drumm C F (1996), 'The thermodynamic and economic modeling of a novel magnesium process', in Lorrimer G W, *Proceedings of the Third International Magnesium Conference*, Institute of Metals, Manchester, UK, 7–18.

Cameron A M (1997), 'Advances in thermal reduction technology for magnesium', in Bickert C M and Guthrie R L, 'Light Metals 1997', Metaux Legers, Montreal, *Proceedings of the International Symposium*, The Metallurgical Society of C.I.M., 579–602.

Campbell M and Hatton H (1951), *Herbert H. Dow, Pioneer in Creative Chemistry*, New York, Appleton-Century-Crofts, Inc.

Christensen J, Creber D, Holywell G, Yamaguchi M and Moria (1997), 'The Evolution of Alcan's MPC Technology', *Proceedings of the First Israeli Conference on Magnesium Science and Technology*, Magnesium Research Institute (MRI) Ltd., Israel.

Christini R A and Ballain M D, 1988, 'Magnetherm atmospheric pressure operation: aluminum reactivity in a silicate slag' in 'Light Metals 1988', TMS Annual Meeting, 1189–1196.

Couling S L and Leontis T E (1980), 'Improved protection of molten magnesium with air/CO2/SF6 gas mixtures' in McMinn C J, 'Light Metals 1980', 109th Annual AIME Meeting, Las Vegas, NV, 997–1007.

Dean K C, Elkins D A and Hussey S J (1967), 'An economic and technical evaluation of magnesium production methods', USBM RI 6946.

Dean K C, Elkins D A and Hussey S J (1965), 'Economic and technical evaluation of magnesium production methods – Part 1: metallo thermic', USBM R.I. 6656.

Dean L G, Olstowski F and Posey K (1959), *Electrolytic production of magnesium metal*, U.S. Patent, 2880151, 1959, March 31.

Dean L G and McCutchen C W (1960), *Electrolytic production of magnesium metal*, U.S. Patent, 2950236, 1960, August 23.

Ditze A and Scharf C (2008), *Recycling of Magnesium*, Clausthal-Zellerfeld, Papierflieger Verlag, Germany.

Eklund H, Engseth P B, Langseth B, Mellerud T, and Wallevik O (2002), 'An improved process for the production of magnesium', in Kaplan H I, 'Magnesium Technology 2002,' TMS Annual Meeting, 9–12.

Elkins A, Dean K C and Rosenbaum J B (1968), 'Economic aspects of magnesium production', in Henrie T S and Baker Jr D H, 'Electrometallurgy', *Proceedings of the Extractive Metallurgy Division Symposium on Electrometallurgy*, AIME, Cleveland, OH, 173–184.

Emeley E F (1966), *Principles of Magnesium Technology*, London, Pergamon.

Evans J (2007), 'The evolution of technology for light metals over the last 50 years: Al, Mg, and Li', *J. Metals*, **59**, 30–38.

Faure C and Marchal J (1964), 'Magnesium by the Magnetherm process', *J. Metals*, **9**, 721–723.

Froats A (1980), 'Pidgeon silicothermic process in the 70's', in McMinn C J, 'Light Metals 1980', 109th Annual AIME Meeting, Las Vegas, NV, 969–979.

Gann J A (1935), 'Magnesium in modern uses of non-ferrous metals', in Mathewson, C H, Modern Uses of Non-Ferrous Metals, *Proceedings of American Institute of Mining and Metallurgy*, 1st, ed., 170–190.

Gossan Concludes Phase III Zuliani Magnesium Process Bench Scale Tests with Excellent Results. Available through http://gossan.ca/news/archive/2010/100427.pdf, viewed on August 17, 2010.

Gwynne B (2010), 'Magnesium alloys in aerospace applications – flammability testing and results', in Agnew S, Sillekens W, Nyberg E and Neelameggham R, 'Magnesium Technology 2010', Seattle, WA, TMS Annual Meeting.

Habashi F (2006), 'A history of magnesium', in Pekguleryuz M O and Mackenzie L W F, *Magnesium Technology in the Global Age*, Montreal, Montreal conference of Metallurgists 2006, 31–42.

Hansgirg (1932), *Carbothermal production of magnesium*, U.S. Patent, 1884993, 1932 October.

Haughton J L and Prytherch W E (1938), *Magnesium and Its alloys*, Chemical Publishing Co., New York, 170.

Hock A L, (1953), 'Magnesium extraction from fused salts', in *Magnesium Review*, Magnesium Elektron Ltd., vol. IX, No. 1, 1–34.

Hopkins D W (1954), 'Significance of the value of Delta G (free energy of reaction)' in *Physical Chemistry and Metal Extraction*, London, J. Garnet Miller Ltd., 76–78.

Jaber N K (2006), 'Silicothermic extraction of magnesium from dolomite by induction heating', in Pekguleryuz M O and Mackenzie L W F, *Magnesium Technology in the Global Age*, Montreal, Montreal conference of Metallurgists, 55–66.

Jackman J R (2004), *Process for magnesium granules*, U.S.Patent 6770115, 2004 April.

Jarrett N (1981), Advances in the smelting of magnesium, in *Metallurgical Treatises* American Institute of Mining and Metallurgy, Warrendale, PA, 159–169.

Kashani-Nejad S, Ng K W and Harris R (2004), 'Effect of impurities on techniques for speciation of magnesium oxides in magnesium production electrolytes' in Luo A A, 'Magnesium Technology 2004', Charlotte, NC, 2004 TMS Annual Meeting, 155–160.

Kelley K K (1945), 'Energy requirements and equilibria in the dehydration, hydrolysis and decomposition of magnesium chloride', *Bulletin of U.S.Bureau of Mines*.

Kipouros G J and Sadoway D R, 1987. 'The chemistry and electrochemistry of magnesium production' in Mamantov G, Mamantov C B and Braunstein, *Advances in molten salt chemistry*, 6th ed. New York, Elsevier, 127–209.

Kipouros G, Medidas H, Vindstad J E, Ostvold T and Tkatcheva 1996, 'Oxide solubilities and phase relations in the system Mg-Nd-O-Cl, in Hale W, 'Light Metals 1996', TMS Annual Meeting, 1123–1128.

Krishnan A, Lu X and Pal U B (2005), 'Solid oxide membrane (SOM) for cost effective and environmentally sound production of magnesium directly from magnesium oxide' in Neelameggham R, Kaplan H I and Powell B R, 'Magnesium Technology 2005', San Francisco, TMS Annual Meeting, 7–15.

Kubaschewski O, Evans E L and Alcock C B (1967), *Metallurgical Thermochemistry*, Pergamon Press, 4th ed., 304–363.

Kubaschewski O, Evans E L, and Alcock C B (1967), *Metallurgical Thermochemistry*, New York, Pergamon Press, 4th ed., 238–240.

Kuchera G H and Saboungi M L (1976) 'Solubility of magnesium oxide in calcium oxide – calcium chloride mixtures', *Metall Trans. B Y* **7**, 213–215.

Legge M H, Clarkson J F and Bachman W D (1983), *Salt coated magnesium granules*, U.S. Patent, 4457775, 1983 May.

Lu H, Jia W, Liao C, Ma R and Yuan W (2005), 'Pilot Experiments of magnesia direct electrolysis in a 5 KA magnesium reduction cell', in Neelameggham R, Kaplan H I and Powell B R, 'Magnesium Technology 2005', San Francisco, CA, TMS Annual Meeting 2005, 23–27.

Lloyd M. Pidgeon Memorial Session, in Hryn J, 'Magnesium Technology 2001', New Orleans, LA, TMS Annual Meeting, 3–20.

Mantell C L (1950), 'Flow sheet of Basic Magnesium Inc.', in *Industrial Electrochemistry*, New York, McGraw Hill Book Company, 521–526.

Mantell C L (1950), 'Continuous chloride process', in *Industrial Electrochemistry*, New York, McGraw Hill Book Company, 509–522.

Mayer A (1944), *Plant for Production of Magnesium by the Ferrosilicon Process*, AIME Presentation, New York.

Mei X, Yu A, Shang S, and Zhu T, (2001), 'Vertical larger diameter vacuum retort magnesium reduction furnace', in Hryn J, 'Magnesium Technology 2001', TMS Annual Meeting, 13–15.

Movsesov E, Sedova L and Drobniy V (1997), 'Surplus chlorine utilization in magnesium production', in Aghion E and Eliezer D, 'Magnesium '97', *Proceedings of the First Israeli Conference on Magnesium Science & Technology*, 25–32.

Muzhzhavlev K D, Lebedev O A, Frantas' ev N A, Olyunin G V, Sheka T S, and Dolgikh T K (1965), 'Results of tests on individual elements in the design of magnesium electrolyzers', *Tsvetn. Metal.*, **38**(5), 57–60.

Muzhzhavlev K D, Kostrev S P, Schegolev V I, Ivanov A B, Romanenko O N, Yazev V D and Vasiliev A V (1977), *Diaphragmless electrolyzer for producing magnesium and chlorine*, U.S. Patent 4058448, 1977, November 15.

Neelameggham R and Priscu J C, (1985), 'Energy reduction approaches in magnesium production' in Bautista R G and Wesley R J, *Proceedings of Energy Reduction Techniques in Metal Electrochemical Processes*, New York, 1985 TMS Annual Meeting, 445–452.

Odle R, *Nozzle based carbo thermic route to magnesium*, Available from www.metallurgicalviability.com/NBCMg.htm (viewed on Sept. 11, 2010).

Olyunin G V, Muzhzhavlev K D, Ivanyushkina L A and Yuzhaninov (1971), 'Heat balance of a large-capacity diaphragm-type magnesium electrolysis cell with overhead current supply', *The Soviet J of non-ferrous metals*, **12**(3), 67–69.

Pidgeon L M and Alexander W A (1944), 'Thermal production of magnesium – pilot plant studied on the retort ferrosilicon process,' *Transactions American Institute of Mining and Materials Engineers*, **159**, 315–352.

Powell A (2010), 'Low-cost zero-emission primary magnesium production by solid oxide membrane (SOM) electrolysis', paper presented in Agnew S, Nyberg E, Sillekens W and Neelameggham R, 'Magnesium Technology 2010', Seattle, TMS Annual Meeting.

Protsenko V M (2000), 'Development of the process for magnesium production by electrolysis from oxygen containing magnesium raw materials', in Aghion E and Eliezer D, 'Magnesium 2000', *Proceedings of the Second Israeli Conference on Magnesium Science & Technology*, 62–65.

Ramakrishnan S and Koltun P (2004), 'A comparison of the greenhouse impacts of magnesium produced by electrolytic and Pidgeon process', in Alan Luo, 'Magnesium Technology 2004', Charlotte, NC, TMS Annual Meeting, 173–178.

Ravelli S E, Bettanini C, Zanier S, Enrici M (1981), *Extraction furnace*, U.S. Patent, 4264778, 1981, April 28.

Ray H S, Sridhar R and Abraham K P (1985), 'Pidgeon process', in *Extraction of Non-ferrous Metals*, New Delhi, Affiliated East-West Press Pvt. Ltd, 285–288.

Skach, Jr E Cobel G B (1983), *Process of making salt-coated magnesium granules*, U.S. Patent 4384887, 1983 May.

Schambra W P (1945), 'The Dow magnesium process at Freeport, Texas', *Trans. Am. Inst. Chem. Eng.*, **41**, 35.

Sharma R A (1996), 'A new electrolytic magnesium production process', *J. Metals*, **48**(10), 39–43.

Shekhovtsov G, Shchegolev V, Devyatkin V, Tatakin A and Zabelin, 2000, 'Magnesium electrolytic production process', in Aghion E and Eliezer D, 'Magnesium 2000', *Proceedings of the Second Israeli Conference on Magnesium Science & Technology*, 57–61.

Shukun M, Xiuming W and Jinxiang X (2010), 'China magnesium development report' in 2009, in *Conference Proceedings, IMA 67th Annual World Magnesium Conference*, Hong Kong, 2010 May, 3–10.

Sivilotti O (1997), 'Advances in electrolytic production of magnesium', in Bickert C M and Guthrie R L, 'Light Metals 1997', Metaux Legers, Montreal, *Proceedings of the International Symposium*, The Metallurgical Society of C.I.M, 543–555.

Sivilotti O G, Vandermulen M, Iseki J and Izumi T (1976), 'Recent developments in the Alcan-type magnesium', 'Light Metals', Vol. **1**, TMS-AIME, 437–456.

Sivilotti O (1988), 'Operating performance of the Alcan multi-polar magnesium cell', in *Reactive Metals symposium III,* 'Light Metals 1988', TMS Annual Meeting, 817–821.

Strelets Kh L (1977), 'Electrolytic production of magnesium', Translated from Russian by J. Schmorak, Israel Program for scientific Translations, 1977, available through NTIS, Springfield, VA.

Thayer R L and Neelameggham R (2001), 'Improving the electrolytic process for magnesium production', *J. Metals*, **53**, 15–17 Available at http://iweb.tms.org/Mg/JOM-0108–15.pdf.

Toomey R D, Davis B R, Neelameggham R and Darlington R K (1981), *Chlorination of impure magnesium chloride melt*, U.S. Patent 4248839, 1981, February.

Toomey R D (1976), *Process for purifying molten magnesium chloride*, U.S. Patent 3953574, 1976 April.

Wadsley M W (2000), 'Magnesium by the Heggie – Iolaire process', in Kaplan H I, Hryn J and Clow B, 'Magnesium Technology 2000', Nashville, TN, TMS Annual Meeting, 65–70.

Wallevik O, Amundsen K, Faucher A and Mellerud T (2000), Magnesium electrolysis – monopolar viewpoint', in Kaplan H I, Hryn J and Clow, 'Magnesium Technology 2000', Nashville, TN, TMS Annual Meeting, 13–16.

Watson K, Ficara P, Charron M, Peacey J, Chin E W and Bishop G (2000), 'The Magnola demonstration plant', in Kaplan H I, Hryn J and Clow B, 'Magnesium Technology 2000', Nashville, TN, TMS Annual Meeting, 27–30.

Woolley D E, Pal U and Kenney G B, 'Solid oxide oxygen-ion-conducting membrane (SOM) technology for production of magnesium metal by direct reduction of magnesium oxide' in Kaplan H I, Hryn J and Clow B, 'Magnesium Technology 2000', Nashville, TN, TMS Annual Meeting, 35–36.

Wynnyckyj J R and Tackie E N (1988), 'Research scale investigation of magnesium winning', in 'Light Metals 1988', TMS Annual Meeting, 807–815.

Zang J C and Ding W, Pidgeon process in China and its future, in Hryn J, 'Magnesium Technology 2001', New Orleans, LA, TMS Annual Meeting, 7–10.

2
Physical metallurgy of magnesium

A. A. KAYA, Mugla University, Turkey

DOI: 10.1533/9780857097293.33

Abstract: Key features in the deformation behaviour of magnesium have been introduced in terms of the consequences of hexagonal crystal structure, dislocation core width and stacking fault (SF) energy concepts. Elastic and plastic deformation behaviour of magnesium has been addressed in relation to critical resolved shear stress, slip and twinning. The anomalies during plastic deformation, fatigue, creep, recrystallization and grain growth in magnesium and its alloys have been pointed out and discussed under individual headings. Future trends in research and use of magnesium alloys have been indicated.

Key words: magnesium alloys, critical resolved shear stress, dislocation core width, stacking fault energy, anomalies in magnesium, slip and twinning in magnesium, fatigue of magnesium, creep of magnesium, recrystallization in magnesium.

2.1 Introduction

Magnesium, with its atomic number 12 and hexagonal close-packed (HCP) crystal structure, is an interesting example of an element of true metallic character, as well as for its crystal class. Its electron configuration ends with s^2 making it an excellent example of a true metallic-bond, possessing a relatively more homogeneous non-localized free electron cloud, at least when in the pure state. No doubt this changes when magnesium makes solute solutions with other metals. However, even at such an early stage of our discussion, solely based on this fact we can infer that the less effective strengthening contribution of the solute alloying elements in magnesium as compared, for example, to the case of its rival light metal, aluminium, may be attributed partly to magnesium's true metallic-bond character. A relatively recent study revealed that some directionality in the bond structure in pure aluminium possibly exists, explaining its abnormal intrinsic stacking fault energy (SFE).[1] Such directionality in bond structure indicates localization in free electron density. The superior strengthening response of aluminium to alloying, even as a dilute solute solution, may thus be due to an additional effect to the already existing directionality in the bond structure.

In magnesium, however, solutes are unable to create sufficient perturbation in the truly delocalized bonding among magnesium atoms. SFE, as a related property to free electron density distribution, is thus given special attention below.

Plastic deformation of magnesium invokes several puzzling questions. These anomalies are pointed out in this chapter, and an attempt has been made to gather from the existing literature the possible accounts of each phenomenon. The reader is also referred to the other chapter related to deformation, Chapter 6 'Forming of magnesium and its alloys', in this book. Topics like fatigue, creep and grain size related phenomena have hopefully been compressed into a comprehensible size, highlighting both more interesting as well as mainstream concepts without greatly compromising the meaning of these otherwise vast topics. The title of 'Physical metallurgy' (arguably 'elastic') had to be limited here to a reasonable size for a book chapter. Needless to say, the chapter may fall short of being an exhaustive coverage of the topic. Finally, a few directions for further research have been pointed out in the closing section.

2.2 Crystal structure and its consequences

HCP crystal lattice and major planes of magnesium are shown in Fig. 2.1. With lattice parameters $a = 3.18$ Å and $c = 5.19$ Å, slightly less than the ideal c/a ratio of 1.62354 (at 25°C) of Mg crystal, appears important in explaining some fundamental characteristics of the metal.[2] A comparison of c/a ratios as well as critical resolved shear stress (CRSS) (to be discussed later) for basal planes of different HCP metals is given in Table 2.1. The c/a ratio of

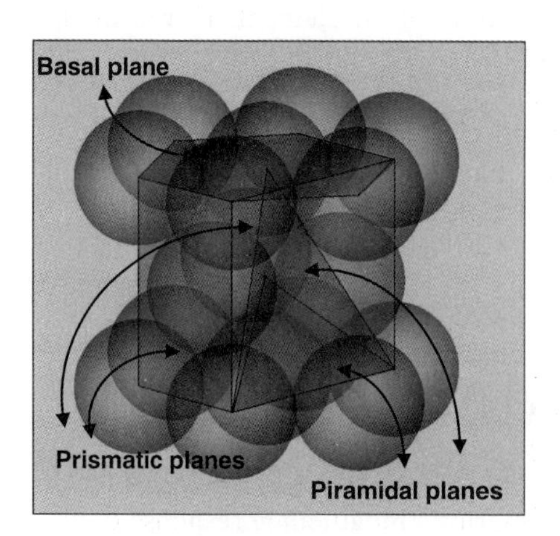

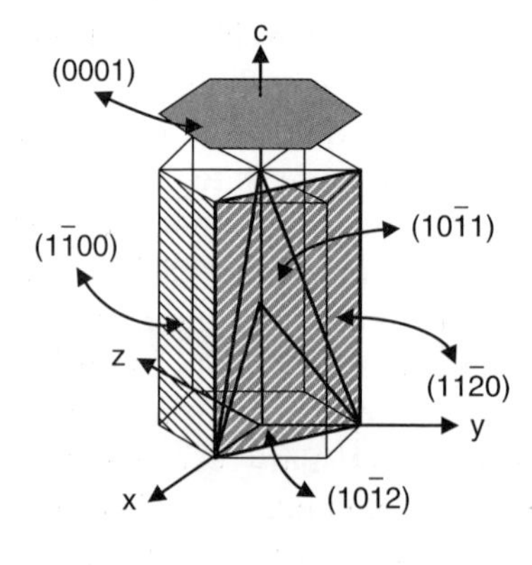

2.1 Schematic description of hexagonal close-packed (HCP) crystal lattice and major planes of magnesium.

Table 2.1 A comparison of CRSS levels for basal planes and *c/a* ratios of different HCP metals (25°C)[2]

Metal	CRSS (psi)	*c/a*
Mg	63	1.624
Cd	82	1.886
Zn	26	1.856
Ti	16000	1.588
Be	5700	1.586

Mg, in turn, results in a somewhat larger primitive cell volume compared to other HCP metals, leading to smaller SFE levels, for example, 36 ergs/cm^2 for basal plane of magnesium. Let us remember that aluminium is considered to have an unusually high SFE as an example of FCC metal. While copper has a lowly SFE of 40 ergs/cm^2 that creates a faulted region of only about 35 Å, aluminium, with its ~7 Å wide fault, has 200 ergs/cm^2 SFE.[3] For the sake of further comparison, it may be pointed out that the SFEs based on the first-principles calculations for Mg and Zr on the high SFE {11–22} pyramidal plane associated with the two <11–2–3>/6 partials are 173 and 388 ergs/cm^2, respectively.[4] These values have very important consequences in terms of dislocation core width, mobility and configurations, and repercussions on the mechanical behaviour of the metals.

2.2.1 Stacking fault energy (SFE) of magnesium

There are few morphological studies on the SFs in magnesium. For a transmission electron microscopy (TEM) work on the interaction of dislocations and SFs, the reader is referred to the detailed work by Li *et al.*[5] A TEM image of SFs in magnesium is given in Fig. 2.2. Michiaki *et al.*[6] also conducted an interesting TEM study, showing that alloying elements form atomic layers around SFs in a Mg-Zn-Gd alloy.

SFs can be generated either during growth processes or deformation. These are planar defects bounded by partial dislocations. The equivalent of the Shockley partials of FCC system lies in the primary slip system, (0001) <11–20>, that is, on the basal plane in magnesium:[7]

$$a/3[11\text{–}20] \rightarrow a/3[10\text{–}10] + a/3[01\text{–}10]$$

SFE, though conventionally defined in terms of surface tension and repulsive forces between the partial dislocations constituting the boundaries of the fault region, is not a simple concept when examined in detail. Essentially, FCC and HCP systems differ from each other only by the choice

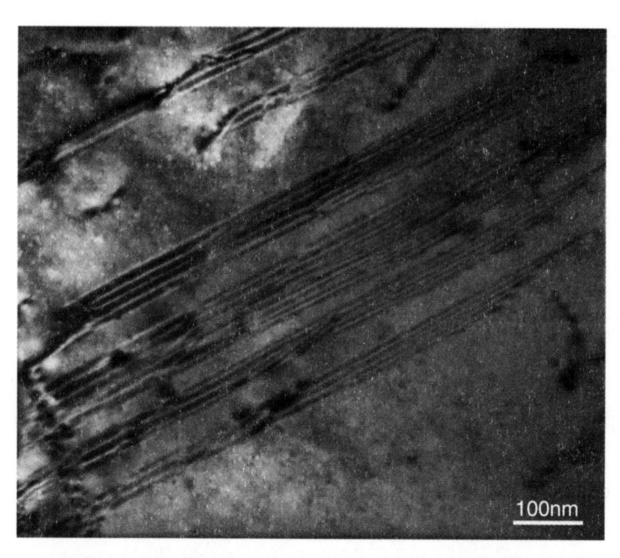

2.2 TEM micrograph of stacking faults in magnesium (courtesy of B. Li *et al.*[5]).

of position for the third layer when stacking the close-packed layers. Thus, an SF in HCP (... ABABAB... stacking) is creating a local FCC stacking (... ABCABCABC... stacking). A distinction was made between growth and deformation faults, and the former was considered to be more important in HCP metals.[8]

Depending on what constitutes the out-of-step stacking, the definitions of 'extrinsic' and 'intrinsic' SFs are introduced – the former being due to an extra layer of atoms (... ABABCABAB... $-E$ stacking) bound by partial dislocations, and the latter due to vacancy condensation. The intrinsic type is further classified into two, known as growth type, which is obtained if a missing A layer is coupled with some shear above it (... ABABCBCB... $-I_1$ stacking), and deformation type (... ABABCACACA... $-I_2$ stacking), which can be directly created by shear on basal plane in magnesium. In both cases, the necessary shear is by 1/3[−1−100]. In addition to these two types, a twin-like sequence is also possible in the basal planes, constituting yet a third type of SF (... ABABCBABABA...$-T$ stacking).[9]

Thus, as can be seen from a brief evaluation of possible stacking orders, several different SFE values must be expected. Since an SFE implies a local FCC magnesium structure, one has to consider its slightly larger crystal dimensions and energy. Table 2.2[9–12] gives a list of energies for different types of SFs, and compares crystal energies for HCP and FCC magnesium. The values given are in agreement with those by others in terms of order scale. As can be deduced from Table 2.2 and the thermodynamic evaluation

Table 2.2 Calculated energies (meV) for different types of SFs on basal planes, and crystal energies of HCP and FCC magnesium

Reference \ Fault type/System	I_1	I_2	E	T_2	HCP	FCC
Chetty and Weitner[9]	11	23	36	27	0	15
A.E. Smith[10]	10	20	32	22	0	12
L. Wen *et al.*[11]	16	34	59	38	–	–
Hui-Yuan Wang *et al.*[12]	–	140	–	–	–	–

Note: The figures of each reference are based on different approximations, and the last line is based on a supercell model of 144 atoms with the composition of Mg_{139} Al_4 Sn_{1}..

Table 2.3 Calculated stacking fault energies (mJ/m²) for (0001) basal (I_2 type fault), (10–10) and (11–20) prismatic planes, and (10–11) and (11–22) pyramidal planes

Planes \ Reference	N. Chetty and M. Weinert[9]	A.E. Smith[10]	L. Wen *et al.*[11]	Hui-Yuan Wang *et al.*[12]	T. Uesugi *et al.*[13]	D. K. Sastry *et al.*[14]	A. Couret and D. Caillard[15]
Basal	44	36	34	140	32	78	<50
Prismatic (10–10)	–	265	354	–	255	–	–
Prismatic (11–20)	–	–	1224	250	–	–	–
Pyramidal (10–11)	–	344	496	–	–	–	–
Pyramidal (11–22)	–	–	452	221	–	–	–

Note: The last column is based on a supercell model of 144 atom with the composition of $Mg_{139}Al_4Sn_1$.

given by the same authors, well separated I_1 type SFs are most likely to be seen in magnesium. A comparison of stable SFEs is given in Table 2.3.[9–15]

Effective SFE, being more complicated, would also depend on local dislocation configurations, the presence of other elements or local defects, and even on grain size and, therefore, is not considered to be an intrinsic property. As such, it cannot be expected to be uniform even on a single plane. For example, SFE decreases in the vicinity of a boundary, for example a twin. In copper, SFE at a coherent twin boundary was reported to be about 2% of the bulk SFE.[16] SFE value at boundaries is therefore particularly important as it is related to the ease of grain boundary sliding, dislocation emission, and formation of microcracks. On the other hand, SFE value, in general, controls twinning, ease of climb and cross-slip of dislocations and, in turn, work hardening. The implication of this is that activation energies for these processes increase with decreasing fault energy, that is, SFE.[17] SFE also controls the formation of dislocation cell-walls, their formation being facilitated by higher SFEs. Whereas, in metals possessing lower SFE value, dislocations tend to remain in coplanar arrays and cell formation is restricted. Yet, one other highly important property that is governed by SFE is the dislocation core width and, through this, dislocation movement – high SFE values leading to smaller dislocation-cores, facilitating dislocation motion, and vice versa.[18]

The discussion given above on SFE is, admittedly, not fundamental enough. A more fundamental approach must relate SFE to the free electron density distribution on the fault plane, or rather to the change in it, due to the creation of the fault.[19,20] It should be stated that as the charge difference on the fault plane increases SFE increases.[21] It is this perspective that has recently explained the interesting behaviour of some magnesium alloy systems, for example, Mg-Zn-Y alloy, at a very fundamental level.[22] This perspective, however, requires a different definition of SFE, based on the electronic structure. A new SFE can thus be defined, corresponding to the energy variation across a cut-plane in the crystal and translation of one part over the other by a fault vector. It is obvious that such a translation has to be made only over a unit cell, and that the fault vector is only a fraction of the lattice translation vector. If this procedure is carried out for the slip plane and expressed for a unit area, then it will define what is known as a 'generalized stacking fault' (GSF)[23] the value of which can be obtained using first-principles calculations and an SFE surface can be plotted.[24]

When the result of this procedure is plotted as a function of 'energy change with respect to the unfaulted atomic stacking' versus 'displacement along a fault vector', for example, {10–10} 1/3<11–20> or {10–12} <10–1–1> (see the schematic plots given in Fig. 2.3[12]), one or two humps may be observed. The first maximum corresponds to the unstable stacking fault (USF), that is, energy barrier to the formation of a partial dislocation, and the following minimum to the stable SF, that is, intrinsic type. If a second maximum is observed, then this can be interpreted as the unstable twin stacking (UT), while the following minimum as corresponding to the stable twin stacking (T) along the fault vector of concern. Thus, the ratio of stable and unstable SFE level, γ_{SF}/γ_{USF}, would indicate ease of slip, while twinning tendency would diminish as the ratio of γ_{UT}/γ_{USF} increases.[12,25] Many studies exist reporting, through first-principles calculation, how the energy barriers for stable and unstable SFs change on certain crystal planes and directions in magnesium and its alloys.[12,26–29]

According to this analysis, for example, it was reported that Mg–Sn and Mg–Pb alloy systems have γ_{SF}/γ_{USF} ratios for basal and pyramidal systems that are much lower than those in pure Mg, with an expected facilitation of

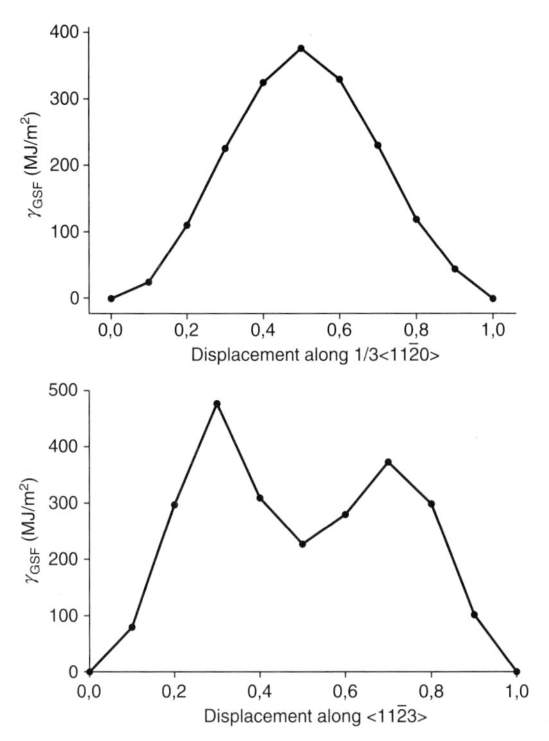

2.3 Plots of 'γ_{SFE}' vs. 'displacement along a fault vector' (adopted from H.Y. Wang *et al.*[12]).

slip.[25] Whereas, the γ_{UT}/γ_{USF} ratios for both alloy systems are higher than that for pure Mg, leading to lower tendency for twinning.

Based on the first-principles calculations by Wang *et al.*,[12] in the basal {0001} <11–20> slip system, the change due to Al and Sn addition was reported to be $\Delta\gamma_{USF} = -29$ mJm^{-2}. This drop can be interpreted as an improvement to the slip activity in this system, since $\Delta\gamma_{USF}$ represents the energy barrier to the formation of partial dislocations rather than the actual value of SFE on a particular atomic plane; whereas for the prismatic {10–10} <11–20> slip system, γ_{USF} appears to increase from 356 to 373 mJm^{-2}. Interestingly, as a result of this change in the presence of Al and Sn, pyramidal {10–11} <11–20> system, and prismatic {10–10} <11–20> slip system change their ranks, and in contrast to the case in pure magnesium, the former becomes more active than the latter owing to a lower γ_{USF}. The drop in γ_{USF} and γ_{SF} values due to Al and Sn were found to be to a lesser extent for the prismatic {11–20} <10–11> and the pyramidal {10–12} <10–1–1> and {11–22} <11–23> slip systems.

To the best knowledge of the author, the experimental studies on Mg–Pb and Mg–Sn systems, despite their large number especially on Mg–Sn, unfortunately, have not paid attention to the fundamental connection between SFE and mechanical properties. Those studies also lack microstructural evaluation in terms of the occurrence and types of twins. Therefore, one can only deduce from some of those studies that apparently twinning, if it exists, is inhibited, and the ultimate tensile strength of Mg–Sn binary alloys increases more rapidly than the proof stress, indicating a strong deformation hardening due to low SFE.[30,31]

It is obvious that any attempt to impart ductility or strength to magnesium through alloying has to involve considerations of SFE and dislocation core properties. Such a critical evaluation should focus on the changes in the ratios of γ_{SF}/γ_{USF} and γ_{UT}/γ_{USF} on particular atomic planes to reveal whether the birth of dislocations or twinning is facilitated by the changes generated in the free electron density distribution due to the presence of a particular alloying element.

The SFE values based on experiments and first-principles calculations are in disparity. However, studies based on, *ab initio* method, that is, density functional theory (DFT),[32,33] and an empirical method, that is, embedded atom model (EAM),[34–37] though reporting marginally different SFE levels, appear to give self-consistent and more reliable results enabling us to compare the behaviour of magnesium with other metals, and to explain the general mechanical behaviour of magnesium based on this fundamental parameter. Further fundamental studies using these methods are obviously needed, for a more complete understanding and better interpretation, especially of the effects of individual alloying elements on the behaviour of crystal defects in magnesium. Such studies have begun to emerge in recent years, explaining individual effects of a number of alloying elements. In a recent report based on *ab initio* calculations, Muzyk *et al.*,[25] gave a list (Table 2.1 of the paper by Muzyk *et al.*) of stable and unstable SFE values for binary magnesium alloys containing Ag, Al, Cu, Fe, Li, Mn, Ni, Pb, Sn, Ti, Y, Zn and Zr. As can be expected, similar calculations covering a wider range of alloying elements, and if possible for more complex alloys, would be highly useful for better alloy design.

2.2.2 Elastic deformation of magnesium

The elastic properties of magnesium, including the atomistic inelastic interaction mechanisms that lead to attenuation of sound waves in metallic materials, have been discussed in detail by Mordike and Lukáč.[38] Magnesium, in contrast to its strong plastic anisotropy, is claimed to be unique among HCP metals in that its elastic constants and thermal elastic coefficients are nearly isotropic, and that elastic compatibility stresses in the vicinity of grain boundaries can be ignored.[39–41] This may be taken to reflect the true metallic character of the interatomic bonding in magnesium, and the Brillouin zone shape which may be compensating for the low symmetry of the hexagonal crystal. With a modulus of elasticity of 45 GPa and a Poisson's ratio of 3.5, magnesium alloys display generally lower values than their competitor alloys aluminium and steel.[38]

An industrially important and relevant property, namely damping, is worth mentioning when discussing the elastic behaviour of magnesium. As a general rule, we can state that lower strength, annealed conditions, larger grain sizes and lower solute contents would result in better damping properties.[42] It is generally cited that magnesium has better damping characteristics than its competitor light weight metals aluminium and titanium.[43] Different magnesium alloy systems may display good damping properties at a variety of temperatures, in many cases outside the range of current interest in magnesium alloys. For automotive power-train applications, for example, only pure magnesium appears to promise a high damping peak for the relevant temperature range, which is of little use as pure metal is not useable in those applications.[44,45] It should be borne in mind that the choice of alloy in terms of damping capacity should be assessed based on the frequency and amplitude of the vibration or noise, and the temperature involved in the particular application of concern. Furthermore, it should be taken into account that there exist attenuation processes related to the defect types and their densities that may influence the outcome of damping measurements. Although damping is considered an elastic response of the material, notably, the ease of dislocation movement and twinning–untwinning determine the level of attenuation.[46,47]

2.3 Plastic deformation behaviour of magnesium and its alloys

Magnesium can be plastically deformed, about 10% elongation to failure, at room temperature despite the fact that as an HCP metal it does not meet the requirement of the von Mises criterion, that is, the necessity of a minimum of five independent slip systems to maintain integrity along grain boundaries during plastic deformation. An inherent additional shortcoming of the HCP crystal system of magnesium is that its low symmetry hinders plastic compatibility, inhibiting transfer of slip across the grain boundaries.[41] Among many differences with its competitor, aluminium, a larger stress exponent (n) variation of magnesium between 1.5 and 5 as compared to the range, 2.7–4, of aluminium with strain rate is worth mentioning here.[48] This feature of magnesium has important implications in terms of plastic deformation and recrystallization, as will be discussed later.

A magnesium single crystal possesses different yield points depending on whether the stresses acting on its c-axis are compressing or extending it, a feature that is essentially different from Bauschinger phenomenon. Likewise, the preferred orientation of wrought polycrystalline magnesium imposes further disadvantages for plastic deformation. As a result, in addition to the in-plane anisotropy, as can be observed in any sheet metal, magnesium and its alloys, when textured due to prior deformation, exhibit additional mechanical anisotropy depending on the direction of the applied stress with respect to the basal plane normal, that is, fibre/texture direction. However, it is said that if estimations are based on CRSS values obtained from single crystals this would lead to anticipation of a much stronger mechanical anisotropy as compared to that resulting from the existing texture.[49] Magnesium alloys also show, in a similar manner based on creep stress direction, a creep anisotropy, which will be discussed below (Section 2.6).

Moreover, a pseudoelastic behaviour appears in magnesium and its alloys upon reversal of loading. The underlying reasons for these counterintuitive features of magnesium and its alloys with respect to its plastic deformation behaviour have been detailed through various studies.[50,51] The information given below is an attempt to highlight those reasons, based on the existing literature.

2.4 Critical resolved shear stress (CRSS), slip and twinning

2.4.1 Critical resolved shear stress (CRSS)

Magnesium compensates for the shortness in the number of slip systems with its easily activating twin systems that provide deformation capacity to the crystal along its *c*-axis until its prismatic and pyramidal slip systems are activated. Although it was proposed[52] that a favourable *c*/*a* ratio of magnesium crystal facilitates the operation of non-basal slip systems more easily than those in other HCP metals with a higher ratio, this view was later disputed based on a more rigorous analysis of the energies of SFs created by dissociated screw dislocations on basal and prismatic planes. The low basal SFE of about 10 mJm^{-2}, being about seven times smaller than that of prismatic SFE, accounts for the occurrence of cross-slip in magnesium only at high temperatures (above 225°C) or the ease of it as compared to other HCP metals.[53] The activity of cross-slip increases with increasing temperature. At this point it should be pointed out that the SFE comparison at room temperature is misleading if the SFE change with temperature is ignored.

CRSS level is particular to a mechanical deformation system, as implicitly expressed above in the discussion on the general features of plastic deformation regarding SFE. It indicates the shear stress level that is resolved onto that specific slip/twin system through Schmid's law, at which level dislocations or twins on that system are mobilized. Since it is also related to dislocation nucleation, comparative evaluation of CRSS values may be considered more meaningful, rather than as absolute quantifications. Schmid's law is expressed as follows:[54]

$$\tau_{CRSS} = \sigma \cos \lambda \cos \phi$$

where σ is the applied stress, λ the angle between the slip direction and the force direction, and ϕ is the angle between force and slip plane normal directions; the cross product $\cos \lambda \cos \phi$ is denoted as m, the Schmid factor.

As is the case for other metals, the ease with which a non-diffusional deformation mechanism, that is, slip or twinning, in magnesium is activated depends on the CRSS. Gharghouri was the first to indicate that twinning in magnesium obeyed Schmid's law.[40] The mechanical deformation systems having the highest Schmid factor yield first, as a result of their angular position with respect to the acting stress. Thus, it is not expected that if one of the deformation systems is activated somewhere in the bulk, the same ones are all necessarily triggered in every grain of the polycrystalline material. When the geometric orientation of each plane and direction of the slip or twinning systems in the bulk is considered with respect to an applied stress of known direction, then Schmid's law determines whether or not the required CRSS is achieved for an individual system. Strongly textured magnesium is thus easier to interpret in this regard, and there exist numerous studies based on the neutron diffraction method where precise angular information is indeed available regarding the directions between applied stress and crystallographic planes together with the knowledge of whether or not the activation is taking place on particular crystallographic indices.[55–61] Such studies can give quite specific and reliable quantification of the CRSS values of individual mechanical deformation mechanisms, albeit to a lesser degree compared to those obtained from single crystals.

Many researchers,[38,62–71] have pointed out the anomalies related to the changes in CRSS; for example, its anomalous dependence on temperature for the second order pyramidal system, much different reduction rate with temperature in the CRSS values of the basal and prismatic planes, and non-monotonic change of the CRSS of the prismatic plane with temperature and concentration in some alloy systems.

Experimentally determined CRSS values for basal and non-basal slip are in the range of 0.5 and 55 MPa, respectively.[72–79] Single crystal experiments revealed CRSS values to be in the range of 5 MPa for basal slip, 10 MPa for extension twinning, 20 MPa for prismatic slip, 40 MPa for pyramidal slip and 70–80 MPa for contraction twinning.[80] Slip on prismatic and pyramidal planes is said to occur by cross-slip of screw dislocations onto those planes. Although commonly believed that non-basal slip systems are only activated at higher temperatures, Obara *et al.*,[63] in an earlier study indicated that even the second order pyramidal slip is operative from room temperature up to 500°C, providing a very high work hardening up to 200°C, and that screw dislocations are more mobile than the edge type. Large Burger's vector, in other words, large dislocation core of dislocations on these non-basal slip planes may explain such observations. It is also known that CRSS values do not change in the same fashion for basal and non-basal planes, the gap between them reducing rapidly as the temperatures increase above 200°C. At this point, let us remember that slip is not the only active deformation system in magnesium.

2.4.2 Slip and twinning

Table 2.4 gives a list of the crystallographic indices of the slip and twinning planes and directions.[2] Balogh *et al.*,[64] in a recent study, showed that $<c + a>$ type dislocations increase while $<a>$ type dislocations and twinning declines during deformation starting from ~100°C. A formal description of the selection of twining plane and direction can be found in studies by Hosford[65] and Christian and Mahajan.[66] In their work on a crystal plasticity modelling Cipoletti *et al.*[67] reported that the relative contribution to total strain by individual slip systems at high temperatures is weakly dependent on grain size and strain rate, and that the difference almost disappears at high strain rates. Thus, even at first glance, mechanical deformation of magnesium involves sequential events, and the answers to the puzzles magnesium presents lie in its deformation mechanisms.

The deformation of magnesium at room temperature starts initially with basal slip and then, as the close CRSS values given above also indicate, at an as little as 0.1% strain level, twinning sets in and deformation commences by the simultaneous action of slip and twinning.[55,57,58,81,82] On this point, Koike *et al.*[83] suggest that basal slip precedes and saturates quickly, thus giving way to twinning. This view additionally leads to the interpretation that, until what appears to be the macro-yielding, actually a strain hardening is taking place. Furthermore, Koike[84] claims that when the Schmid factor for non-basal slip reaches the range of 1.5–2.0 times that of basal planes due to local texture, both types of slip operate collectively. It has been shown that about 40% of dislocation segments were of the non-basal type in a severely deformed AZ31 sample having 8 μ grain size. Thus the non-basal dislocations proliferate due to plastic incompatibility in the vicinity of grain boundaries, leading to grain boundary sliding (GBS) up to 8 pct of total strain at room temperature.[41]

Brown *et al.*[55] defined a 'twinning regime', reporting that the twinning activity in magnesium provided for about 42% of the deformation by the

Table 2.4 Crystallographic indices of the slip and twinning planes and directions of magnesium

Planes	Directions	Type	Slip direction	Number of independent systems
$\{0001\}$	$\langle 11\bar{2}0\rangle$	Basal	*a*	2
$\{10\bar{1}0\}$	$\langle 11\bar{2}0\rangle$	Prismatic	*a*	2
$\{11\bar{2}1\}$	$\langle 11\bar{2}0\rangle$	Pyramidal	$c + a$	4
$\{10\bar{1}2\}$	$\langle 10\bar{1}1\rangle$	Pyramidal (Twinning)	$c + a$*	3
$\{10\bar{1}1\}$	$\langle 10\bar{1}\bar{2}\rangle$	Pyramidal (Twinning)	$c + a$*	3
$\{11\bar{2}2\}$	$\langle 11\bar{2}3\rangle$	Pyramidal (Twinning)	$c + a$	4
$\{11\bar{2}1\}$	$\langle 11\bar{2}\bar{6}\rangle$	Pyramidal (Twinning)	$c + a$	4

*Frequent ones.

time the deformation in the parent structure reached 14%, during which process the contribution of twinning was linear up to 6% deformation of the parent material, peaking at 8% and diminishing thereafter. When the matrix was deformed by 14% the volume per cent of the twinned region comprised about 80%. The same study also describes the strain partitioning between the twinned and untwined regions of the parent structure, pointing out that the matrix relaxes within the twinned area as the new orientation corresponds to 'hard' orientations and therefore starts to only elastically strain for the continuation of the deformation process. The saturation of twinning by about 8% matrix deformation was also confirmed by others.[56,58,85] For the continuation of deformation, the matrix grains became further divided, as if into sub-grains, as the extension twins rotated the twinned regions by 86.3°, and in the case of compression twinning by 56°.[86] Once the twinning is saturated, the slip systems requiring higher CRSS, that is, prismatic and pyramidal planes, take on the deformation.[57,58] However, as a result of plastic incompatibility, large stresses accumulate in the vicinity of grain boundaries during plastic deformation, leading to activation of non-basal slip in these regions at an earlier stage as compared to the grain interiors.[41]

When the value of *c/a* ratio is equal to the ideal value of $\sqrt{3}$, the {0002} plane makes a 43.15° angle with the {10–12} plane, inevitably assigning a high Schmid factor to the {10–12} <10–11> twinning system, thus leading to its low CRSS value when the crystal is in tension parallel to – or, in compression, perpendicular to – the *c*-axis. Hence, the slightly less-than-ideal *c/a* ratio of magnesium, explaining the ease of the well-known deformation twin system of magnesium and its alloys.

Increased temperature, for example, above 250°C, leads to an overlap of first and second Brillouin zones, which in turn, allows for an elastic increase in the *c*-axis.[87] This effect of temperature is akin to stretching the *c*-axis with externally applied stresses. From a mechanical point of view, changes of *c*/a ratio thus achieved may be interpreted in terms of changes in the Schmid factor assignable to the twinning and non-basal slip planes, leading to facilitation or hindering of their activation. However, this consideration alone cannot explain all observations. The underlying effect ought to be more subtle and be related to the changes in free electron density distribution and its consequences in terms of dislocation behaviour. Especially the influence of alloying elements on *c/a* ratio appears to be more complex, necessitating considerations on the relative changes in *c*- and *a*-dimensions, and the resulting changes in Peierls stresses. A noteworthy example is the addition of Li and In. These elements are said to lower the *c/a* ratio yet promote non-basal slip activity, while indium addition beyond 2.2 at% increases the axial ratio.[87,88] For a more detailed discussion on the effects of alloying the reader is referred to Chapter 5 in this book.

The ratio of CRSS values of the basal slip and extension twinning in magnesium is given as 1:3.5.[89] Barnett reported a CRSS ratio of 1:0.7:2:15 for basal, twinning, and slip on prismatic and pyramidal planes, respectively.[50] The ratio of CRSS values given by Jonas *et al.*, which is in general agreement with others, as 1:9:9:11:4:30:30 for basal glide, prismatic glide, pyramidal glide, extension twin nucleation, extension twin growth, contraction–twin nucleation and contraction–twin growth, respectively.[90] These ratios show concisely the sequential nature and relative ease of deformation events in magnesium.

The amount of shear provided by common twining in magnesium is 0.13. Twinning, unlike slip, occurs on non-close-packed pyramidal planes in magnesium on the following planes and directions: {10–11}<10–1–2>, {10–12}<10–11>, {11–21}<11–2–6> and {11–22}<11–23>.[66] It should be noted that in other HCP metals additional twinning systems may operate, totalling seven modes of twinning.[91] The CRSS values for the most frequent ones in magnesium, that is, {10–12}<10–1–1> and {10–11}<10–1–2> twins are about 2 MPa, and in the range of 65–153 MPa, respectively, indicating why the former, known as the 'extension twinning', occurs much more easily than the latter, the so-called 'contraction twin'.[40,92–95] The {10–12} twinning mode, in addition to a low shear stress, requires a simpler atomic shuffling,[92] and as a consequence it occurs more frequently than those of any other twinning modes.

The name 'extension twin' somewhat misleadingly refers to the case when the applied stress is compressive and parallel to – or a tensile stress perpendicular to – basal planes, the resulting twin actually extending the crystal along the *c*-axis. 'Contraction twin' applies to the case when the applied stress is compressive and perpendicular to the basal plane, the resulting twin shortening the crystal in the *c*-axis direction.[50,51] The morphology of contraction twins appears to be more of a lens shape, facilitating their identification via microscopy.[96] There exist studies reporting the formation of the more difficult twin mode, that is, contraction mode, even when the stress conditions are set for the more common type of twinning, that is, extension mode. If deformation by the extension mode is saturated, or the grain size is fine enough, some contraction twins are also observable in the same structure.[90,97] Though the two types of twins require different initial stress conditions, there are observations on the formation of contraction twins following saturation of extension twins at 8% deformation of the parent material.[55,58,81,96,98] Twins and non-basal slip, that is, slip through the more-difficult-to-activate <*c* + *a*> dislocations, provide means for deformation along the *c*-axis of magnesium crystal.[99]

Double twinning, in the sequence of {10.1}–{10.2}, can also create co-existence of the two modes of twinning, the latter being inside the former one. This is the situation when a heavily textured magnesium with basal planes oriented parallel to the process deformation direction, as in extrudites or rolled sheet, is subjected to tension parallel to the original processing direction. Double twinning, simply by the rotation of the twinned region for a second time, brings the twinned region into suitable orientation for slip, that is, the *c*-axis parallel to the loading direction. Since the structure is saturated for twinning by two consecutive twinning actions, further deformation becomes only possible by slip modes. Such a twinning sequence is apparently only possible in coarse-grained structures and at large strains prior to failure.[51,57,82,84,100–102] It has been pointed out that inside the double twinned region the crystal becomes suitably oriented for continuation of deformation by slip modes. As a consequence of this, the deformation is said to be localized within that space, leading to diffuse necking and failure therein.[51,103,104]

Twins are compositionally homogeneous interfaces within a matrix. Their formation is, as commonly known, affected by grain size, deformation rate and temperature. Deformation by twinning requires increasing stresses, that is, becomes more difficult, with decreasing grain size.[82,105–107] Increasing temperatures and decreasing strain rates also diminish deformation twinning.[108] Twins, as they offer an additional deformation mechanism in magnesium, starting from their nucleation, act as 'softeners', not because the resulting structure becomes easier to deform by slip but, simply, until saturation of twinning, twins do not create a substantial hardening effect.[56]

Twinning activity is closely related to dislocations, both at the nucleation and growth stages. The Burgers vector of these theoretical elementary twinning dislocations is very small and depends on *c*/*a* ratio, for example, for the {10–12} <10–1–1> twinning in Mg is only 0.024 nm.[109] Nucleation stage, almost exclusively, initiates from grain boundaries, where it is said, in some cases, to be facilitated by the impinging twins on the nucleation site in the neighbouring grains.[81] Since the low symmetry of HCP system does not permit the twinning reaction to be of a simple shear, as would be the case in high symmetry cubic systems, the twin interfaces are said to contain complex, so-called zonal, dislocation structures. These zonal dislocations, arise from having to accommodate the necessary non-shear type slight displacements of atoms at the twin boundary so that the mirror symmetry is accomplished.[90] Several proposals, based on TEM observations and modelling, exist describing these twin interface zonal dislocations.[109–112]

Twinning in HCP crystals is closely related to the grain size, in the sense that below an apparent threshold grain size twinning does not take place. The limiting grain size for magnesium is about 1 μ. What is being less emphasized is the relationship between twin formation, and SFE and dislocation core properties, both of which are, in turn, related to free electron density distribution. Dependence of CRSS on temperature for prismatic slip planes depends on the alloy system, and is usually complex.[38] While this relationship is linear in some alloys, it is rather inexplicable in others. However, the root cause of such non-linear behaviour may be sought in the relationship between the electron density function and temperature. Free electron density distribution around each atom beyond Brillouin zones is not homogeneous. Furthermore, this distribution must be subject to changes with temperature as well as to the existing crystal defect configurations. These factors altogether may render the observations less discernible. CRSS, in other words, homogeneous flow through activation of an independent slip system, should be related to dislocation core properties/SFE levels and, in turn, to the free electron density distribution. Fortunately, there have been an increasing number of investigations focusing on this issue, and presenting us with some progressively refined modelling studies over the past decade (see the discussion on SFE).

A final point of interest regarding the mechanical behaviour of magnesium and its alloys is the pseudoelastic behaviour due to twinning–detwinning upon reversal of the applied stress direction.[113] {10–12}<10–11> extension twinning is mostly responsible for this pseudoelasticity. However, such reversibility cannot be sustained indefinitely, due to the involvement of dislocations in twinning reactions, and therefore the phenomenon is highly important in fatigue cycles. This behaviour is more readily observed in pure Mg,[114] and small-grained material, at least in AZ91.[115]

Magnesium, having an inherently anisotropic crystal, due to its low symmetry, is particularly inclined to evolve texture by slip as well as twinning. Therefore, starting materials that are the products of processes involving deformation present us with a variety of mechanical behaviours depending on the path of loading during the final shaping. Likewise, the finished products of processes involving deformation exhibit mechanical anisotropy under service conditions.

Twinning must be regarded as both a deformation mechanism in itself, and an aid facilitating further plastic deformation by slip due to reorientation of the basal planes into favourable directions as a result of it. Thus, magnesium avoids otherwise inevitable loss of integrity along grain boundaries as an expected outcome of the lack of a sufficient number of, that is 5, independent slip systems to satisfy von Mises criterion.

2.5 Fatigue behaviour

Fatigue is a failure mechanism which, in addition to the level of cycling stress resulting in a net tensile stress and its frequency, is highly sensitive to discontinuities and inhomogeneties present in materials. Such pre-existing adverse features can facilitate both crack initiation and crack propagation. However, the absence of such undesirable constituents, it though would prolong fatigue life, does not prevent fatigue completely. Some fatigue initiation sites are born during service life as well. Having this perspective, magnesium alloys should be considered in terms of both the potential fatigue initiators present prior to service of the part and the discontinuities/inhomogenities that can come into life during service due to the dynamic nature of fatigue conditions. Furthermore, the fatigue strength of magnesium is linearly related to the tensile strength, the slope being 0.4 in wrought alloys and 0.3 in castings.

General design precautions against fatigue, such as avoiding sharp corners, better surface finish, employing welded joints instead of methods involving drilling, etc., will not be considered here. The topic will be limited to the issues related to more inherent features of magnesium alloys. For more comprehensive studies on fatigue of magnesium alloys, the reader may also refer to recent studies in literature.[96,116]

Yet another omission from the discussion here is the effect of temperature and cycling stress level, for which the reader is referred to the data given elsewhere.[117] Related to the effect of temperature, we will only mention that crack propagation rate appears to accelerate with temperature.[118] Regarding the level and frequency, it may be worthwhile also to mention that both high-cycle and low-cycle fatigue conditions are of interest due to the potential applications for which magnesium alloys are considered. However, there are studies indicating that, at least for high pressure die-casting samples, frequency did not seem to be a major influence.[119]

Potential fatigue initiators in magnesium alloys, as for other metals, can be broadly categorised into two groups, those related to the production methods of the subject parts, and those related to the magnesium system, that is, those related to the alloy chemistry and/or to the behaviour of the hexagonal crystal. Needless to say, that the size of each defect type is highly important in terms of fatigue life. However, the scope here will be limited to the qualitative evaluation of the potential fatigue initiators.

In the first category of crack initiators, that is, those related to the manufacturing methods, the following can be listed: porosity,[119] inclusions,[119,120] formation of large grains,[121] and surface discontinuities.[122] The second category involves: twinning related crack initiation,[51,116,121,123] intermetallics of the alloy system,[120,124] and corrosion related surface defects.[125,126] It should be pointed out that in some cases features that appear in the first category as potential initiators may also trigger a mechanism related to the second category. For example, production method is one of the dominant factors determining the grain size, through which it could also influence the ease of twin formation. Likewise, a production method can also be suitable for a particular alloy, and the alloy type of magnesium, through the types, morphology and distribution of second phase particles, may have an important effect on fatigue life.[119]

One of the most attractive mass production methods for magnesium parts is high pressure die-casting (HPDC). This method, by its design and practice, inevitably mixes air into the liquid metal when forcing it through the shot sleeve into the die cavity. As a result of this, the product contains porosities that practically prohibit even post-manufacturing heat treatment (see Fig. 2.4). Such pores strongly act as fatigue initiators, and the fatigue life only corresponds to the crack propagation stage of fatigue. Since the effectiveness of pores in causing fatigue failure depends on their size and amount, fatigue data for cast materials may show great scatter.[119,127]

Wrought products, by the nature of their manufacturing method, are much leaner compared to cast products in terms of microstructural defects, most importantly, of porosity. This is not to imply that microstructure is not important in terms of the fatigue behaviour of wrought products. Extrusions are relatively more sensitive to strain rates than cast materials during fatigue.[128] It should be noted that wrought magnesium alloys appear to be more predictable compared to cast ones. However, studies exist, showing that the fatigue life of both cast and wrought Mg alloys, AZ31 and AZ91 fits into Manson–Coffin relationship when tested under controlled stress conditions rather than under strain control.[116,129]

Plots of strain amplitude versus number of cycles for magnesium alloys (S/N curves) show a transition 'knee'. This transition in strain-life behaviour, interpretable as a clear transition from a low-cycle regime to a high-cycle regime, corresponds to around 0.5% strain levels.[116,130] This apparent transition was attributed, as can also be inferred from the discussion on twinning behaviour, to the change in the deformation behaviour in magnesium at the atomic level. When the strain amplitude was above this threshold value, fatigue cracking was associated with a greater amount of twinning and shear deformation, and below, with more dislocation movement.[131]

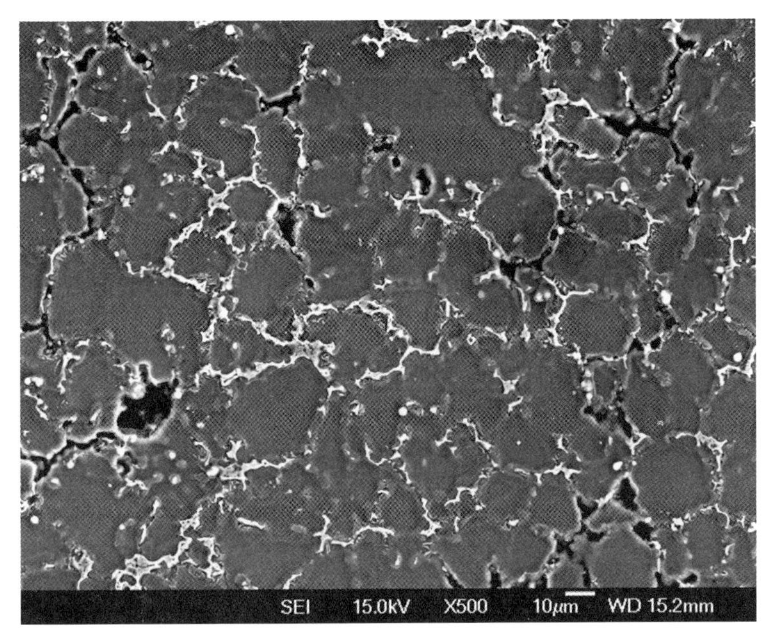

2.4 Porosity in high pressure die-cast AZ31 alloy microstructure.

It has been proposed that when grain size is smaller than the plastic-zone size ahead of a fatigue crack tip, the reversal of strain becomes more difficult, and thus fatigue crack propagation is facilitated.[128] On the other hand, it is also known that twinning is more favourable in large grains. Therefore, twinning–detwinning activity would be expected to dominate with large grains (see the general discussion on twinning above). Twinning and detwinning activity during cycling loading have been widely studied, including experiments involving *in situ* diffraction techniques.[59,130,132,133] It has been shown that the asymmetry of the hysteresis loops increases with the applied strain or stress amplitude.[96] Even when porosity is absent in the initial microstructure, eventual formation of microcracks parallel to contraction twin paths revealed the close relationship between the twinning and fatigue failure.[51]

In the light of the brief overview given above it may be clear that the non-reversing portion of the strain, that is, accumulation of damage to lead to fatigue failure, may be facilitated by the action of the consecutive deformation mechanisms operating in magnesium. Each of the three mechanisms, that is, slip and almost simultaneous action of tensile twinning followed by compressive twinning, could be preventing the reversal of the plastic movement created by the other. Twins, in addition were observed to be closely associated with formation of microcracks which may, in turn, be related to stress concentrations at the tip of twins.[51]

Due to the scatter in fatigue test results of magnesium, a set of tests should be taken to represent the batch of materials that comes from the manufacturing route the samples have experienced. If a general comparison with other light metals is to be made, then it is fair to say that Mg alloys appear to exhibit faster fatigue crack propagation rates than aluminium alloys, such as 6063 and 7075 and pure titanium.[126] In a similar trend to other metals, improvements were also achieved on the fatigue resistance of magnesium alloys by treatments that produced compressive stresses on the surfaces of parts.[134,135]

2.6 Creep behaviour

Creep is a time-dependent deformation process that manifest itself even when applied stress is constant and below the yield strength of material. While in an ordinary deformation the movement of dislocations, on an atomic scale, can cover large distances at each step, in thermally activated deformation this movement occurs due to diffusion of atoms and therefore is restricted to a few atomic distances at each step. In such a process, dislocations attempt to pass the obstacles on their path individually by mobilizing their pertinent segment locally. Hence, while the ordinary long-range deformation involves the movement of long segments of sequential dislocations, the thermally activated deformation component is associated with the movement of small sections of a dislocation, for example, dislocation jogs, or the cutting of a dislocation by another. This nevertheless causes slow and permanent deformation, called creep. Since atoms vibrate anywhere above absolute zero, a thermal component always exists and creep is possible even at very low temperatures, given sufficient time.[136]

The basic mathematical expression of the dominant mechanism at high stresses, that is, 'dislocation creep', assumes the form of an Arrhenius type relationship in which strain rate has a power law dependence on stress:[137]

$$\dot{\varepsilon}_{ss} = \mathrm{A}\sigma \exp\left(\frac{-Q}{RT}\right) \tag{2.1}$$

'Diffusional creep', on the other hand, operates at lower stress levels and at high temperatures with a linear dependence of strain rate on stress:[138]

$$\dot{\varepsilon}_{ss} = \mathrm{B}\sigma \exp\left(\frac{-Q}{RT}\right) \tag{2.2}$$

Since the diffusion is affected by grain size, a term, d, representing it may also be incorporated into this expression:[137]

$$\dot{\varepsilon}_{ss} = \mathrm{C}\frac{1}{d^2}\sigma \exp\left(\frac{-Q}{RT}\right) \tag{2.3}$$

In these equations A, B and C are material constants, and d is grain size; R is the gas constant and T is the absolute temperature. Q represents the activation energy for creep, and n is known as the stress exponent.

There exist several modified versions of the classical Equation [2.1] relating the steady-state creep rate $\dot{\varepsilon}$ to the stress exponent n. One of them is the theoretical creep rate equation developed by Weertman:[139]

$$\dot{\varepsilon} = a\left(\frac{D_l}{b^{3.5}M^{0.5}}\right)\left(\frac{G\Omega}{kT}\right)\left(\frac{\sigma}{G}\right)^n \quad [2.5]$$

where D_l is the coefficient for lattice self-diffusion, M is the number of dislocation sources per unit volume (equivalent to $0.27\rho^{1.5}$, ρ being the dislocation density), Ω is the atomic volume, and α is a constant in the range $0.015 < \alpha < 0.33$.

Garofalo's[140] steady-state creep rate equation (sin hyperbolic equation), compared to the other equations, is applicable to a wider creep temperature and strain ranges, and not necessitating a steady-state creep condition unlike the others, is:[141]

$$\dot{\varepsilon} = A[\sinh(\alpha\sigma)]^n \exp\left(\frac{-Q}{RT}\right) \quad [2.6]$$

Norton equation, being another alternative equation, can be given as:

$$\dot{\varepsilon} = \left(\frac{ADGb}{kT}\right)\left(\frac{\sigma}{G}\right)^n \quad [2.7]$$

where $\dot{\varepsilon}$ is steady-state creep rate, σ creep stress, α and A are dimensionless materials constants, D the diffusion coefficient, Qc the activation energy for creep, R the gas constant and T the absolute temperature, G the shear modulus, b the magnitude of the Burgers vector and k the Boltzmann constant.

Yet another modified rate equation as suggested by Mohamed and Langdon[142] is in the form:

$$\dot{\varepsilon} = A'\left(\frac{\sigma}{G}\right)^n \exp\left(\frac{-Q}{RT}\right) \quad [2.8]$$

where A' is a material parameter, being proportional to γ^3 (γ being SFE).

Equations [2.7] and [2.8] further have also modified versions.[143–145] Some workers when employing the modified versions of the Equations [2.7] and [2.8] adapted a reduction scheme in the stress level, essentially in order to reduce the unusually high n values otherwise obtained. While doing so, a threshold σ_0 was subtracted from σ (i.e., $\sigma-\sigma_0$ is used instead of σ in Equations [2.7] and [2.8]) so that only that portion of the stress that serves purely dislocation glide could be taken into account for the observed creep rates.[143,144]

Despite their inherent shortcoming in assuming homogeneous materials, these alternative equations can serve well for determining the stress exponent n using the slope of a plot of log $\dot{\varepsilon}$ against log σ, and some of these expressions are based on somewhat more practical parameters. However, it should be borne in mind that, though representing the same parameter, the stress exponent n may not be identical in value in all of these constitutive equations, and instead its values can have a statistical correlation with each other.

Although a clear-cut distinction between various creep modes, and specifically between possible rate-controlling micro-mechanisms, is often difficult, values of Q (activation energy for an individual creep mechanism)[146] and n (the stress exponent in the expressions of dislocation creep)[147] are being used to establish the operating creep mechanism and, especially due to the difficulty in singularly resolving Q parameter, leading to contradictory interpretations among authors.

In addition to the interpretations of 'Q' and 'n', there also exist indications in strain-time creep diagrams that can be related to creep mechanisms. Hence, the lack of significant instantaneous strain upon loading in the primary region, and an inverted primary stage were related to a viscous glide process of dislocations, while the opposite of these conclusions were attributed to dislocation climb or cross-slip of dislocations.[148]

2.6.1 Creep behaviour of magnesium

Earlier interest on the creep behaviour of magnesium stemmed from its use in nuclear canning applications.[148–150] The current interest, however, arises from the expected economic and environmental benefits of the use of magnesium alloys, particularly in the automobile's power-train applications requiring creep resistance at temperatures up to 200°C.

Creep properties of magnesium alloys are of great importance as they are considered by automotive industry for power-train applications where service conditions impose temperatures that are high enough for these alloys to undergo creep damage. Although the subject temperatures (~150ºC) may appear rather low, due to the low melting temperature of magnesium, and its hexagonal crystal structure that allows for relatively rapid diffusion rates, creep becomes a dominant design criterion especially for applications such as engine block and transmission case. It appears that the expansion of the current use of Mg casting alloys hinges, to a large extend, on the development of die-castable and economically competitive an alloy with good creep strength up to 200°C. Current studies show that magnesium alloys can indeed be promising towards this goal.[151–154]

Magnesium is inherently prone to creep deformation due to high diffusion rates in its lattice ($D_{L,Mg}/D_{L,Al}$ and $D_{gb,Mg}/D_{gb,Al}$ ratios are greater than unity over a wide temperature range and the ratio of D_{gb}/D_L for magnesium is five times higher than that for aluminium at 0.67 T_m).[140] Furthermore, as we have previously discussed, between approximately 200°C and 250°C depending on the alloy, additional pyramidal and prismatic slip planes become operative, lowering the creep resistance.

Creep is a complex phenomenon in many engineering applications, and it is perhaps more so in the case of magnesium alloys. Although there appear to be disagreements in the literature regarding its severity even for the same alloy (e.g., high pressure die-cast[155] and squeeze-cast[144] AE42) a different type of tensile-compression anisotropy of magnesium manifests itself once again. In the absence of texture, the better compressive creep resistance,[144,156] than that under tensile above a certain stress level, of the cast magnesium alloys cannot be related to the inherent mechanical anisotropy of the hexagonal crystal. Instead, this type of anisotropy was, in aluminium containing alloys, related to the suppression by compression type stresses, or facilitation otherwise, of the dynamic precipitation of the β-$Mg_{17}Al_{12}$ phase,[157] and to changes in twinning behaviour.[144] However, this creep anisotropy was also observed in an alloy (Mg-1.92 RE-0.33Zn-0.26Mn, wt%) that does not form the β-$Mg_{17}Al_{12}$ phase.[155,158] Moreover, Xu *et al.*[158] showed that the effect of the stress type was reversed in AM50, AE44 and AJ62A alloys. AM50, among these three alloys, was further complicated in that low pressure die-cast samples tested below a low transition stress showed no anomalous behaviour.

The reason for this creep anisotropy can be understood by the fact that twinning is no longer in play at the temperature ranges of creep tests, except for a minor contribution.[158] Dieringa *et al.* claimed that, while the stress level increases the disparity between compressive and tensile stresses diverges, and at sufficiently low stresses where twinning is negligible, the difference disappears, reversing in behaviour below that level (see Fig. 2.5).[144] It would be interesting to see the effect of crystal anisotropy on creep strength, if a patient worker was to, only for the sake of curiosity, creep test magnesium at the sufficiently low temperatures such that extensive twinning is possible. To the best knowledge of the author, no conclusive argument explaining exclusively the stress direction anomaly in creep has been provided to date.

Yet another interesting point is that creep resistance in interrupted creep tests is inferior and creep life is shorter compared to that in continuous creep, indicating that the designs based on continuous creep test data might be misleading in certain applications.[157,158] In these cases, creep–fatigue

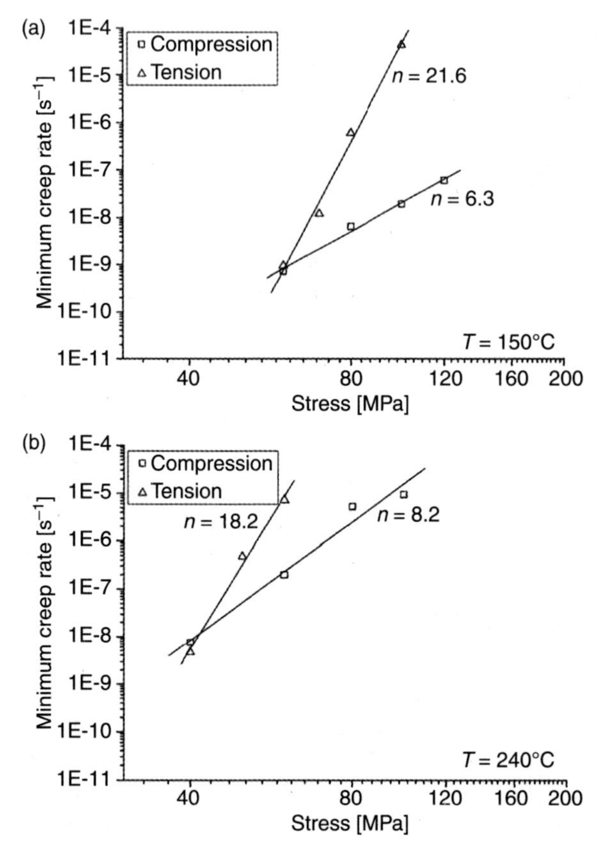

2.5 Norton plots for AE42 alloy showing the divergence of strain rates under compression and tension at two different temperatures (adapted from H. Dieringa *et al.*[144]).

interactions can come into play and different tests tailored to the specific stress conditions need to be conducted. The service conditions exerted on a magnesium alloy, for example, in an engine block application, involve multi-axial and cyclic stress conditions, corrosive medium, inhomogeneous and cyclic temperatures. Mondal *et al.*[157] attributed the higher strain levels in cyclic tests to the reappearance of the primary creep at the beginning of each cycle.

Furthermore, a severe microstructural instability of the cast structure also plays a major role in determining the creep response of magnesium alloys.[155] Dynamic precipitation reactions often come into play, further complicating the evaluation of creep properties of magnesium alloys. Thus, the already questionable assumptions of conventional creep tests such as homogenous and constant stress-temperature distributions, uniaxially applied stress, non-corrosive test atmosphere may be even less representative of the conditions that apply to the case of magnesium applications due to the additional complexity brought about by the inherent microstructural instability and lack of a self-healing surface oxide in magnesium. This instability problem will be further discussed later.

When assessing magnesium and its alloys in terms of creep mechanisms, particular emphasis must be placed upon the creep modes corresponding to medium to high stresses, and low to moderate temperatures. This selective attention is mainly due to the service conditions imposed by the perceived use of magnesium alloys in automobile applications, where such stress and temperature conditions prevail. However, it should be borne in mind that some magnesium alloys also serve in more severe conditions, at least in terms of temperature, such as nuclear canning applications.[149]

Despite the fact that magnesium alloys are generally recognized as materials with relatively low creep resistance, one magnesium alloy in the magnesium–thorium system, namely HZ22, shows the highest ratio of service temperature (350°C) over melting temperature compared to any material except some superalloys.[159] Although this alloy system is no longer exploited due to the radioactivity of thorium, the example it sets may be taken to indicate the true potential of magnesium alloys for future applications if a proper alloy design is achieved.

2.6.2 Low stress creep regimes

A deformation mechanism map for pure magnesium was first established by Frost and Ashby.[160] Watanabe *et al.*[161] showed in their work on deformation mechanism maps of fine-grained Mg alloys the existence of a critical grain size above and below which lattice diffusion and grain boundary diffusion, respectively, govern GBS. Kim *et al.*[162] further developed deformation mechanism maps for AZ61 and AZ31 alloys by normalizing the stress levels for high temperatures.

Three different creep mechanisms, although not specific to magnesium, may be encountered depending on the material and test conditions in the low stress region. These mechanisms are diffusion creep, Harper–Dorn creep,[163] and GBS.[164,165] Diffusion creep is further subdivided into Nabarro–Herring creep operating through bulk diffusion at relatively higher temperatures[166,167] and Coble creep operating through boundary diffusion at relatively lower temperatures.[168]

Vacancies diffuse from grain boundaries subjected to tensile stress to boundaries subjected to compression with a corresponding counter flow of atoms. This mechanism, independent of the alloy type, forms the basis of *diffusion creep*, leading to elongation of grains in the direction of the applied stress. When this flow occurs through the grain interiors, that is, lattice diffusion, the case corresponds to Nabarro–Herring creep, and if through grain boundaries then to Coble creep. If diffusion creep is operating, little or no primary creep is observed, and this mode is not associated with metallographic evidence such as slip lines.[150]

Although the grains retain their relative positions during diffusional creep, it is recognized that some sliding movement is necessary to maintain the integrity of the polycrystalline structure, without an eventual increase in the number of grains along the applied stress direction.[169] This accommodation process was first observed by Lifshitz[170] and later by others.[171,172]

As a result of diffusional activity, particle-denuded zones can form in certain alloy systems, such as Mg–Zr (Mg-0.5 wt% Zr alloy, ZA)[173,174] and Mg–Mn[175] along grain boundaries perpendicular to the stress axis during creep. Originally, these denuded zones were attributed to diffusional creep.[174–178] It has been shown that such denuded zones, though not so pronounced and characteristic in terms of their location as in creep, may also be observed after annealing.[179–181] Dissolution of precipitates at migrating grain boundaries via GBS is an alternative explanation for the formation of these denuded zones.[182,183]

Harper–Dorn creep[163] occurs at even higher temperatures, that is, close to melting point, leading to much faster creep rates than those expected in true diffusion creep.[184,185] Therefore, it is explained in terms of a viscous flow of dislocations. A change in the shape of the grains is again inevitable in this mode of creep as well. However, the change in the shape of the grains is not accompanied by a displacement of grains.[169] This mode of creep has, to the best knowledge of the author, not been reported for magnesium alloys.

Creep via grain boundary sliding

It has long been known that grain boundary sliding (GBS) is a major creep mechanism for magnesium alloys,[186] especially at the temperature and stress conditions they are currently being considered for, that is, under-bonnet automotive applications. Consequently, great research effort has been devoted to strengthening the grain boundaries in magnesium alloys by employing suitable precipitates (see Chapter 7). Under constant-creep conditions the ratio of strain due to GBS to the total strain remains constant, and this ratio increases with decreasing stress.[187] The process is attributed to a crystallographic deformation mechanism in the vicinity of the grain boundary, rather than a viscous flow of grains over each other. Koike *et al.*[188] suggested a tentative demarcation between the slip induced GBS and pure GBS in terms of temperature to be 150°C, above which grain boundary diffusion with activation energy of 80 kJ/mol operated. The movement of grains over each other leads to an increase in their number along the applied stress direction without a considerable elongation of the grains.[169] Such a displacement of grains is termed Rachinger sliding[189] and may be attributed to GBS in the absence of a considerable change in the grain aspect ratio during creep. This mechanism is akin to the superplastic behaviour of the fine-grained metals, which is well known in magnesium alloys.[190,191] It has been reported that the transition from GBS-controlled deformation to dislocation creep occurs at higher strain rates compared to aluminium alloys of equivalent grain size, approximately at 2×10^{-3} sn^{-1} and 5×10^{-4} sn^{-1}, respectively.[192]

Even though finer grain sizes are desired for magnesium alloys, to enhance many other mechanical properties, this leads to conditions that possibly facilitate grain boundary creep unless the grain boundaries are strengthened or anchored by intergranular precipitates. Casting techniques, such as HPDC, produce fine grain sizes. This process inherently produces bimodal grain size distributions as the solidification prematurely starts even before the liquid metal enters the mould[193] and grain sizes as different as 0.5–50 μ can exist in the same die-cast structure.[194] Furthermore, as a result of rapid solidification rates, metastable microstructures are produced. Due to this, together with the large freezing range of alloys like AZ91, microstructural instability becomes an inherent problem of die-cast alloys, complicating the creep behaviour of the material and its interpretation. Microstructural instability manifests as the occurrence of a discontinuous precipitation reaction of $Mg_{17}Al_{12}$. Dargush *et al.*[194,195] have suggested that the low resistance to GBS is related to the discontinuous precipitation of $Mg_{17}Al_{12}$ in the Al-enriched peripheries of the grains. Since this reaction requires the movement of high-angle boundaries it is reasonable to assume that the reaction also facilitates grain sliding and migration during creep. It should also be borne in mind that, even at moderate temperatures, such regions of the microstructure operate at relatively higher homologous temperatures compared to the grain interiors, thus leading to more diffusion in these regions. In support of this view, indeed, abundant occurrence of the discontinuous reaction in such regions has been observed after creep. Therefore, its association with GBS during creep should be regarded as highly possible.[196]

The information given above regarding the GBS, on the other hand, does not mean that grain boundaries should constitute the sole concern for the alloy designers. It has been shown that strengthening the α-Mg matrix by solid solution and/or precipitation can be more important than grain boundary reinforcement by intermetallic phases for potentially good creep-resistant alloys containing RE elements.[155,197]

2.6.3 Effect of precipitates

The major second phase in the earlier popular Mg–Al based alloys, such as AZ91, is the β-($Mg_{17}Al_{12}$) phase that exists in several distinct morphologies (Fig. 2.6). These are grain boundary blocky particles, lamellae of the colonies of discontinuous precipitation, irregular ones in the α constituent of the cellular nodules, and two types of intragranular continuous precipitates, that is, plates and plaques.[198]

The role of the β-$Mg_{17}Al_{12}$ in creep has been discussed since the 1950s.[186,199] It seems reasonable to assume that the β-phase softens easily, due to its low melting point (~460°C) and its partly metallic bonding, and, therefore, cannot serve to inhibit GBS. Contrary to this view, it was found that thinner

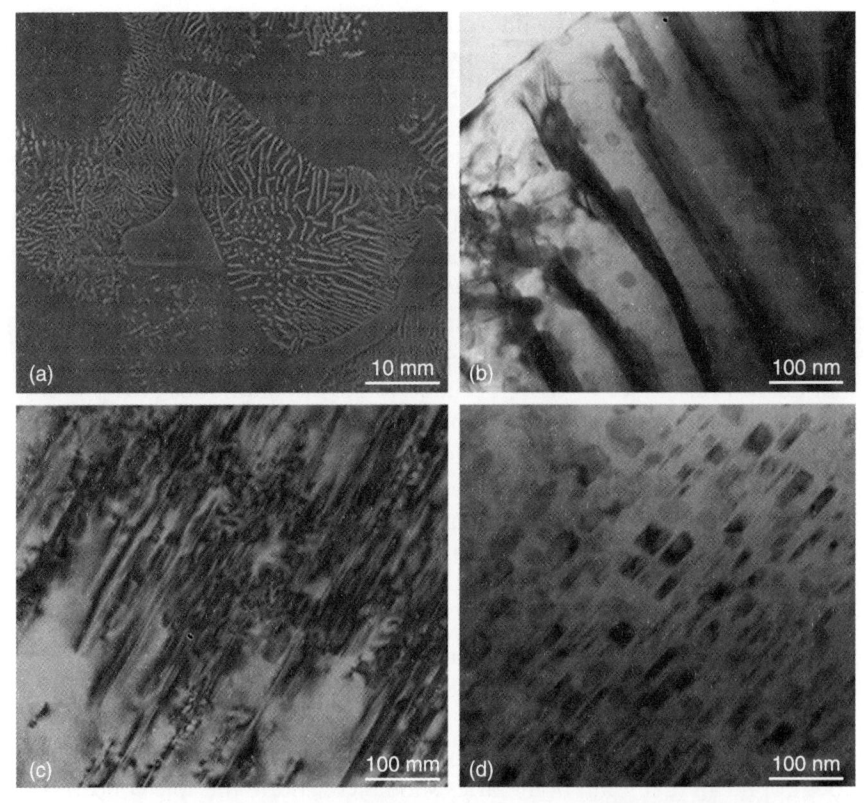

2.6 The major second phase in Mg–Al alloys, β-($Mg_{17}Al_{12}$), exists in several distinct morphologies in AZ91: (a) grain boundary blocky particles, (b) the lamellar colonies of discontinuous precipitation, and the irregular ones in the α_{Mg} constituent of these cellular nodules; and two types of intragranular continuous precipitates, (c) plates and (d) plaques.[198]

sectioned die-cast specimens containing finer and more β-phase showed greater creep strength. Furthermore, heat treatment promoting β precipitation prior to creep seemed to improve creep strength.[195] Although, it has been shown[200] on bulk $Mg_{17}Al_{12}$ that β-phase, due to its strong covalent bonding, would not soften markedly below 260°C, $Mg_{17}Al_{12}$ has very complex bonding showing a mixture of homopolar, heteropolar and normal metallic bonding, where softening of the compound cannot be completely excluded.[186,201] Hence precipitation of the $Mg_{17}Al_{12}$ phase, as a process causing diffusion, during creep and not its existence in the pre-creep structure, seems to be the main factor in creep deformation of Mg–Al alloys.

The alloys that have been developed more recently do not rely on the questionable precipitates for elevated temperature strengthening. Instead, many of those alloys derive their strength from Ca, RE, Y, Sn and Sr containing intermetallic precipitates. Since a chapter of this book has been devoted to the topic of the alloying behaviour of magnesium, our discussion of the topic is hereby concluded.

2.6.4 Dislocation creep and its interpretation based on *Q* and *n*

It is known that at sufficiently high stresses dislocation glide takes place regardless of the temperature. However, stress levels that would not cause ordinary dislocation glide but are still higher than those required by the above-mentioned modes lead to a different dominant mechanism known as dislocation creep. At such stress levels, if temperatures are sufficiently high (at about 1/3 of T_m), dislocations are no longer confined to glide motion only in their original slip planes. Creep rates are essentially controlled by diffusion, and, hence, vacancy creation in this temperature range. Although creep deformation essentially takes places through dislocation movement on basal or non-basal planes, this mode of creep involves different rate-controlling micro-mechanisms, since the dislocations surmount the barriers on their path via short-range diffusion at each step. Due to thermal activation (creation of vacancies) they can climb out of their slip planes via vacancy diffusion or cross-slip, and thus overcome obstacles. Therefore, dislocation creep may also be described as thermally activated non-conservative motion of dislocations. The climb motion mentioned may be achieved via two different vacancy diffusion mechanisms at the same stress level, namely volume diffusion at higher temperatures and dislocation core diffusion at lower temperatures.[202] This allows the dissipation of interlocked dislocation networks as a recovery process. The deformation process is, in this case, a combination of climb and glide motions, glide being responsible for strain.[203] Although it is assumed that the contribution of climb to the steady-state creep rate is zero, the deformation rate may not be solely due to glide, as some energy may also be released by climb movement of dislocation segments[204] and the activation energy (e.g., involving the vacancy creation and the movement of a jog) would reflect this combined mechanism. It should further be pointed out that the steady-state creep rate is affected by stress through increasing dislocation velocity[205] and dislocation density.[52]

In such successive processes, namely dislocation climb and viscous glide (also referred to as recovery), the slower one controls the steady-state creep rate, giving rise to the different creep behaviour of pure metals and alloys.[148,150] In pure metals, as a result of recovery process, an internal sub-grain structure forms. Alloys, on the one the hand, are classified as Class-I and Class-II. The creep behaviour of the Class-I alloys is similar to that of pure metals. In Class-II alloys, however, some form of viscous drag process is assumed to operate on dislocations, due to the presence of solute atoms, with the result that the rate of glide motion is slower than the rate of climb. Glide motion, therefore, becomes the rate-controlling mechanism with an activation energy equal to that of solute diffusion,[206] and in such cases sub-grain formation is generally not observed. The rate-controlling mechanism of dislocation creep is thus understood as the dragging of solute atmosphere around a dislocation in the lower stress range and, with the climb of dislocations, in the higher stress range.[148]

Operation of a particular rate-controlling mechanism of dislocation creep can vary depending on the alloy, heat treatment, stress and temperature. Positive identification of a specific rate-controlling mechanism, though often difficult, is essentially made on the basis of Q and n. The stress exponent, n, may be more indicative of a particular rate-controlling mechanism within the stress–temperature domain of dislocation creep,[196] provided that the creep test stress is below that of yield and that long-range plastic deformation is absent. Although there appears to be disagreement in the literature on the values of the activation energy and stress exponent[144,155] these parameters are often taken to indicate a predominant rate-controlling creep mechanism.

Proliferation of Q and n values reported in the literature may be due to the sample or test conditions. Activation energy calculations can be complicated in the case of thermally unstable systems as is generally the case for cast or precipitation hardened magnesium alloys usually suffering from non-equilibrium solidification and metastable phase formation. It is conceivable that microstructural instabilities may induce or remove barriers to dislocation movement, for example, formation or coarsening of precipitates, thus changing the creep strain rate.[207] At low stresses and high temperatures creep rates can decrease due to grain coarsening during the test, and stress exponent may slightly increases due to the same reason. Some experimental scatter in the values of Q may also be attributable to concurrent grain growth.[150] Furthermore, there may exist decreasing levels of contribution to the total creep strain by GBS with increasing stress levels in the dislocation creep region.[187]

Activation energy is a structure sensitive property and therefore it is assumed that its value for dilute alloys approaches that of elements. Furthermore, the activation energy is assumed to be insensitive to changes in temperature and stress within the stress-temperature domain of a particular micro-mechanism controlling the creep rate.[186] However, when considering a wider span of stress and temperature, for a given temperature, the activation energy Q has a strong stress dependency, decreasing with increasing stress.[52,148,150] The general trend in HCP metals is that activation energy increases from values close to that of self-diffusion at ~0.5–0.6 T_m to higher values at ~0.7–0.8 T_m, T_m being the melting point in Kelvin.[150,208] For the same change in temperature range HCP metals also show a decrease in the stress exponent.

Dislocation creep in magnesium

Since dislocation creep is the main rate-controlling process under the service conditions (T, σ) of current magnesium applications, this section focuses specifically on this mechanism.

Pure magnesium shows little creep strain in the primary stage under low stress levels or at high temperatures.[150] The activation energy for the creep of pure magnesium was reported by Shi and Northwood[209] to be 106 $kJmol^{-1}$, in the temperature range of 150–250°C and at a stress range of 20–50 Mpa which is above the yield stress. Although this value is smaller than the activation energy for self-diffusion of magnesium, ~135 $kJmol^{-1}$, similar values of activation energies for the same temperature range were also reported earlier by others.[208] The difficulty in the interpretation of such figures may also be attributed to the seemingly high stress range that is above the yield stress of pure Mg in the temperature range.

Activation energies in the range of lattice self-diffusion of magnesium, that is, 135 ± 10 $kJmol^{-1}$, have also been reported in the literature for the temperature range of ~200–477°C.[150] Based on metallographic observations of straight slip lines across the grains, it was suggested that this high stress–temperature region corresponded to a deformation mechanism via basal slip. Although basal slip, based on an activation energy value of ~135 $kJmol^{-1}$ that was independent of stress, can be said to be controlled by lattice self-diffusion of magnesium, the micro-mechanism of creep still need be determined. In this regard, provided that the stress is below yield, the stress exponent n may be taken as an indication to distinguish between the two micro-mechanisms, that is, glide or climb-controlled basal slip. Since the process of glide-controlled dislocation creep is associated with $n = 3$,[206] the conclusion was made based on a higher stress exponent value of ~5, thus the specific rate-controlling process was identified as climb of dislocations, while Q is equal to the activation energy for solute diffusion.[150,208] This was further confirmed using the Equation [2.5] for dislocation climb.[150]

The stress exponent value n was also assessed by Jones and Harris[210] in terms of strain rate, and a lower value of 4.5 was determined for low strain rates ($\leq 10^{-6}$ s^{-1}) where dislocation climb is the rate-controlling mechanism. For higher strain rates where non-basal slip predominates, n was found to increase with increasing temperature, reaching a value as high as 8.

At temperatures above about 0.6 T_m, however, the creep behaviour changes and the rate-controlling mechanism is understood to be the cross-slip of dislocations as the apparent activation energy of creep becomes larger than that for diffusion. At this high temperature range and under high stress levels, an activation energy level of 220 ± 10 $kJmol^{-1}$ was found corresponding to cross-slip of dislocations from basal to prismatic planes, that is, non-basal slip.[148,150,208] Thus, it may be concluded for pure magnesium that, compared to cross-slip, climb occurs at lower temperatures and requires lower activation energies for a given stress level.

Admittedly, much of the reported data in literature appear to have been taken at stress levels that are above the yield stress of the subject magnesium alloys. This point should be taken into consideration when interpreting the results.

Dislocation creep in magnesium alloys

The overall creep behaviour of a dilute Mg alloy may be divided into low and high temperature ranges to facilitate understanding of the operating mechanisms.[148]

i. In the low temperature range, two independent mechanisms have been shown to operate in an Mg-0.8Al alloy, the transition taking place between 327°C and 477°C and at lower temperatures as the stress increased.[148] The change of mechanism from glide to climb-controlled creep within the lower temperature range manifests itself with a change in the associated stress exponent from about $n = 3$ at lower stresses to $n = 6$ at higher stress.
ii. At temperatures between 200°C and 477°C and at stresses below 10 MPa Class-II behaviour, that is, uniform distribution of dislocations and absence of sub-grain formation associated with a viscous glide process is observed. Extensive basal slip that may be observed metallographically, an activation energy of 140 $kJmol^{-1}$ that matches well the value for interdiffusivity of aluminium (143 $kJmol^{-1}$) in magnesium, a low stress exponent of 3, lack of significant instantaneous strain upon loading, and

a brief normal or inverted primary stage are all associated features of creep in this stress-temperature range.[209,211]

The experimental creep rates, however, were more than an order of magnitude slower than those predicted by the existing three theoretical models to calculate the creep rate based on a glide-controlled mechanism, that is, dislocation-solute interaction.[212–214]

Above the stress levels associated with $n = 3$ the activation energy remains the same, while the creep behaviour changes sharply. The unchanging value of activation energy while the creep mechanism is changing from viscous glide to climb can be rationalized via the small difference between the activation energies involved in each case, the interdiffusion of aluminium in magnesium (143 $kJmol^{-1}$)[211] when glide is operative, and self-diffusion of Mg atoms (135 $kJmol^{-1}$)[215] when climb is operative.

A higher stress exponent, $n = 6$, is attributed to dislocation climb.[148] This interpretation is supported by the observed creep rate, measurable instantaneous strain, extensive normal primary stage, formation of dislocation substructure and the agreement between the experimentally estimated and observed values of binding energy (~0.14 eV) between a dislocation and a solute atom.[148]

The high temperature behaviour in Mg-0.8Al alloy above 600–750 K is similar to pure magnesium. Extensive non-basal slip and an activation energy value that is significantly higher than those of lattice self-diffusion of Mg or interdiffusion of Al are observed depending on the applied stress level.[148]

As a rate-controlling creep mechanism at high temperatures, glide on non-basal planes is excluded since the existing theoretical models for this mechanism that are based on a single[216] or double[212] kink formation do not produce results in agreement with the experimental observations. On the other hand, cross-slip of screw dislocations from basal to prismatic planes according to Friedel's model[214] seems to be in excellent agreement with the experimental estimations in terms of the constriction energy which is equal to 160 $kJmol^{-1}$. A stress exponent of 4, measurable instantaneous strain, a normal primary stage, extensive non-basal slip and presence of free dislocations and some well-defined sub-boundaries altogether provided further evidence for the operation of cross-slip mechanism at high temperatures.[148]

A similar trend also seems to be valid for non-dilute magnesium alloys. It was shown for Mg–Y–Nd–Zr and Mg–Y–Nd alloys that, compared to cross-slip, climb occurs at lower temperatures requiring lower activation energies for a given stress level.[207,217,218]

The effects of dispersoids on creep behaviour are also interesting and complex.[52] Oxide dispersions in Mg powder metallurgy samples exhibit self-diffusion controlled basal slip at low temperatures (144–253°C) as the rate-controlling mechanism. This corresponds to the case where the stress exponent n shows mainly dependence on temperature, also indicating dislocation climb within low temperature range, with a value ranging from 7.5–10.[219] When the rate-controlling mechanism shifts to cross-slip at higher temperatures, n shows dependence on both stress and temperature with its value varying drastically, first decreasing with increasing stress from 15 to 7.5 and then increasing again to 18. This may be attributed to the fact that n has two contributing terms, one due to the activation energy Q which decreases with increasing stress, and a second one due to moving dislocation density which increases with increasing stress.[52]

2.7 Recrystallization and grain growth

Grain refinement and deformation are the two most effective strengthening mechanisms for magnesium alloys. However, magnesium presents us with yet another peculiarity related to grain size. Contrary to the behaviour of many other metals,[220] when strengthened via grain refinement, magnesium displays an associated ductility at sub-micron grain sizes due to room temperature GBS.[83] As discussed earlier in this chapter, grain size can also be the determining factor for the onset and/or extent of twinning as an important deformation mechanism, and in turn, for the evolution of deformation textures. The success of shaping processes that subject pre-deformed starting materials to further hot deformation intimately depends on a thorough understanding of grain size related phenomenon, that is, recrystallization, which would determine the crystallographic texture and mechanical anisotropy of the product. Recrystallization is an important mechanism for reducing/eliminating texture created by deformation processes. A creep mechanism that is often encountered, GBS, is also primarily related to the grain size of magnesium-based materials. Finally, magnesium and its alloys exhibit superplasticity, also depending on the strain rate, below a grain size of approximately 8 μ size. Thus, it seems that there are many reasons to make an attempt to discuss the static or dynamic processes involved in changing the grain size.

As in other metals, magnesium also obeys the Hall–Petch (H–P) type strengthening by grain refinement. However, there are several grain size related phenomena that are particularly important in the case of magnesium and its alloys. In conventional H–P behaviour strength and grain size are related through the following equation:

$$\sigma = \sigma_0 + kd^{-\alpha}$$

where σ is the yield strength, σ_0 and k are material constants independent of grain size, and the exponent α is typically between 0.5 and 1.[221,222] Of the two competitive deformation mechanisms, twinning, being suppressed as the temperature increases, is more dependent on grain size than slip. Therefore, two separate k factors may be defined. Thus, it is possible to observe a transition point in the strength versus $d^{-1/2}$ plot where twinning is suppressed due to grain refinement.[82,221,223]

Refinement of grain size requires deformation followed by recovery and recrystallization. The recovery stage, however, is not inevitable in all metals; in its absence the process is defined as 'discontinuous recrystallization'. If recrystallization is preceded by recovery then the whole process is known as 'continuous recrystallization', and if not 'discontinuous recrystallization'.

During the precursor to the recrystallization, that is, the recovery process, it is assumed that the high-angle boundaries do not move while dislocations are annealed out and/or rearranged into low-angle boundaries with an accompanying softening. This stage is conventionally termed as polygonization. However, this change in the defect structure depends very much on the SFE level of the system. In the case of low SFE metals like magnesium, grouping of dislocations in this manner is difficult due to large dislocation core structures and associated strain fields. Since recovery is a difficult process in magnesium because of this, the stored energy release is left to the operation of recrystallization stage. If this effect is considered to couple with the high self-diffusion of magnesium, fast dynamic recrystallization rates of magnesium systems can be understood.[224–227]

Discontinuous recrystallization, on the other hand, is defined as a nucleation and a growth process. The second stage, by definition, occurs by migration of high-angle boundaries. However, during the recrystallization stage in some magnesium alloys, movement of low-angle twin boundaries has also been observed.[228] This was interpreted as the equivalent of polygonized grain boundary movement observed in continuous recrystallization, and therefore the whole recrystallization process was somewhat erroneously considered as CDRX in some studies.[224,228–230] The CDRX process may be realized if the SFE becomes sufficiently high, thus enabling dislocations of smaller core width to constitute low-angle boundaries at the polygonization stage.

The processes of deformation and recovery/recrystallization can be conducted as separate steps. If the route of separate stages is pursued, the process is termed as 'static recrystallization' (SRX). On the other hand, the whole sequence of these processes can occur simultaneously during hot deformation. Then the grain refinement takes place through what is known as 'dynamic' (DRX) or 'metadynamic' (MDRX) recrystallization, the latter occurring during post-hot work. If appropriate, a prefix of 'continuous' or 'discontinuous' can also be used before these terms. The desired result, always, is to obtain an isotropic material with fine equiaxed grain morphology.

Since the deformation of magnesium involves sequential mechanisms in terms of both temperature and accumulation of strain, it follows that the rate or ease of recrystallization and its mechanism should change accordingly. We can thus infer that until a pronounced non-basal slip is activated, that is, about at 200°C, the mechanism of recrystallization should be dependent on the mechanisms that produce the strain up to that point, that is, basal slip and twinning. When the activity of <*c* + *a*> is also added to the former mechanisms, that is, above 200°C, at least the rate, if not the mechanism, of recrystallization may be expected to change. The results of the study by Ion *et al.*[226] may thus be interpreted as new grain formations along the former grain boundaries, which may be named as 'necklace' formation (further discussed below) or as 'rotation crystallization'. Thus, coupled with the increased diffusion rate as the temperature is increased, the rate and mechanism of recrystallization may again be expected to change in relation to the facilitated creation/movement of <*c* + *a*> dislocations as temperatures approach 300°C. Yet, one other consequence of increased temperature in precipitate forming alloys of magnesium would be due to the function of precipitates as nucleation sites for the formation of new grains (also see below).

Where dislocations increase in number, that is, near parent grain boundaries, or at the twin boundaries, such defect substructures are, by definition, potential nucleation sites for new grains. As a major factor in magnesium deformation, twins, the formation of which is extended up to 450°C,[231] are expected to be involved in recrystallization processes. We will now go into further details on these points below.

As discussed earlier, the basal, prismatic and pyramidal planes of magnesium have very different SFE values from one another, not to mention that prismatic and pyramidal planes have SFE values that are essentially higher than that of aluminium. As we have already associated the SFE of a metal to the stages of a full recrystallization process, we may then differentiate between the behaviour the low SFE planes and the high SFE planes. Indeed, Somjeet *et al.*[232] suggested that during the high temperature deformation, while basal planes presented DDRX, the prismatic and pyramidal planes expectedly promoted CDRX.

Enhanced slip activity would be expected to accumulate a greater amount of internal energy due to an increase in dislocation density. Given such a condition, the resulting DRX would be expected to be faster. Indeed, it has been established that under conditions that all three slip mechanisms, basal, prismatic and pyramidal, are activated, DRX is promoted.[230] Barnett *et al.*[233] indirectly supported the same conclusion in their study on a textured AZ31 material, where prismatic slip was dominant, resulting in delayed DRX. While some other studies[234,235] showed that, when extensive twinning took place, recrystallization was further promoted, as the twin boundaries

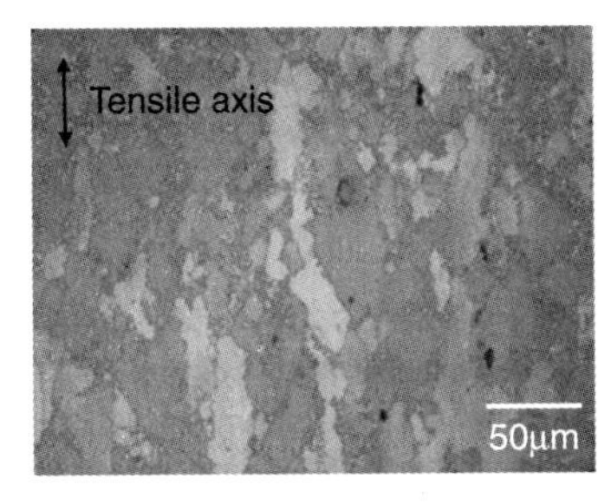

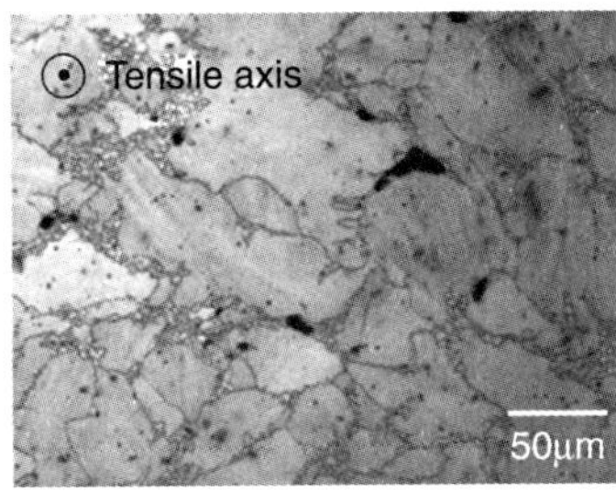

2.7 Recrystallization in the form of 'necklace' formation along the grain boundaries in AZ31rod stretched 30% at 300°C.

provided nucleation sites for the newly forming grains. The same authors also showed that the material was more resistant to DRX when high strain rates were employed. They attributed this observation to inhibition of the associated diffusion, as well as to less pronounced decrease of CRSS for pyramidal slip.

An interesting feature associated with recrystallization of certain magnesium alloys is that a fine recrystallized grain structure called 'necklace' forms along grain boundaries and, to a lesser degree, at twin boundaries (Fig. 2.7).[98,236] As mentioned earlier, plastic compatibility stresses creates non-basal dislocation activity in the vicinity of grain boundaries and twin boundaries, within several microns.[237] Thus, the 'necklace' morphology can be considered as an outcome of the plastic anisotropy within especially large grains and the plastic incompatibility at grain boundaries. Since necklace structure does not expand well into the interiors of grains, it creates inhomogeneity in grain morphology. This affects further refinement of grains and also plastic forming of the alloy. Necklace regions are said to facilitate deformation, based on the observed ductile shear zones in these locations.[226,227] It may also be speculated that if a well-developed necklace structure exist, it may provide additional deformation mechanism for large grained magnesium-based materials due to grain sliding.[224]

Another important issue in recrystallization is that the DRX grain size dependence on deformation temperature, rate and initial texture. It was found that DRX grain size of the AZ31 alloy was more sensitive to the deformation temperature and deformation rate rather than initial crystallographic texture, higher temperatures and deformation rates giving larger grain sizes.[98] However, this is not to say that initial texture does not exert a pronounced influence on DRX. Wang *et al.*[238] compared the development of DRX in AZ31 having two different textures, and showed that when only {10–12} extension twins exist the process was delayed, while double {10–11}–{10–12} twinning greatly promoted it. The same study also showed that accelerated DRX in case of double twinned structure was attributed to further accumulation of stored energy due to additional dislocation glide within the double twinned regions.

DRX is also an inevitable phenomenon in twin-roll casting of magnesium alloy sheet material, which is a promising popular industrial process. The process creates somewhat deformed cast structures, and in some recent studies, Masoumi *et al.*[106,107] reported weakened texture in twin-roll cast AZ31 sheet material due to the inherent non-equilibrium chemical composition of the cast structure.

Texture developed due to the alloying elements that strongly influence it, also have been shown to affect recrystallization. In a comparative study[239] on binary magnesium alloys containing Y and Zn, DRX was suppressed due to yttrium during rolling at 350°C. On the other hand, despite a similar prior texture structures, only Y-containing alloy exhibited texture weakening during SRX at 400°C with a much finer final grain size for equal annealing times, with Mg–Zn giving practically equal grain size as pure Mg. This may be interpreted on the basis of reduced SFE values in the presence of Y and Zn. It should be born in mind that SFE values differ depending on the plane of interest and present element, as revealed with the calculated values by Muzyk *et al.*[25] While γ_{USF} is reduced by both Y and Zn, γ_{SF} remains unchanged with Zn, with the result that Y facilitates slip on basal planes to a greater extent. Thus, the equal grain sizes attained via SRX in pure Mg and Mg–Zn, and smaller in Mg–Y can be rationalized. Likewise, twin formation is easier in Y-containing binary alloy as compared to Mg–Zn, while non-basal slip is promoted more with Zn. Consequently, it can be inferred that, due to kinetic reasons, DRX is delayed in Y-containing alloy, despite an expectedly higher twinning and dislocation content. Whereas, as a result of accumulated energy and availability of more nucleation sites during SRX, the initial texture was weakened.

The Zener–Holloman parameter is important in relating the strain rate and temperature to the development of grain size. Practically, this parameter combines the two independent parameters, that is, temperature and strain rate, into one, and thus facilitates comparative interpretations by using a unified term, that is, a single Z value. Although, modified expressions of it can be found in literature,[240,241] the original expression is in the form:

$$Z = \dot{\varepsilon} \exp\left(\frac{Q}{RT}\right)$$

where $\dot{\varepsilon}$ is strain rate, Q activation energy, which may be taken as the activation energy for the diffusion of the rate-controlling species, that is, self-diffusion in magnesium or that of the solute atom, R the gas constant (8.318 J/mol K) and T the deformation temperature. Thus, given the conditions of low Z value, it was found that pure magnesium develops larger grain sizes compared to that in alloys, for example, AZ31. The shortcoming of the evaluation solely based on *Z* parameter is that when precipitates exert their pinning effect on the grain boundary movement, or act as nucleation sites for recrystallized grains, then the expression is rendered inefficient.[228,242] When precipitation reactions are taking place in parallel to recrystallization,[243,244] as is the case of magnesium alloys containing precipitate forming elements such as Al, Ca, RE and Y, then the number of precipitates that form, depending on the temperature, deformation ratio and deformation rate, can be expected to influence the recrystallized grain size towards its reduction. Xu *et al.* reported that lesser amount of precipitation resulting from higher temperatures, lower strain rates and ratios led to coarser recrystallized grain sizes in several magnesium alloys during dynamic recrystallization.[245] Precipitates are also known to promote nucleation of new grains during static recrystallization as well, and to reduce pre-existing texture, leading to improvement of formability.[246,247] This phenomenon is known as particle-stimulated nucleation (PSN) in the recrystallization process.[248]

Twin boundaries behave in an interesting way during dynamic recrystallization for two reasons. Studies on both precipitate-free and precipitate-containing magnesium alloys have established that the twin boundaries were favourable sites for DRX. It is known that compression twin boundaries are sinks for basal dislocations, whereas tension twins repel basal dislocations and thus leading to pile up of dislocations at tension twin boundaries, eventually to recovery and recrystallization especially if the grain size is large. A micrograph from the report by Al-Samman and Gottstein[235] showing on-twin recrystallized grains is given in Fig. 2.8, confirming the similar observations by others.[236,245] It was also shown that the effect of twinning had a great span of temperature as revealed by DRX on {10–12} <10–1–1> twinning that formed in AZ61 extruded at 450°C.[231]

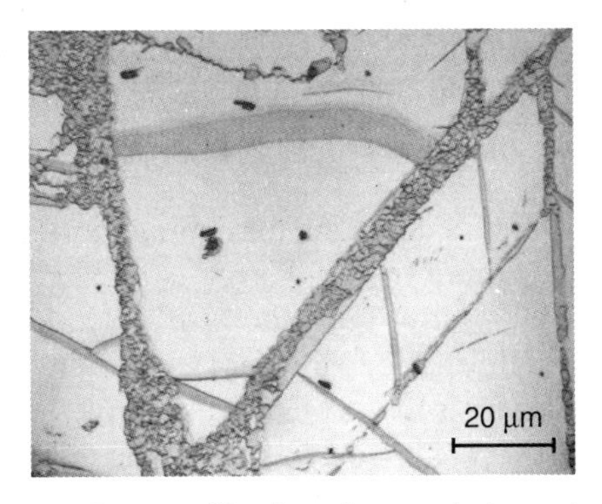

2.8 Recrystallization along twin boundaries in AZ31 deformed at 200°C/$10^{-2}s^{-1}$ up to $\varepsilon = -1.2$ (courtesy of T. Al-Samman and G. Gottstein[235]).

The second interesting point regarding especially the extension twin boundaries is that they are also mobile at the temperatures of recrystallization process. It is claimed that the immediate vicinity of twin boundaries on the un-twinned matrix side are associated with greater amount of dislocations, and this provides the boundary with the necessary driving force to propagate, and expand the twinned area. Furthermore, the formation of new extension twins is also possible.[249]

Texture prior to and following recrystallization in magnesium are, therefore, worth considering. It is reasonable to assume that extension twinning and dynamic recrystallization are, on the one hand, competing processes during high temperature deformation. On the other hand, twinning is also effective as a cooperative process to partition the parent grains into smaller sections. Albeit, both processes are not very effective to work against texture formation at least in wrought magnesium alloys that cannot form precipitates,[250] with the exception of yttrium-containing alloys.[251,252] It is particularly well known that the prior crystallographic texture does not markedly change during recrystallization of magnesium,[253] rendering it different from its rival light metal, aluminium.

Ideally, a recrystallization process is expected to give not only a small grain size with weakened texture but also a homogeneous size distribution. Samman and Gottstein[235] reported that the grain size and texture evolution at higher temperatures (400°C) were more strain-rate sensitive than at lower ones (200°C). The same authors pointed out the difficulty in analysing the recrystallization texture in terms of macro texture. However, *in situ* methods may be employed to elucidate whether recrystallized grain orientations obey the local texture, and if so, to what extent.

On the other hand, when grain size distribution is considered, Wang *et al.*[254] reported that it gets worse with increasing recrystallization temperature and annealing time. The same authors, based on a comparison of their modelling and experimental results, suggested two critical temperatures at least for AZ31 alloy, 250°C as the boundary where activation energy for grain boundary migration changes, and 300°C, above which variation of grain size becomes severe. Such an effect may not be observed in alloy systems of magnesium containing precipitates.

In an attempt to relate the grain refinement and hardening through deformation to the fundamental properties of metals, Kaveh *et al.*,[255] also based on the work of others, reported that steady-state grain size and hardness are achieved beyond which further deformation does not create any change. They have evaluated the threshold levels based on atomic bond type/energy, specific heat capacity, activation energy for self-diffusion, and SFE, using a large collection of related physical parameters of, as many as 22, pure metals gathered from the literature (see Table 2.1 of Reference 255). Their claim is that the threshold grain size apparently is not related to SFE but to the other parameters mentioned. Nevertheless, their approach to determination of grain sizes on various metals may warrant criticism.

A final point of interest regarding recrystallization of magnesium and its alloys is a novel process defined as the electropulsing effect. Xu *et al.*[256] showed electropulsing employed during recrystallization dramatically accelerated DRX even at a relatively low temperatures and high strain rates in AZ31 compared to a conventional process. However, industrial applicability of such a high energy process is debatable.

2.8 Future trends

In the light of the above, it seems reasonable to suggest that great effort should be devoted to alloy development to meet the demands of various potential applications, be it transport industry, aerospace, sporting goods, household appliances, etc. However, to speed up the process and to achieve specific goals within a reasonable time span, alloy development must rely more on the prior computer modelling approach, which focuses attention on estimating SFE and dislocation core structures via *ab initio* methods. Further development of such computer programs may just be possible and eventually lead to the development of more complicated, yet far more effective, alloys for future applications. On the other hand, while pursuing the conventional routes to developing new magnesium alloys, nitride or boride forming compositions may also be worth exploring in addition to the more conventional intermetallic formers. Needless to say, the more recent forming techniques as well as the conventional ones should be systematically studied further in order to determine the working envelope, that is, the boundaries of processing parameters for different magnesium alloys.

2.9 References

1. J. Li, MRS Bulletin, v.32, 2007, 151–159. www.mrs.org/bulletin.
2. M. Pekguleryuz, M. Celikin, M. Hoseini, A. Becerra, L. Mackenzie, 'Study on edge cracking and texture evolution during 150 C rolling of magnesium alloys: the effects of axial ratio and grain size', *J. Compd.*, **510**, 2012, 15–25.
3. M.J. Whelan, P.B. Hirsch, R.W. Horne, W. Bollmann, 'Dislocations and stacking faults in stainless steels', *Proc. Royal Soc.*, **240**(1223), 1957, 524–529.
4. J.R. Morris, J. Scharff, K.M. Ho, D.E. Turner, Y.Y. Ye, M.H. Yoo, *Phil. Mag. A* **76**, 1997, 1065–1072.
5. B. Li, P.F. Yan, M.L. Sui, E. Ma, 'Transmission electron microscopy study of stacking faults and their interaction with pyramidal dislocations in deformed Mg', *Acta Materialia*, **58**, 2010, 173–179.
6. M. Yamasaki, M. Sasaki, M. Nishijima, K. Hiraga, Y. Kawamura, 'Formation of 14H long period stacking ordered structure and profuse stacking faults

in Mg–Zn–Gd alloys during isothermal aging at high temperature', *Acta Materialia*, **55**, 2007, 6798–6805.
7. J.P. Hirth, J. Lothe, *Theory of dislocations*, Wiley Publishing, London, 1982.
8. J.W. Christian, 'A note on deformation stacking faults in hexagonal close packed Lattices', *Acta Cryst.*, **7**, 1954, 415–421.
9. N. Chetty, M. Weinert, 'Stacking faults in magnesium', *Physical Review B*, **56(17)**, 1997, 844–851.
10. A.E. Smith, 'Surface, interface and stacking fault energies of magnesium from first principles calculations', *Surface Science*, **601**, 2007, 5762–5765.
11. L. Wen, P. Chen, Z.-F. Tong, B.-Y. Tang, L.-M. Peng, W.-J. Ding, 'A systematic investigation of stacking faults in magnesium via first-principles calculation', *Eur. Phys. J. B*, **72**, 2009, 397–403.
12. H.Y. Wang, N. Zhang, C. Wang, Q.C. Jiang, 'First-principles study of the generalized stacking fault energy in Mg–3Al–3Sn alloy', *Scripta Materialia*, **65**, 2011, 723–726.
13. T. Uesugi, M. Kohyama, M. Kohzu, M. Higashi, 'Generalized stacking fault energy and dislocation properties for various slip systems in magnesium: a first-principles study', *Materials Science Forum*, **419–422**, 2003, 225–230.
14. D. K. Sastry, Y.V.R.K. Prasad, K.I. Vasu, 'On the stacking fault energies of some cph Metals', *Scripta Met.*, **3**, 1969, 927–933.
15. A. Couret, D. Caillard, 'An in situ study of prismatic glide in magnesium. II. Microscopic activation parameters', *Acta Metall.*, **33**, 1985, 1455–1462.
16. H. Gleiter, H.P. Klein, 'The stacking fault energy in the vicinity of a coherent twin boundary', *Philos. Mag.*, **27**(5), 1973, 1009–1026.
17. J.S. Hirschhorn, 'Stacking faults in the refractory metals and alloys – A review', *J. Less Common Metals*, **5**(6), 1963, 493–509.
18. M.E. Eberhart, *Acta Materialia.*, **44**, 1996, 2495–2501.
19. N. Kioussis, M. Herbranson, E. Collins, M.E. Eberhart, *Phys. Rev. Lett.*, **88**, 2002, 1255–12601.
20. S. Ogata, J. Li, S. Yip, *Science*, **298**, 2002, 807–812.
21. Q. Yue, K. M. Raja, 'Ab initio study of the effect of solute atoms on the SFE energy in aluminium', *Phys. Rev*. B, **75**, 2007, 224105.
22. Aditi Datta, U.V. Waghmare, U. Ramamurty, 'Structure and stacking faults in layered Mg–Zn–Y alloys: A first-principles study', *Acta Materialia*, **56**, 2008, 2531–2539.
23. V. Vitek, *Philos. Mag.*, **18**, 1968, 773–780.
24. U.V. Waghmare, E. Kaxirus, M.S. Duesbery, *Phys. Status Solid* B, **217**, 2000, 545–552.
25. M. Muzyk, Z. Pakiela and K.J. Kurzydlowski, 'Generalized stacking fault energy in magnesium alloys: Density functional theory calculations', *Scripta Materialia*, **66**, 2012, 219–222.
26. P. Tang, L. Wen, Z. Tong, B. Tang, L. Peng, W. Ding, 'Stacking faults in B2-structured magnesium alloys from first principles calculations', *Comp. Mater. Sci.*, **50**, 2011, 3198–3207.
27. S. Kibey, J.B. Liu, D.D. Johnson, H. Sehitoglu, *Appl. Phys. Lett.*, **89**, 2006, 1911–1916.
28. S. Kibey, J.B. Liu, M.J. Curtis, D.D. Johnson, H. Sehitoglu, *Acta Materialia.*, **54**, 2006, 2991–2996.

29. N.I. Medvedeva, O.N. Mryasov, Y.N. Gornostyrev, D.L. Novikov, A.J. Freeman, *Phys. Rev.* B, **54**, 1996, 13506.
30. B.Q. Shi, R.S. Chen, W. Ke, 'Solid solution strengthening in polycrystals of Mg–Sn binary alloys', *J. Alloy Compd.*, **509**, 2011, 3357–3362.
31. H. Liu, Y. Chen, Y. Tang, S. Wei, G. Niu, 'The microstructure, tensile properties, and creep behavior of as-cast Mg–(1–10)%Sn alloys', *J. Alloy Compd.*, **440**, 2007, 122–126.
32. R. Wang, S.F. Wang, X. Z. Wu, Q. YiWei, 'First-principles determination of dislocation properties in magnesium based on the improved Peierls–Nabarro equation', *Phys. Scr.*, **81**, 2010, 065601–07.
33. J.A. Yasi, T. Nogaret, D.R. Trinkle, Y. Qi, L.G. Jr. Hector, W.A. Curtin, *Modelling Simul. Mater. Sci. Eng.*, **17**, 2009, 055012.
34. Z.S. Basinski, M.S. Duesbery, R. Taylor, *Can. J. Phys.*, **48**, 1970, 1480–86.
35. M.H. Liang and D.J. Bacon, *Phil. Mag.* A, **53**, 1986, 163–181.
36. A. Serra, R.C. Pond, D.J. Bacon, *Acta Metall.*, **39**, 1991, 1469–1475.
37. M. Igarashi, M. Khantha, V. Vitek, *Phil. Mag.* B, **63**, 1991, 603–610.
38. B.L. Mordike, P. Lukáč, 'Physical metallurgy', in *Magnesium Technology, Metallurgy, Design, Data, Applications*, H.E. Friedrich, B.L. Mordike (eds), Springer-Verlag Berlin Heidelberg, ISBN-10 3–540–20599–3, 2006, 63–105.
39. A. Kelly, G.W. Groves, *Crystallography and Crystal Defects*, Addison-Wesley, Reading, MA, 1970, 163.
40. M. Gharghouri, G. Weatherly, J. Embury, J. Root, *Phil. Mag.* A, **79**(7), 1999, 1671–1695.
41. J. Koike, T. Kobayashi, T. Mukai, H. Watanabe, M. Suzuki, K. Maruyama, K. Higashi, 'The activity of non-basal slip systems and dynamic recovery at room temperature in fine-grained AZ31B magnesium alloys', *Acta Materialia*, **51**, 2003, 2055–2065.
42. J.W. Jensen, *Metalscope,* 1965, 7–10.
43. J.C. Kaufman, 'Damping of light metals', *Materials in Design Engineering*, **56**(2), 1962, 104–112.
44. A. A. Baikov, 'Institute of Metallurgy. Translated from Metallovedenie i Termicheskaya Obrabotka', *Metallov*, **11**, 1970, 48–51.
45. G. Fantozzi, C. Esnouf, S.M. Seyed Reihani, G. Revel, *Acta Metall.*, **32**, 1984, 2175–1281.
46. A. Granato, K.Lücke, *J. Appl. Phys.*, **27**, 1956, 583–587.
47. A.E. Schwaneke, R. W. Nash, 'Effect of preferred orientation on the damping capacity of magnesium alloys', *Metallurgical Transactions*, **2**, 1971, 3453–3460.
48. M.A. Kulas, W.P. Green, E.M. Taleff, P.E. Krajewski, T.R. McNelley, *Met. Mat. Trans.* A, **36**, 2005, 1249–1255.
49. W.B. Hutchinson, M.R. Barnett, 'Effective values of critical resolved shear stress for slip in polycrystalline magnesium and other HCP metals', *Scripta Materialia*, **63**, 2010, 737–740.
50. M.R. Barnett, 'Twinning and the ductility of magnesium alloys: Part I: Tension twins', *Mater. Sci. Eng. A*, **1**–2(464), 2007, 1–7.
51. M.R. Barnett, 'Twinning and the ductility of magnesium alloys: Part II.: Contraction twins'. *Mater. Sci. Eng.* A, **464**(1–2), 2007, 8–16.
52. K. Milicka, J. Cadek, P. Rys, 'High temperature creep mechanisms in magnesium', *Acta Metal.*, **18**, 1970, 1071–1082.

53. G. Neite, K. Kubota, K. Higashi, F. Hehmann, 'Magnesium-based alloys', vol. **8**, Chp. 4 in 'Structure and properties of nonferrous alloys', volume ed. K.H. Matucha, *Materials Science and Technology series*, R.W. Cahn, P. Haasen, E.J. Kramer (eds), VCH, 1996, 113–213.
54. E. Schmid, W. Boas, *Plasticity of Crystals*, Chapman & Hall Ltd., London., 1968, 55–76.
55. D.W. Brown, S.R. Agnew, M.A.M. Bourke, T.M. Holden, S.C. Vogel, C.N. Tomé, 'Internal strain and texture evolution during deformation twinning in magnesium', *Mater. Sci. Eng. A*, **399**(1–2), 2005, 1–12.
56. O. Muránsky, D.G. Carr, P. Šittner, E.C. Oliver, 'In situ neutron diffraction investigation of deformation twinning and pseudoelastic-like behaviour of extruded AZ31 magnesium alloy', *Int. J. Plasticity*, **25**, 2009, 1107–1127.
57. S.R. Agnew, D.W. Brown, C.N. Tomé, 'Validating a polycrystal model for the elasto-plastic response of magnesium alloy AZ31 using in-situ neutron diffraction', *Acta Materialia*, **54**(18), 2006, 4841–4852.
58. S.R. Agnew, C.N. Tomé, D.W. Brown, T.M. Holden, S.C. Vogel, 'Study of slip mechanisms in a magnesium alloy by neutron diffraction and modelling', *Scripta Materialia*, **48**(8), 2003, 1003–1008.
59. L. Wu, A. Jain, D.W. Brown, G.M. Stoica, S.R. Agnew, B. Clausen, D.E. Fielden, P.K. Liaw, 'Twinning–detwinning behavior during the strain-controlled low-cycle fatigue testing of a wrought magnesium alloy, ZK60A', *Acta Materialia*, **56**, 2008, 688–695.
60. L. Wu, S.R. Agnew, Y. Ren, D.W. Brown, B. Clausen, G.M. Stoica, H.R. Wenk, P.K. Liaw, 'The effects of texture and extension twinning on the low-cycle fatigue behaviour of a rolled magnesium alloy, AZ31B', *Mater. Sci. Eng. A*, **527**, 2010, 7057–7067.
61. B. Clausen, C.N. Tomé, D.W. Brown, S.R. Agnew, 'Reorientation and stress relaxation due to twinning: Modeling and experimental characterization for Mg', *Acta Materialia*, **56**, 2008, 2456–2468.
62. J.F. Stohr, J. Poirier, *Phil.Mag.*, **25**, 1972, 1313–1319.
63. T. Obara, H. Yoshinaga, S. Morozumi, '{11–22} <-1–123> slip system in magnesium', *Acta Metall.*, **21**, 1973, 845–853.
64. L. Balogh, G. Tichy and T. Ungár, 'Twinning on pyramidal planes in hexagonal close packed crystals determined along with other defects by X-ray line profile analysis', *J. Appl. Cryst.*, **42**, 2009, 580–591.
65. W.F. Hosford, *The Mechanics of Crystals and Textured Polycrystals*, Oxford University Press, Oxford, 1993, 172.
66. J.W. Christian, S. Mahajan, *Prog. Mater. Sci.*, **39**, 1995, 1–9.
67. D.E. Cipoletti, A.F. Bower, P.E. Krajewski, 'A microstructure-based model of the magnesium alloy', *Scripta Materialia*, **64,** 2011, 931–934.
68. S. Ando, K. Nakamura, K. Takashima, H. Tonda, *J. Japan Institute Light Metals*, **42**, 1962, 765–772.
69. P.W. Flynn, J. Mote, J.E. Dorn, *Trans. AIME*, **221**, 1961, 1148–1155.
70. A. Akhtar, E. Teghtsoonian, *Acta Metall.*, **17**, 1969, 1351–1358.
71. C. Escaravage, P. Bach, G. Champier, in: *Proc. 2nd International Conference on the Strength of Metals and Alloys*. ASM, Pacific Grove, 1979, 299–303.
72. A. Urakami, A. Meshii, M.F. Fine, in: *Proc. 2nd International Conference on the Strength of Metals and Alloys*. ASM, Pacific Grove, 1970, 272–276.
73. A. Ahmadieh, J. Mitchell, J.E. Dorn, *Trans. AIME*, **233**, 1965, 1130–1137.

74. S. Ando, H. Tonda, *Mater. Sci. Forum*, **43**, 2000, 350–351.
75. S. Ando, H. Tonda, *Mater. Trans. JIM*, **41**, 2000, 1188–1194.
76. M.R. Barnett, Z. Keshavarz, X. Ma, *Metall. Mater. Trans. A*, **37**, 2006, 2283–2290.
77. G. Proust, C.N. Tomé, A. Jain, S.R. Agnew, *Int. J. Plasticity*, **25**, 2009, 861–868.
78. T. Mayama, T. Ohashi, K. Higashida, Y. Kawamura, 'Crystal plasticity analysis on compressive loading of magnesium with suppression of twinning', *Magnesium Technology* 2011, W.H. Sillekens, S.R. Agnew, N.R Neelameggham and S.N. Mathaudhu (eds), TMS, 2011, 273–278.
79. S. Hong, S.H. Park, C.S. Lee, 'Strain path dependence of (10–12) twinning activity in a polycrystalline magnesium alloy', *Scripta Materialia*, **64**, 2011, 145–148.
80. A. Chapuis, J.H. Driver, *Acta Materialia*, **59**, 2011, 1986–1993.
81. O. Muránsky, M.R. Barnett, D.G. Carr, *Acta Materialia*, **58**, 2010, 1503–1517.
82. M.R. Barnett, Z. Keshavarz, A.G. Beer, D. Atwell, 'Influence of grain size on the compressive deformation of wrought Mg–3Al–1Zn', *Acta Materialia*, **52**, 2004, 5093–5103.
83. J. Koike, R. Ohyama, T. Kobayashi, M. Suzuki, K. Maruyama, *Mater. Trans.*, **44**, 2003, 1–7.
84. J. Koike, 'Enhanced deformation mechanisms by anisotropic plasticity in polycrystalline Mg alloys at room temperature', *Metall. Mater. Trans. A*, **36**(7), 2005, 1689–1696.
85. P. Yang, Y. Yu, L. Chen, W. Mao, 'Experimental determination and theoretical prediction of twin orientations in magnesium alloy AZ31', *Scripta Materialia*, **50**, 2004, 1163–1168.
86. Y.N. Wang, J.C. Huang, *Acta Materialia*, **55**, 2007, 897–905.
87. A. Becerra, M. Pekguleryuz, *J. Mater. Res.*, **23**(12), 2008, 3379–3386.
88. F.E. Hauser, P.r. Landon, J.E. Dorn, 'Deformation and fracture of alpha solid solutions of lithium in magnesium', *Trans. Am. Soc. Metals*, **50**, 1958, 856–863.
89. H. Li, E. Hsu, J. Szpunar, R. Verma, J. T. Carter, 'Determination of active slip/twinning modes in AZ31 Mg alloy near room temperature', *JMEPEG*, **16**, 2007, 321–326.
90. J. J. Jonas, S. Mu, T. Al-Samman, G. Gottstein, L. Jiang, E. Martin, 'The role of strain accommodation during the variant selection of primary twins in magnesium', *Acta Materialia*, **59**, 2011, 2046–2056.
91. J. Wang, J.P. Hirth, C.N. Tomé, '(-1012) Twinning nucleation mechanisms in hexagonal-close-packed crystals', *Acta Materialia*, **57**, 2009, 5521–5530.
92. M. Yoo, ' Slip, twinning and fracture in hexagonal-close packed metals', *Met. Trans. A*, **12**(3), 1981, 409–418.
93. H. Yoshinaga, R. Horiuchi, *Trans. JIM*, **4**, 1963, 1–8.
94. B.C. Wonsiewicz, W.A. Backofen, *Trans. Metall. Soc. AIME*, **239**, 1967, 422–431.
95. R.E. Reed-Hill, W.D. Robertson, *Acta Metall.*, **5**, 1957, 717–727.
96. M. Huppmann, M. Lentz, K. Brömmelhoff, W. Reimers, 'Fatigue properties of the hot extruded magnesium alloy AZ31', *Mater. Sci. Eng. A*, **527**, 2010, 5514–5521.
97. M. Huppmann, M. Lentz, S. Chedid, W. Reimers, 'Analysis of deformation twinning in the extruded magnesium alloy AZ31 after compressive and cycling loading', *J. Mater. Sci.*, **46**, 2011, 938–950.
98. T. Al-Samman, G. Gottstein, *Mater. Sci. Forum*, **539–543**, 2007, 3401–3406.

99. S.R. Agnew, O. Duygulu, 'Plastic anisotropy and the role of non-basal slip in magnesium alloy AZ31B', *Int. J. Plasticity*, **21**, 2005, 1161–1193.
100. Z. Keshavarz, M.R. Barnett, 'EBSD analysis of deformation modes in Mg–3Al–1Zn', *Scripta Materialia*, **55**, 2006, 915–918.
101. L. Jiang, J.J. Kpmas, A.A. Luo, A.K. Sachdev, S. Godet, 'Influence of {10–12} extension twinning on the flow behaviour of AZ31 Mg alloy', *Mater. Sci. Eng. A*, **452**, 2007, 302–309.
102. O. Muránsky, D.G. Carr, M.R. Barnett, E.C. Oliver, P.Šittner, 'Investigation of deformation mechanisms involved in the plasticity of AZ31 Mg alloy in situ neutron diffraction and EPSC modelling', *Mater. Sci. Eng. A*, **334**, 2008, 8–15.
103. A.A. Luo, A.K. Sachdev, in: A.A. Luo (Ed.), *Magnesium Technology 2004*, TMS, Warrendale, PA, 2004, 79–84.
104. M.R. Barnett, S. Jacob, B.F. Gerard and J.G. Mullins, 'Necking and failure at low strains in a coarse-grained wrought Mg alloy', *Scripta Materialia*, **59**, 2008, 1035–1038.
105. M. A. Meyers, O. Vohringer, V. A. Lubarda, *Acta. Materialia*, **49**, 2001, 4025–4032.
106. M. Masoumi, F. Zarandi, M. Pekguleryuz, *Scripta Materialia.*, **62**(11), 2010, 823–826.
107. M. Masoumi, F. Zarandi, M. Pekguleryuz, *Mater. Sci. Eng.* A, **528**, 2011, 1268–1279.
108. C.N. Tomé, S.R. Agnew, W.R. Blumenthal, M.A.M. Bourke, D.W. Brown, G.C. Kaschner, *Mater. Sci. Forum*, **408–412**, 2002, 263–268.
109. N. Thompson, D.J. Millard, *Philos. Mag.*, **43**, 1952, 422–429.
110. B. Li, E. Ma, 'Zonal dislocations mediating {10–11}<10–1–2> twinning in magnesium', *Acta Materialia*, **57**, 2009, 1734–1743.
111. D.G. Westlake, *Acta Metal*, **9**, 1961, 327–334.
112. S. Vaidya, S. Mahajan, *Acta Metal.*, **28**, 1980, 1123–1130.
113. G.E. Mann, T. Sumitomo, C.H.Cáceres, J.R. Griffiths, 'Reversible plastic strain during cyclic loading–unloading of Mg and Mg–Zn alloys', *Mater. Sci. Eng. A*, **456**, 2007, 138–146.
114. T. Sumitomo, C.H. Cáceres, M. Veidt, in: A.K. Dahle (Ed.), *First International Conference on Light Metals Technology*, CAST Centre Pty. Ltd., Brisbane, Australia, 2003, 349–351.
115. C.H.Cáceres, T. Sumitomo, M. Veidt, *Acta Materialia*, **51**, 2003, 6211–6218.
116. M. Matsuzuki, S. Horibe, 'Analysis of fatigue damage process in magnesium alloy AZ31', *Mater. Sci. Eng A*, **504**, 2009, 169–174.
117. S.D. Henry, G.M. Davidson, S.R. Lampman, F. Reidenbach, R.L. Boring, W.W. Scott, editors. *Fatigue data book: Light structural alloys*, Materials Park, OH., ASM International, 1995.
118. X.S. Wang, J.H. Fan, 'An evaluation on the growth rate of small fatigue cracks in cast AM50 magnesium alloy at different temperatures in vacuum conditions', *International Journal of Fatigue*, **28**, 2006, 79–86.
119. H. Mayer, M. Papakyriacou, B. Zettl, S.E. Stanzl-Tschegg, 'Influence of porosity on the fatigue limit of die cast magnesium and aluminium alloys', *International Journal of Fatigue*, **25**, 2003, 245–256.
120. T.S. Shih, W.S. Liu, Y.J. Chen, 'Fatigue of as-extruded AZ61A magnesium alloy', *Mater. Sci Eng.* A, **325**, 2002, 152–62.

121. S. Begum, D.L. Chen, S. Xu, A.A. Luo, *Int. J. Fatigue*, **31**, 2009, 726–735.
122. X.S. Wang, X. Lu, D.H. Wang, 'Investigation of surface fatigue microcrack growth behavior of cast Mg-Al alloy', *Mater. Sci. Eng. A*, **364**, 2004, 11–16.
123. J. Koike, N. Fujiyama, D. Ando, Y. Sutou, 'Roles of deformation twinning and dislocation slip in the fatigue failure mechanism of AZ31 Mg alloys', *Scripta Materialia* **63,** 2010, 747–750.
124. Y. Uematsu, K. Tokaji, M. Matsumoto, 'Effect of aging treatment on fatigue behaviour in extruded AZ61 and AZ80 magnesium alloys', *Mater. Sci. Eng. A*, **517**, 2009, 138–145.
125. Z.Y. Nan, S. Ishihara, T. Goshima, 'Corrosion fatigue behavior of extruded magnesium alloy AZ31 in sodium chloride solution', *Int. J. Fatigue*, **30**, 2008, 1181–8.
126. K. Tokaji, M. Nakajima, Y. Uematsu, 'Fatigue crack propagation and fracture mechanisms of wrought magnesium alloys in different environments', *Int J. Fatigue*, **31**, 2009, 1137–1143.
127. C.J. Bettles, M.A. Gibson, 'Current wrought magnesium alloys: strengths and weaknesses', *J. Miner Met. Mater. Soc.*, **57**, 2005, 46–49.
128. R. Zeng, E. Han, W. Ke, W. Dietzel, K.U. Kainer, A. Atrens, 'Influence of microstructure on tensile properties and fatigue crack growth in extruded magnesium alloy AM60', *Int. J. Fatigue*, **32**, 2010, 411–419.
129. S. Hasegawa, Y. Tsuchida, H. Yano, M. Matsui, 'Evaluation of low cycle fatigue life in AZ31 magnesium alloy', *Int. J. Fatigue*, **29**, 2007, 1839–45.
130. C.L. Fan, D.L. Chen, A.A. Luo,' Dependence of the distribution of deformation twins on strain amplitudes in an extruded magnesium alloy after cyclic deformation', *Mater. Sci. Eng. A*, **519**, 2009, 38–45.
131. Q. Li, Q. Yu, J. Zhang, Y. Jiang, 'Effect of strain amplitude on tension–compression fatigue behavior of extruded Mg6Al1ZnA magnesium alloy', *Scripta Materalia*, **62**, 2001, 778–81.
132. M. Huppmann, S. Stark, W. Reimers, *Mater. Sci. Forum*, **638–642**, 2010, 2411–2416.
133. D.W. Brown, A. Jain, S.R. Agnew, B. Clausen, 'Twinning and detwinning during cyclic deformation of Mg alloy AZ31B', *Mater. Sci Forum*, **539**–543, 2007, 3407–3413.
134. P. Zhang, J. Lindemann, *Scr. Mater.*, **52**, 2005, 485–490.
135. W. Liu, J. Dong, P. Zhang, C. Zhai, W.J. Ding, *Mater. Trans.*, **50**, 2009, 791–798.
136. R. E. Reed-Hill, *Physical Metallurgy Principles*, 2nd Edition, D. Van Nostrand, New York, 1973.
137. M. F. Ashby, D.R.H. Jones, *Engineering Materials I*, Pergamon Press, Oxford, 1980.
138. F.R.N. Nabarro, H.L., DeVilliers, *The Physics of Creep*, London, Taylor and Francis, 1995.
139. J. Weertman, 'Rate Processes in plastic deformation of materials', J.C.M. Li and A.K. Mukherjee (eds), ASM, Metals Park, Ohio, 1975, 315.
140. G. Garofalo, *Trans. AIME*, **227**, 1963, 351.
141. I. Rieriro, V. Gutiérrez, J. Castellanos, J. Muñoz, M. Carsí, M.T. Larrea, O.A. Ruano, *Metall. Mater. Trans.* A, 2010, doi10.1007/s11661–010–0259–6.
142. F.A. Mohamed, T.G. Langdon, *Acta Metall.*, **22**, 1974, 779–788.
143. Y. Li, T.G. Langdon, *Scripta Materalia*, **36**(12), 1997, 1457–1460.

144. H. Dieringa, N. Hort, K.U. Kainer, 'Investigation of minimum creep rates and stress exponents calculated from tensile and compressive creep data of magnesium alloy AE42', *Mater. Sci. Eng. A*, **510–511**, 2009, 382–386.
145. S. Spigarelli, M. El Mehtedi, 'Microstructure-related equations for the constitutive analysis of creep in magnesium alloys', *Scripta Materialia*, **61**, 2009, 729–732.
146. M. El Mehtedi, S. Spigarelli, E. Evangelista, G. Rosen, 'Creep behaviour of the ZM21 wrought magnesium alloy', *Mater. Sci. Eng. A*, **510–511**, 2009, 403–406.
147. C.J. Boehlert, K. Knittel, *Mater. Sci. Eng. A*, **417**, 2006, 315–321.
148. S.S. Vagarali, T.G. Langdon, 'Deformation mechanism in H.C.P. metals at elevated temperatures-II creep behaviour of a Mg-0.8%Al solid solution alloy', *Acta Metal.*, **29**, 1982, 1157–1170.
149. C. Brown, *J. Nuclear Mater.*, 12(2), 1964, 243–247.
150. S. S. Vagarali, T. Langdon, 'Deformation mechanism in H.C.P. metals at elevated temperatures-I creep behaviour of magnesium', *Acta Metal.*, **29**, 1981, 1969–1982.
151. M. Pekguleryuz, M. Celikin, 'Creep mechanisms in magnesium alloys', *Int. Mater. Rev.*, **55**(4), 2010, 197–217.
152. M.O. Pekguleryuz, A.A. Kaya, 'Creep resistant magnesium alloys for powertrain applications', *Adv. Eng. Mater.*, **5**, 2003, 866–878.
153. Y. Terada, D. Itoh, T. Sato, 'Dislocation movements during creep in a die-cast AM50 magnesium alloy', *Mater. Chem. Phys.*, **117**, 2009, 331–334.
154. A. Srinivasan, J. Swaminathan, M.K. Gunjan, U.T.S. Pillai, B.C. Pai, 'Effect of intermetallic phases on the creep behavior of AZ91 magnesium alloy', *Mater. Sci. Eng. A*, **527**, 2010, 1395–1403.
155. I.P. Moreno, T.K. Nandy, J.W. Jones, J.E. Allison, T.M. Pollock, 'Microstructural stability and creep of rare-earth containing magnesium alloys', *Scripta Materialia*, **48**, 2003, 1029–1034.
156. B. R. Powell, A. A. Luo, V. Rezhets, J. J. Bommarito, B.L. Tiwari, *SAE Tech. Paper Ser.* 2001-01-0422, 2001.
157. A.K. Mondal, D. Fechner, S. Kumar, H. Dieringa, P. Maier, K.U. Kainer, 'Interrupted creep behaviour of Mg alloys developed for powertrain applications', *Mater. Sci. Eng. A*, **527**, 2010, 2289–2296.
158. S. Xu, M.A. Gharghouri and M. Sahoo, 'Tensile-compressive creep asymmetry of recent die cast magnesium alloys', *Adv. Eng. Mater.*, **9**(9), 2007, 807–812.
159. B.L. Mordike, T. Ebert, *Mater. Sci. Eng. A*, **302**, 2001, 37–45.
160. H.J. Frost, M.F. Ashby, *Deformation Mechanism Maps*, Pergamon Press, Oxford, 1982.
161. H. Watanabe, T. Mukai, M. Kohzu, S. Tanabe, K. Higashi, *Acta. Mater.*, 1999, **47**, 1999, 3753–3759.
162. W. J. Kim, S. W. Chung, C. S. Chung, D. Kum, 'Superplasticity in thin magnesium alloy sheets and deformation mechanism maps for magnesium alloys at elevated temperatures', *Acta Materialia*, **49**, 2001, 3337–3345.
163. J. Harper, J.E. Dorn, *Acta Metall.*, **5**, 1957, 654–660.
164. H. Luthy, R.A. White and O.D. Sherby, *Mater. Sci. Eng.*, **39**, 1979, 211–217.
165. O.A. Ruano and O.D. Sherby, *Mater. Sci. Eng.*, **51**, 1981, 9–15.

166. F.R.N. Nabarro, in: *Report of a Conference on Strength of Solids*, The Physical Society of London, London, 1948, 75–80.
167. C. Herring, *J. Appl. Phys.*, **21**, 1950, 437–441.
168. R.L. Coble, *J. Appl. Phys.*, **34**, 1963, 1679–1684.
169. T.G. Langdon, *Mater. Sci. Eng. A*, **283**, 2000, 266–273.
170. I.M. Lifshitz, *Soviet Phys. JETP*, **17**, 1963, 909–914.
171. R. Raj, M.F. Ashby, *Metall. Trans.*, **2**, 1971, 1113–1119.
172. W.R. Cannon, *Phil. Mag.*, **25**, 1972, 1489–1495.
173. R. L. Squires, R. T. Weiner, M. Phillips, *J. Nucl. Mat.*, **8**, 1963, 77–82.
174. J. E. Harris, R. B. Jones, *J. Nucl. Mat.*, **10**, 1963, 360–366.
175. V. J. Haddrell, *J. Nucl. Mat.*, **18**, 1966, 231–235.
176. F.R.N. Nabarro, in: Mishra RS, Mukherjee AK, Murty KL, editors. *Proceedings of Symposium on Creep Behaviour of Advanced Materials for the 21st Century.* Warrendale: TMS, 1999, 391–395.
177. J.E. Harris, *J. Metal Sci.*, **7**, 1973, 1–6.
178. R.C. Gifkins, T.G. Langdon, *Scripta Metall.*, **4**, 1970, 563–569.
179. E.H. Aigeltinger, R.C. Gifkins, *J. Mater. Sci.*, **10**, 1975, 1889–1894.
180. B.W. Pickles, J. Inst . *Metals*, **95**, 1967, 333–339.
181. W. Vickers, P. Greenfield, *J. Nucl . Mater.*, 24, 1967, 249–260.
182. L. E. Rarety, *J. Nucl. Mater.*, **20**, 1966, 344–350.
183. J. Wadsworth, O.A. Ruano, O.D. Sherby, *Metall. Mater. Trans. A*, **33**, 2002, 219–229.
184. C.R. Barrett, E.C. Muehleisen, W.D. Nix, *Mater. Sci. Eng.*, **10**, 1972, 33–38.
185. A.J. Ardell, S.S. Lee, *Acta Metall.*, **34**, 1986, 2411–2416.
186. G.V. Raynor, *The Physical Metallurgy of Mg and its Alloys*, Pergamon Press, London, UK, 1959.
187. J. E. Dorn, 'Some fundamental experiments on high temperature creep', *J. Mechanics Phys. Solids*, **8**, 1954, 85–116.
188. J. Koike, R. Ohyama, T. Kobayashi, M. Suzuki, K. Maruyama, 'Grain-boundary sliding in AZ31 magnesium alloys at room temperature to 523K', *Mater. Trans.*, **44**, 2003, 445–451.
189. W.A. Rachinger, *J. Inst. Metals*, **81**, 1952, 33–38.
190. M. Mabuchi, H. Iwasaki, K. Yanase, K. Higashi, 'Low temperature superplasticity in an AZ91 magnesium alloy processed by ECAE', *Scripta Materialia*, **36**(6), 1997, 681–686.
191. H. Watanabe, T. Mukai, K. Ishikawa, K. Higashi, 'Low temperature superplasticity of a fine-grained ZK60 magnesium alloy processed by equal-channel-angular extrusion', *Scripta Materialia*, **46**, 2002, 851–856.
192. S. Agarwal, C.L. Briant, P.E. Krajewski, A.F. Bower, E.M. Taleff, *J. Mater. Eng. Perf.*, **16**, 2007, 170–176.
193. W. Sequeira, M.T. Murray, G.L. Dunlop G.L., 'Effect of Section Thickness and Microstructure on the Mechanical Properties of High Pressure Die Cast AZ91D', *Proceedings of the Third International Magnesium Conference, London, Institute of Materials,* 1997, 63–67.
194. M.S. Dargush, G.L. Dunlop, K. Pettersen, in 'Magnesium alloys and their applications', Proc. Volume sponsored by Volswagen AG, eds. B.L. Mordike and K.U. Kainer, Werkstoff-Informationsgesellschaft, Frankfurt, Germany, 1998, 277.

195. M.S. Dargush, M. Hisa, C.H. Caceres, G.L. Dunlop, in *Proc. of the Third International Magnesium Conf.*, ed. G.W. Lorimer, Manchester, UK, The Institute of Materials, 10–12 April 1996, 153–165.
196. F. von Buch, B.L. Mordike, in *Magnesium Alloys and Technologies*, ed. K.U. Kainer, 106–129.
197. S.M. Zhu, M.A. Gibson, M.A. Easton, J.F. Nie, 'The relationship between microstructure and creep resistance in die-cast magnesium–rare earth alloys', *Scripta Materialia*, **63**, 2010, 698–703.
198. A. A. Kaya, P. Uzan, D. Eliezer, E. Aghion, An Electron Microscopy Investigation on As-Cast AZ91D Alloy, *J. Mater. Sci. Technol.*, **16**, 2000, 1001–1011.
199. I.J. Polmear, Overview on 'Magnesium alloys and applications', *Mater. Sci. Tech.*, **10**, 1995, 1–16.
200. M. Fukuchi, F. Watanabe, *J. Jpn Institute of Metals*, **39**(5), 1980, 253–257.
201. A. Luo and M.O. Pekguleryuz, 'Cast Magnesium alloys for elevated temperature applications', *J. Materials Sci.*, **29**, 1994, 5259–5271.
202. V. Lupinc, in *Creep and Fatigue in High Temperature Alloys*, ed. J. Bressers, Applied Science Publishers, London, 1981, 7–40.
203. R. Lagneborg, in *'Creep and Fatigue in High Temperature Alloys'*, ed. J. Bressers, Applied Science Publishers, London, 1981, 41–71.
204. L. Shi and D.O. Northwood, 'A dislocation network model for creep', *Scripta Metal.*, **26**, 1992, 777–780.
205. J.C.M. Li, *Dislocation dynamics*, ed. A.R. Rosenfield, G.T. Hahn, A.L. Bement and R.I. Jaffee, Mc Graw Hill, 1968, 87.
206. O.D. Sherby, M. Burke, 'Mechanical behaviour of crystalline solids at elevated temperature', *Prog. Mater. Sci.*, **13**, 1968, 323–389.
207. B.L. Mordike, 'Creep-resistant magnesium alloys', *Mater. Sci. Eng.* A, **324**, 2002, 103–112.
208. W.J.G. Tegart, *Acta Met.*, **9**, 1961, 614–617.
209. L. Shi, D.O. Northwood, *Acta Metall. Mater.*, **42**, 1994, 871–877.
210. R.B. Jones, J.E. Harris, *Proc. of the Joint Int. Conf. on Creep*, Inst. Mech. Eng. Proc., London, **1**, 1963, 1–110.
211. G. Moreau, J.A. Cornet, D. Calais, *J. Nucl. Mater.*, **38**, 1971, 197–202.
212. J. Weertman, *J. Appl. Phys.*, **28**, 1957, 1185–1191.
213. S. Takeuchi, A.S. Argon, *Acta Metall.*, **24**, 1976, 883–889.
214. J. Friedel, *Dislocations*, Pergamon Press, Oxford, 1964.
215. P.G. Shewmon, F.N. Rhines, Trans. Am. Inst. Min. Engrs. V.200, 1954, 1021–1026.
216. J.J. Gilman, *J. Appl. Phys.*, **36**, 1965, 3195–3200.
217. B.L. Mordike, P. Lukac, in: G.W. Lorimer (Ed.), *Third International Magnesium Conference*, Pub. Inst. of Materials, London, 1997, 419–429.
218. B.L. Mordike, I. Stulikova, *Proc. of the Int. Conf. on Metallic Light Alloys*, Institution of Metallurgists, London, 1983, 146–153.
219. S.L. Robinson, O.D. Sherby, *Acta Materialia*, **17**, 1969, 109–125.
220. H. Miura, G. Yu, X. Yang, 'Multi-directional forging of AZ61Mg alloy under decreasing temperature conditions and improvement of its mechanical properties', *Mater. Sci. Eng. A*, **528**, 2011, 6981–6992.
221. M.R. Barnett, *Scripta Materalia*, **59**, 2008, 696–698.

222. N. Hansen, *Scripta Materalia*, **51**, 2004, 801–806.
223. H.Y. Wang, E.S. Xue, W. Xiao, Z. Liu, J.B. Li, Q.C. Jiang, 'Influence of grain size on deformation mechanisms in rolled Mg–3Al–3Sn alloy at room temperature', *Mater. Sci. Eng. A*, v.**528**, 2011, 8790– 8794.
224. J.C. Tan and M.J. Tan, *Mater. Sci. Eng. A*, **339**, 2003, 124–132.
225. M.R. Barnett, A.G. Beer, D. Atwell, A. Oudin, *Scripta Materialia*, **51**, 2004, 19–24.
226. S.E. Ion, F.J. Humphreys, S.H. White, *Acta Metall.*, **30**, 1982, 1909–1919.
227. M.T. Perez-Prado, J.A. del Valle, J.M. Contreras, O.A. Ruano, *Scripta Materialia*, **50**, 2004, 661–665.
228. S.W. Xu, N. Matsumoto, S. Kamado, T. Honma, Y. Kojima, *Scripta Materalia*, **61**, 2009, 249–255.
229. T. Al-Samman, X. Li, S.G. Chowdury, *Mater. Sci. Eng. A*, **527**, 2010, 3450–3456.
230. J.A. Del Valle, M.T.Pérez-Prado, O.A. Ruano, *Mater. Sci. Eng. A*, **355**, 2003, 68–74.
231. Q. Ma, B. Li, E.B. Marin, S.J. Horstemeyer, 'Twinning-induced dynamic recrystallization in a magnesium alloy extruded at 450°C', *Scripta Materialia*, **65**, 2011, 823–826.
232. S. Biswas, S.S. Dhinwal, S. Suwas, 'Room-temperature equal channel angular extrusion of pure magnesium', *Acta Materialia*, **58**, 2010, 3247–3261.
233. M.R. Barnett, *J. Light Met.*, **1**, 2001, 11–17.
234. N.V. Dudamell, I. Ulacia, F.Gálvez, S. Yi, J. Bohlen, D. Letzig, I. Hurtado, M.T.Pérez-Prado, 'Influence of texture on the recrystallization mechanisms in an AZ31 Mg sheet alloy at dynamic rates', *Mater. Sci. Eng. A*, **532**, 2012, 528–535.
235. T. Al-Samman, G. Gottstein, 'Dynamic recrystallization during high temperature deformation of magnesium', *Mater. Sci. Eng. A*, **490**, 2008, 411–420.
236. M.R. Barnett, *Mater. Trans.*, **44**, 2003, 571–577.
237. T. Kobayashi, J. Koike, Y. Yoshida, S. Kamado, M. Suzuki, K. Maruyama, Y. Kojima, 'Grain size dependence of active slip systems in an AZ31 magnesium alloy', *J. Jpn. Inst. Met.*, **67**, 2003, 149–156.
238. M. Wang, R. Xin, B. Wang, Q. Liu, 'Effect of initial texture on dynamic recrystallization of AZ31 Mg alloy during hot rolling', *Mater. Sci. Eng. A*, **528**, 2011, 2941–2951.
239. S.A. Farzadfar, É. Martin, M. Sanjari, E. Essadiqi, M.A. Well, S. Yue, 'On the deformation, recrystallization and texture of hot-rolled Mg–2.9Y and Mg–2.9Zn solid solution alloys—A comparative study', *Mater. Sci. Eng. A*, 2011, doi:10.1016/j.msea.2011.11.061.
240. H. Watanabe, H. Tsutsui, T. Mukai, K. Ishikawa, Y. Okanda, M. Kohzu, K. Higashi, *Mater. Trans.*, **7**, 2001, 1200–1207.
241. M. Mabuchi, K. Kubota, K. Higashi, *Mater. Trans.*, **36**, 1995, 1249–1255.
242. M. Hakamada, A. Watazu, N. Saito, H. Iwasaki, *J. Mater. Sci.*, **43**, 2008, 2066–2072.
243. F.J. Humphreys, *Acta Metall.*, **25**, 1977, 1323–1330.
244. F.J. Humphreys, *Acta Metall.*, **27**, 1979, 1801–1808.
245. S.W. Xu, S. Kamado, T. Honma, 'Recrystallization mechanism and the relationship between grain size and Zener–Hollomon parameter of Mg–Al–Zn–Ca alloys during hot compression', *Scripta Materialia*, **63**, 2010, 293–296.

246. X. Huang, K. Suzuki and Y. Chino, 'Static recrystallization and mechanical properties of Mg-4Y-3RE magnesium alloy sheet processed by differential speed rolling at 823K', *Mater. Sci. Eng. A*, doi:10.1016/j.msea.2012.01.044.
247. A. Sadeghi, M. Pekguleryuz, 'Recrystallization and texture evolution of Mg–3%Al–1%Zn–(0.4–0.8)%Sr alloys during extrusion', *Mater. Sci. Eng. A*, **528**, 2011, 1678–1685.
248. F.J. Humphreys, M. Hatherly, *'Recrystallization and Related Annealing Phenomena'*, 2nd ed. Pergamon, Oxford, 2004.
249. S.W. Xu, K. Oh-ishi, S. Kamado and T. Honma, 'Twins, recrystallization and texture evolution of a Mg–5.99Zn–1.76Ca–0.35Mn (wt.%) alloy during indirect extrusion process, *Scripta Materialia*, **65**, 2011, 875–878.
250. L.W.F. Mackenzie, M. Pekguleryuz, *Mater. Sci. Eng.* A, **480**, 2008, 189–197.
251. K. Hantzsche, J. Bohlen, J. Wendt, K.U. Kainer, S.B. Yi, D. Letzig, *Scripta Materalia*, **63**, 2010, 725–730.
252. L.W.F. Mackenzie, M. Pekguleryuz, *Scr. Mater.*, **59**, 2008, 665–668.
253. I. Samajdar, R.D. Doherty, *Acta Materialia*, **46**, 1998, 3145–3158.
254. M. Wang, B.Y. Zong, G. Wang, 'Grain growth in AZ31 Mg alloy during recrystallization at different temperatures by phase field simulation', *Comp. Mater. Sci.*, **45**, 2009, 217–222.
255. K. Edalati, Z. Horita, 'High-pressure torsion of pure metals: influence of atomic bond parameters and stacking fault energy on grain size and correlation with hardness', *Acta Materialia*, **59**, 2011, 6831–6836.
256. Q. Xu, G. Tang, Y. Jiang, 'Thermal and electromigration effects of electropulsing on dynamic recrystallization in Mg–3Al–1Zn alloy', *Mater. Sci. Eng. A*, **528**, 2011, 4431–4436.

3 Thermodynamic properties of magnesium alloys

S. L. SHANG and Z. K. LIU, Pennsylvania State University, USA

DOI: 10.1533/9780857097293.85

Abstract: Thermodynamic properties and their determination from first-principles and phonon calculations and CALPHAD (CALculation of PHAse Diagram) modeling are reviewed for Mg-based alloys and compounds, encompassing enthalpy, entropy, Helmholtz energy, Gibbs energy, heat capacity, isothermal and isentropic bulk moduli, anisotropic thermal expansions, and isothermal and isentropic elastic constants as functions of temperature, pressure and composition. Furthermore, various strategies based on first-principles calculations for the treatment of disordered phases are discussed, including the cluster expansion, special quasirandom structure (SQS), and partition function methods. Finally, other capabilities of the first-principles calculations are pointed out, such as: determination of Helmholtz energy for unstable phases; calculation of defect energies, such as vacancy formation energy, stacking fault energy, anti-phase boundary (APB) energy, and surface and interfacial energies; evaluation of diffusion coefficients in solid and liquid phases; assessment of creep properties, tensile and shear strengths, and solute strengthening due to alloying elements.

Key words: magnesium alloy, thermodynamics, first-principles, phonon, elasticity, thermal expansion.

3.1 Introduction

With a density two-thirds that of aluminum and one-quarter of steel, magnesium (Mg) alloys are of growing importance. The development of Mg-based light alloys for vehicle structures has been promoted in the light of rising oil prices and global climate change due to greenhouse gases (Shang *et al.*, 2008). Computational thermodynamics, based on the CALPHAD (CALculation of PHAse Diagram) approach (Saunders and Miodownik, 1998; Lukas *et al.*, 2007; Liu, 2009), is a key enabling method to help accelerate the pace of materials research and development, developing relationships and data that can reduce the design-to-alloy synthesis and production cycle time (Olson, 1997; Allison *et al.*, 2008).

This technology has the capability to model and solve a vast number of materials-related problems related to phase equilibria, phase stability, and phase transformations by using well developed CALPHAD software packages such as Thermo-Calc (Andersson *et al.*, 2002), FactSage (Bale *et al.*, 2009), and PANDAT (Cao *et al.*, 2009), where a modeled thermodynamic database is prerequisite for any thermodynamic analyses (Shang *et al.*, 2010e; Shang *et al.*, 2008). However, measured thermodynamic data from experiments, thermochemical data in particular, are scarce, resulting in large uncertainties in thermodynamic modeling. Fortunately, the advanced computational tools available today, for instance first-principles calculations based on the density functional theory (DFT) (Hohenberg and Kohn, 1964), can provide considerable insight into these basic materials properties. These advances provide the focus of this chapter on the thermodynamic properties of Mg-based alloys and compounds determined by first-principles calculations.

The remainder of this chapter is organized as follows. In Section 3.2, the basic thermodynamic relationships and their derivatives for single crystals and polycrystals are presented, in particular those relations which are relevant to first-principles calculations. Section 3.3 presents the measurements and empirical predictions of thermodynamic properties together with the modeled thermodynamic properties for binary Mg-based alloys and compounds using the CALPHAD approach. Section 3.4 illustrates the capabilities of first-principles calculations in determining the thermodynamic properties and the anisotropic thermal expansions and elastic properties of Mg-based alloys and compounds as functions of temperature. This section also includes details regarding the treatment of disordered structures within first-principles approaches. Finally Section 3.5 shows recent developments of first-principles thermodynamics.

3.2 Fundamentals of thermodynamics

3.2.1 Basic relationships

Based on the combined first and second laws of thermodynamics, the change of internal energy U of a system is given by (Hillert, 2008)

$$\mathrm{d}U = T\mathrm{d}S - P\mathrm{d}V + \sum \mu_i \mathrm{d}N_i - D\mathrm{d}\xi \qquad [3.1]$$

The internal energy U is thus a function of the natural variables S (entropy), V (volume), N_i (amount of independent component i), and ξ (internal process) of the system, that is, $U(S, V, N_i, \xi)$. The other variables T (temperature), P (pressure), μ_i (chemical potential) and D (driving force) are dependent variables and can be represented by partial derivatives of the internal energy with respect to their corresponding natural variable while keeping other natural variables constant. Because the major focus of this chapter is the thermochemical properties of individual phases in a closed system without mass change (i.e., $dN_i = 0$) and at the equilibrium condition (i.e., $Dd\xi = 0$), the variables N_i and ξ are therefore ignored herein unless otherwise mentioned.

In experiments, the variables of temperature T and pressure P are typically controlled. In theoretical calculations, it is more convenient to control temperature T and volume V. For the convenience of both experiments and calculations, thermodynamic variables of enthalpy H, Helmholtz energy F, and Gibbs energy G are defined as follows:

$$H = U + PV \quad [3.2]$$

$$F = U - TS \quad [3.3]$$

$$G = U - TS + PV = H - TS = F + PV \quad [3.4]$$

Other commonly used thermodynamic variables can be obtained with the following derivatives:

$$\alpha_V = \left(\frac{\partial \ln V}{\partial T}\right)_P = \frac{1}{V}\left(\frac{\partial V}{\partial T}\right)_P = \frac{1}{V}\frac{\partial^2 G}{\partial P \partial T} = \frac{1}{B_T}\left(\frac{\partial P}{T}\right)_V = \frac{1}{B_T}\left(\frac{\partial S}{\partial V}\right)_T = 3\alpha_L \quad [3.5]$$

$$B_T = -V\left(\frac{\partial P}{\partial V}\right)_T = V\left(\frac{\partial^2 F}{\partial V^2}\right)_T \quad [3.6]$$

$$B_S = B_T C_P / C_V = B_T + (\alpha_V B_T)^2 TV / C_V \quad [3.7]$$

$$C_V = \left(\frac{\partial U}{\partial T}\right)_V = T\left(\frac{\partial S}{\partial T}\right)_V = -T\left(\frac{\partial^2 F}{\partial T^2}\right)_V \quad [3.8]$$

$$C_P = \left(\frac{\partial H}{\partial T}\right)_P = T\left(\frac{\partial S}{\partial T}\right)_P = -T\left(\frac{\partial^2 G}{\partial T^2}\right)_P = C_V + \alpha_V^2 B_T TV \quad [3.9]$$

where α_V is volume thermal expansion, α_L the linear volume thermal expansion, B_T the isothermal bulk modulus, B_s the isentropic (adiabatic) bulk modulus, C_V the heat capacity at constant volume, and C_P the heat capacity at constant pressure.

The third law of thermodynamics states that entropy is zero at zero Kelvin for an ordered phase, that is, $S_0 = 0$. However, the absolute value of internal energy U is unknown, and we must select a reference state. The same is true for H, F, and G. A widely used reference state in the CALPHAD community for thermodynamic modeling is to set $H_{298.15} = 0$ for pure elements at their respective stable structures at room temperature (298.15 K) and ambient pressure. This is known as the stable element reference (SER). From the viewpoint of theoretical predictions, for example, first-principles calculations, thermodynamic properties can be easily obtained using the above relations once F or G is known.

One common expression for Gibbs energy of pure elements and stoichiometric compounds at finite temperatures and ambient pressure is as follows:

$$G = a + bT + cT\ln T + dT^2 + eT^{-1} + fT^3 \quad [3.10]$$

where $a, b, c, d, e,$ and f are model parameters. The expressions for other thermodynamic properties can be obtained from G, for example (see Equation [3.9])

$$C_P = -c - 2dT - 2eT^{-2} - 6fT^2 \quad [3.11]$$

For solution phases, such as the substitutional solutions of liquid, bcc, fcc, and hcp, the Gibbs energy is written as follows (Saunders and Miodownik, 1998; Lukas *et al.*, 2007):

$$G = \sum_i x_i G_i^0 + RT \sum_i x_i \ln x_i + G^{ex} \quad [3.12]$$

where G_i^0 is the Gibbs energy of the pure element i usually expressed by Equation [3.10] with functions from the SGTE database (Dinsdale, 1991) commonly used, x_i the mole fraction of i, and R the gas constant. G^{ex} is the excess Gibbs energy, expressed using the Redlich-Kister polynomial (Redlich and Kister, 1948):

$$G^{ex} = \sum_i \sum_{<j} x_i x_j \sum_{n=0} L_{ij}^n (x_i - x_j)^n + \sum_i \sum_{<j} \sum_{<k} x_i x_j x_k L_{ijk} + \ldots \quad [3.13]$$

where L_{ij}^n is the nth binary interaction parameter between elements i and j. L_{ijk} represents the ternary interaction parameter among elements i, j, and k, that is, $L_{ijk} = x_i L_{ijk}^0 + x_j L_{ijk}^1 + x_k L_{ijk}^2$. These L parameters may depend on temperature:

$$L = A + BT + CT\ln T \quad [3.14]$$

where A, B, and C are the modeling parameters to be evaluated.

For the convenience of CALPHAD modeling, the Gibbs energy of a stoichiometric compound can be expressed by:

$$G = \sum_i x_i G_i^0 + \Delta G_f \quad [3.15]$$

where G_i^0 is the Gibbs energy of pure element i. The Gibbs energy of formation ΔG_f is temperature dependent and follows the same format as Equation [3.10]:

$$\Delta G_f = a + bT + cT \ln T + dT^2 + eT^{-1} + fT^3 \quad [3.16]$$

For non-stoichiometric compounds, one typically adopts the sublattice model, for details see Saunders and Miodownik (1998) and Lukas *et al.* (2007). The method of Hillert and Jarl (1978) is commonly used by the CALPHAD community to incorporate magnetic contributions into thermodynamic descriptions of phases.

3.2.2 Anisotropic thermal expansion and elasticity

Many properties of single crystals are anisotropic, such as thermal expansion and elasticity. Second-order strain tensors, which have components ε_{ij}, are symmetric, and consequently can be expressed in terms of six independent components ε_i:

$$\varepsilon = \begin{pmatrix} \varepsilon_{11} & \varepsilon_{12} & \varepsilon_{13} \\ \varepsilon_{21} & \varepsilon_{22} & \varepsilon_{23} \\ \varepsilon_{31} & \varepsilon_{32} & \varepsilon_{33} \end{pmatrix} = \begin{pmatrix} \varepsilon_1 & \varepsilon_6/2 & \varepsilon_5/2 \\ \varepsilon_6/2 & \varepsilon_2 & \varepsilon_4/2 \\ \varepsilon_5/2 & \varepsilon_4/2 & \varepsilon_3 \end{pmatrix} \quad [3.17]$$

The reduced representations ε_i are also referred to the engineering strains or the representations by matrices. Table 3.1 shows the relationship between the representations by tensors and matrices (Nye, 1985).

Increasing the temperature results in six independent linear thermal expansions α_{ij} (or α_i in matrix form, see Table 3.1), one corresponding to each ε_{ij} (or ε_i):

$$\alpha_{ij} = \left(\frac{\partial \varepsilon_{ij}}{\partial T}\right)_P = \left(\frac{\partial \varepsilon_i}{\partial T}\right)_P = \alpha_i \quad [3.18]$$

After considering crystal symmetry, the number of independent coefficients of linear thermal expansion is reduced. For example, there exist only two

Table 3.1 Relationship between representations by tensors and by matrices

Representation by tensors	11	22	33	23,32	31,13	12,21
Representation by matrices	1	2	3	4	5	6

Source: Nye (1985).

independent coefficients of linear thermal expansion for hexagonal, trigonal, and tetragonal crystals: $\alpha_L^{\parallel}$ (along the principal axis, that is, the c-axis direction in crystals) and $\alpha_L^{\perp}$ (normal to the principal axis) (Ho and Taylor, 1998). We have:

$$\alpha_L(\theta) = \alpha_L^{\parallel} \cos^2\theta + \alpha_L^{\perp} \sin^2\theta \qquad [3.19]$$

where θ is angle with respect to the principal axis. For volumetric thermal expansion, $\alpha_V = \alpha_L^{//} + 2\alpha_L^{\perp}$ (see also Equation [3.5]).

The next property to consider is elasticity, which is the second derivative of energy with respect to the strains ε_i within the elastic region. By ignoring the higher-order elastic constants, strain energy per volume can be represented by (Nye, 1985):

$$e = \frac{1}{2}\sum_i \sum_j c_{ij}\varepsilon_i\varepsilon_j \qquad [3.20]$$

where $i, j = 1, 2, \ldots, 6$, and c_{ij} are the elastic stiffness constants,

$$\mathbf{C} = \begin{pmatrix} c_{11} & c_{12} & c_{13} & c_{14} & c_{15} & c_{16} \\ & c_{22} & c_{23} & c_{24} & c_{25} & c_{26} \\ & & c_{33} & c_{34} & c_{35} & c_{36} \\ & & & c_{44} & c_{45} & c_{46} \\ & & & & c_{55} & c_{56} \\ & & & & & c_{66} \end{pmatrix} \qquad [3.21]$$

The lower-left corner of **C** is omitted as the elastic constant matrix is symmetric. The reduced indices for c_{ij} are used according to the matrix representations in Table 3.1. These are, in fact, fourth-order tensors. Note that the elastic strain energy e must be positive, and since the strains $\varepsilon_i \geq 0$, the elastic constant matrix **C** must be positive in order for a given structure to be stable. This is known as Born's criteria (Born and Huang, 1998). For example, in a cubic system:

$$c_{12} - |c_{12}| > 0,\ c_{11} + 2c_{12} > 0,\ c_{44} > 0 \qquad [3.22]$$

Table 3.2 Bulk modulus B and shear modulus G in the Voigt and Reuss approaches, estimated by elastic stiffness constants c_{ij} and elastic compliance constants s_{ij}

Voigt approach	Reuss approach
$B_{Voigt} = (A + 2B)/3$	$B_{Reuss} = 1/(3a + 6b)$
$G_{Voigt} = (A - B + 3C)/5$	$G_{Reuss} = 5/(4a - 4b + 3c)$
$A = (c_{11} + c_{22} + c_{33})/3$	$a = (s_{11} + s_{22} + s_{33})/3$
$B = (c_{12} + c_{13} + c_{23})/3$	$b = (s_{12} + s_{13} + s_{23})/3$
$C = (c_{44} + c_{55} + c_{66})/3$	$c = (s_{44} + s_{55} + s_{66})/3$

Note: The s_{ij} matrix is the inverse of the c_{ij} matrix, and *vice versa*.

For a hexagonal system:

$$c_{11} - |c_{12}| > 0,\ (c_{11} + c_{12})c_{33} - 2c_{13}^2 > 0,\ c_{44} > 0 \qquad [3.23]$$

Based on the c_{ij} values obtained for single crystals, polycrystalline aggregate properties such as the bulk modulus (B) and shear modulus (G) are usually estimated by the Voigt–Reuss–Hill approach (Simmons and Wang, 1971). Voigt's approach gives the upper bound of elastic properties in terms of the uniform strain, and Reuss's approach gives the lower bound in terms of the uniform stress. See Table 3.2 for details on how B and G are expressed by the Voigt and Reuss approaches. Hill's approach (Hill, 1952) gives average between Voigt and Reuss.

With the combination of bulk modulus B and shear modulus G, we can judge the ductility and brittleness according to the Pugh criterion (Pugh, 1954), which states the B/G ratio differentiates ductile (> ~1.75) and brittle (< ~1.75) materials. Furthermore, based on the B and G values obtained from the Voigt and Reuss approaches, a universal elastic anisotropy index A^U for a crystal with any symmetry has been proposed (Ranganathan and Ostoja-Starzewski, 2008):

$$A^U = \frac{5G_{Voigt}}{G_{Reuss}} + \frac{B_{Voigt}}{B_{Reuss}} - 6 \geq 0 \qquad [3.24]$$

Note that for locally isotropic single crystals $A^U = 0$, and the departure of A^U from zero defines the extent of single crystal anisotropy and accounts for both the shear and bulk contributions. For a crystal with cubic symmetry, the relationship between A^U and the commonly used Zener anisotropy ratio A is: $A^U = (6/5)(\sqrt{A} - 1/\sqrt{A})^2$, where $A = 2c_{44}/(c_{11} - c_{12})$.

Usually the elastic stiffness constants, especially those at high temperatures, are measured by the resonance method. In this method, the process is considered adiabatic because elastic waves travel faster than heat diffuses, and the deformation due to the elastic waves can be viewed as an isentropic (adiabatic) process (Liu *et al.*, 2010; Shang *et al.*, 2010g). Thus, the isothermal elastic stiffness constants c_{ij}^T are different from the isentropic ones c_{ij}^s, and their thermodynamic relations are given by (Davies, 1974):

$$c_{ij}^s = c_{ij}^T + \frac{TV\lambda_i\lambda_j}{C_V} \quad [3.25]$$

$$\lambda_i = -\sum_{j=1}^{6}\left(\frac{\partial\sigma_i}{\partial\varepsilon_j}\right)_T\left(\frac{\partial\varepsilon_j}{\partial T}\right)_\sigma = -\sum_{j=1}^{6}\alpha_j c_{ij}^T \quad [3.26]$$

where σ_i is stress. Usually, $\lambda_i \le 0$ due to the non-negative values of c_{ij}^T and thermal expansion α_j in most cases, resulting in $c_{ij}^S \ge c_{ij}^T$. For a cubic system, λ_i reduces to $-\alpha_V B_T$ if $i = 1, 2$, or 3, and reduces to 0 if $i = 4, 5$, or 6 (Shang *et al.*, 2010g). For the case of bulk modulus, Equation [3.25] reduces to the commonly used Equation [3.7].

3.3 Thermodynamic properties of Mg alloys and compounds

3.3.1 Measurement and empirical estimation

Details regarding methods for measuring equilibrium data (i.e., phase diagram data) between phases and thermochemical data for individual phases can be found in Kubaschewski *et al.* (1993); Zhao (2007). For instance, phase diagram data can be determined via diffusion couples or multiples, differential thermal analysis (DTA), and heat-flux differential scanning calorimetry (HF-DSC) of metals and alloys. Thermochemical properties, on the other hand, can be determined by calorimetric methods, including DTA and DSC, and electromotive forces (Ipser *et al.*, 2010). The thermochemical data can also be predicted using empirical approaches, that is, the enthalpy of formation estimated by Miedema's method (Miedema *et al.*, 1980), Le Van's method, or Slobodin's method, see Kubaschewski *et al.* (1993) for more details. In this chapter, the thermochemical properties will be predicted by first-principles calculations as detailed in Section 3.4.

3.3.2 Thermodynamic properties evaluated for binary Mg alloys and compounds

With thermodynamic modeling based on the CALPHAD approach (Saunders and Miodownik, 1998; Lukas *et al.*, 2007; Liu, 2009), thermodynamic properties can be estimated; however, some of the thermochemical properties are still assigned arbitrary values due to the lack of experimental data. Table 3.3 summarizes the modeled L parameters (see Equations [3.12–3.14]) for the liquid and hcp phases of Mg-X binary systems from a thermodynamic database developed for Mg-based alloys (Shang *et al.*, 2008), where X is one of the alloying elements Al, Ca, Ce, Cu, Fe, K, La, Li, Mn, Na, Nd, Pr, Si, Sn, Sr, Y, Zn, or Zr. Two or three L parameters are typically used to describe the liquid and hcp phases, and one to three fitting parameters are employed within each L parameter. In order to better illustrate the effect of these parameters, Figure 3.1 compares the liquidus between the hcp and liquid phases predicted for Mg-X systems with up to 0.1 mole fraction of X. In addition, Table 3.4 lists the liquidus and solidus temperatures predicted with mole fractions of X at 0.05 and 0.1 when possible. Figure 3.1 and Table 3.4 show that the addition of alloying elements lowers the melting point of hcp Mg, and the element Li has the smallest effect on both the liquidus and solidus. Alloying with the rare-earth elements La, Ce, Pr, and Nd – and especially the alkaline earth metal Ca – suppresses the liquidus and solidus significantly. Some elements, such as Fe, K, and Zr, are nearly insoluble in hcp Mg.

For each compound presented in the Mg-X binary systems from the aforementioned database, the enthalpy and entropy of formation have been modeled with the CALPHAD method and are given in Table 3.5. These values are at 298.15 K and ambient pressure. The ΔH values predicted by first-

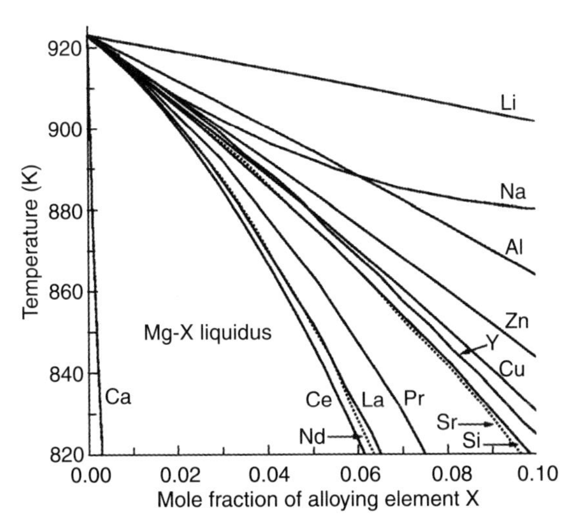

3.1 Liquidus between liquid and hcp phases predicted by CALPHAD modeled thermodynamic data of Mg-X systems as shown in Table 3.3

Table 3.3 Modeled *L* parameters of liquid and hcp phases for Mg-X binary systems, see Equations [3.13] and [3.14]

System and reference	Phase	L0	L1	L2
Mg–Al (Zhong *et al.*, 2005)	Liquid	−9031 + 4.855T	−891 + 1.137T	434
	hcp	4272–2.190T	−1.077–1.015T	−965
Mg–Ca (Zhong *et al.*, 2006a)	Liquid	−32322.4 + 16.721T	60.3 + 6.549T	−5742.3 + 2.760T
	hcp	−9183.2 + 16.981T		
Mg–Ce (Zhang *et al.*, 2008)	Liquid	−36703 + 13.831T	30962–17.297T	−15090
	hcp	−94338 + 79.952T		
Mg–Cu (Zuo and Chang, 1993)	Liquid	−36962.7 + 4.744T	−8182.19	
	hcp	22500–3T		
Mg–Fe (Ansara *et al.*, 1998)	Liquid	61343 + 1.5T	−2700	
	hcp	92400		
Mg–K (Zhang, 2006)	Liquid	37272.2		
	hcp	−1072.6		
Mg–La (Guo and Du, 2004)	Liquid	−32472.5 + 8.367T	35610.1–24.012T	−13162.4
	hcp	−25 000		
Mg–Li (Ansara *et al.*, 1998)	Liquid	−14935+10.371T	−1789 + 1.143T	6533–6.6915T
	hcp	−6856	4000	4000
Mg–Mn (Grobner *et al.*, 2005)	Liquid	+25922.4 + 9.036T	−3470.8	
	hcp	+37148–1.810T		

Mg–Na (Zhang, 2006)	Liquid	26025.779	4509.96384	
	hcp	75698.31		
Mg–Nd (Meng *et al.*, 2007)	Liquid	−43547.1 + 19.415T	−30060.5 + 10.775T	−26879.5 + 22.230T
	hcp	−13200		
Mg–Pr (Guo and Du, 2005)	Liquid	−41498.5 + 13.863T	−38739.5 + 27.386T	
	hcp	−10 000		
Mg–Si (Ansara *et al.*, 1998)[a]	Liquid	−83864.3 + 32.4T	18027.4–19.612T	2486.67–0.311T
	hcp	−7148.79 + 0.894T		
Mg–Sr (Zhong *et al.*, 2006b)	Liquid	−18647.1 + 9.070T	−12831.6 + 7.509T	
	hcp	10 000		
Mg–Y (Fabrichnaya *et al.*, 2003)	Liquid	−41165.3 + 17.564T	−15727.0 + 4.705T	
	hcp	−26612.8 + 13.946T	−2836.2	
Mg–Zn (Agarwal *et al.*, 1992)	Liquid	−81439.7 + 518.25T-64.71Tln(T)	2627.5 + 2.931T	−1673.3
	hcp	−1600.8 + 7.624T	−3823.0 + 8.026T	
Mg–Zr (Arroyave *et al.*, 2005)	Liquid	4961.4 + 38.180T		
	hcp	30384.0 + 13.723T	18588.9–25.175T	

[a]For liquid phase L^3 = 18541.2–2.318T; L^4 = −12338.8 + 1.542T

Table 3.4 Liquidus and solidus temperatures (K) predicted by CALPHAD for Mg-X systems at mole fractions of X at 0.05 and 0.1, when available

System	Liquidus		Solidus	
	0.05	0.1	0.05	0.1
Mg–Al	894.4	863.7	832.6	733.9
Mg–Ca	870.1	798.8	512.9	440.3
Mg–Ce	846.4	711.4	309.3	239.4
Mg–Cu	879.4	830.4	–	–
Mg–Fe	–	–	–	–
Mg–K	–	–	–	–
Mg–La	851.9	731.6	591.5	483.9
Mg–Li	912.5	901.5	909.3	893.8
Mg–Mn	–	–	–	–
Mg–Na	892.2	880.1	–	–
Mg–Nd	851.0	714.9	–	–
Mg–Pr	863.7	765.3	–	–
Mg–Si	875.7	818.2	–	–
Mg–Sr	875.8	815.8	–	–
Mg–Y	879.1	824.7	823.9	776.1
Mg–Zn	884.2	843.5	–	–
Mg–Zr	–	–	–	–

Note: See Table 3.3 for details of thermodynamic parameters.

principles (at 0 K) are also shown in the last column for comparison. Mg_2Si has the largest enthalpy of formation, and the Mg–Al compounds have the smallest enthalpies of formation. These observations have been confirmed by first-principles calculations as shown in Table 3.5, and for most cases, the ΔH values modeled with the CALPHAD method agree reasonably well with those of first-principles predictions, indicating the capabilities of first-principles methodology. For more details regarding these comparisons, see (Zhang *et al.*, 2009).

3.4 First-principles thermodynamics of Mg alloys and compounds

3.4.1 Introduction to first-principles calculations

First-principles calculations, such as the one based on DFT (Hohenberg and Kohn, 1964), only require the knowledge of atomic species and crystal structure to define the energetics of the structure and hence are predictive in nature. In DFT the total energy of a many-electron system in an external potential is expressed by a unique functional of the electron density $\rho(\vec{r})$, and this functional has its minimum at the ground-state electron density

Table 3.5 Enthalpies of formation ΔH (kJ/mol) and entropies of formation ΔS (J/mol-K) at 298.15 K predicted by CALPHAD for Mg-X compounds

System	Compound	x_{Mg}	Struk	Space group	Prototype	$\Delta H^{298.15}$	$\Delta S^{298.15}$	ΔH (F-P)
Mg–Al	$Mg_{89}Al_{140}$	0.3886		$Fd\bar{3}m$	Cd_2Na	–3.533	–0.502	–3.42[a]
	$Mg_{23}Al_{30}$	0.4340		$R\bar{3}$	$Co_5Cr_2Mo_3$	–3.246	0.110	
	$Mg_{17}Al_{12}$	0.5862	A12	$I\bar{4}3m$	α-Mn	–3.585	–0.767	–3.60[a]
Mg–Ca	Mg_2Ca	0.6667	C14	$P6_3/mmc$	$MgZn_2$	–13.795	–5.078	–12.14[b]
Mg–Ce	MgCe	0.5000	B2	$Pm\bar{3}m$	ClCs	–14.075	–10.221	–0.96[c]
	Mg_2Ce	0.6667	C15	$Fd\bar{3}m$	Cu_2Mg	–16.779	–11.317	–13.70[c]
	Mg_3Ce	0.7500	$D0_3$	$Fm\bar{3}m$	BiF_3	–18.092	–12.574	–7.69[c]
	$Mg_{41}Ce_5$	0.8913		$I4/m$	Ce_5Mg_{41}	–18.762	–6.250	
	$Mg_{17}Ce_2$	0.8947		$P6_3/mmc$	$Ni_{17}Th_2$	–14.819	–2.358	
	$Mg_{12}Ce$	0.9231		Im/mmc	$Mg_{12}Ce$	–13.726	–2.201	
Mg–Cu	$MgCu_2$	0.3333	C15	$Fd\bar{3}m$	$MgCu_2$	–10.909	1.239	–4.75[b]
	Mg_2Cu	0.6667		$Fddd$	Mg_2Cu	–9.540	–0.622	
Mg–La	MgLa	0.5000	B2	$Pm\bar{3}m$	ClCs	–16.701	–4.594	–11.64[b]
	Mg_2La	0.6667	C15	$Fd\bar{3}m$	$MgCu_2$	–8.938	3.378	–12.55[b]
	Mg_3La	0.7500	$D0_3$	$Fd\bar{3}m$	BiF_3	–19.698	–7.303	–13.44[b]
	$Mg_{17}La_2$	0.8947		$P6_3/mmc$	$Ni_{17}Th_2$	–8.663	–2.070	–7.70[b]
	$Mg_{12}La$	0.9231		Im/mmc	$Mg_{12}Ce$	–6.398	–1.304	–5.79[b]
Mg–Nd	MgNd	0.5000	B2	$Pm\bar{3}m$	ClCs	–15.658	–3.885	
	Mg_2Nd	0.6667	C15	$Fd\bar{3}m$	$MgCu_2$	–15.110	–2.726	
	Mg_3Nd	0.7500	$D0_3$	$Fm\bar{3}m$	BiF_3	–19.765	–7.558	

(*Continued*)

Table 3.5 (Continued)

System	Compound	x_{Mg}	Struk	Space group	Prototype	$\Delta H^{298.15}$	$\Delta S^{298.15}$	ΔH (F-P)
	$Mg_{41}Nd_5$	0.8913		$I4/m$	$Mg_{41}Ce_5$	−17.104	−13.106	
Mg–Pr	MgPr	0.5000	B2	$Pm\bar{3}m$	ClCs	−15.882	−3.174	
	Mg_2Pr	0.6667	C15	$Fd\bar{3}m$	$MgCu_2$	−11.587	0.846	
	Mg_3Pr	0.7500	$D0_3$	$Fm\bar{3}m$	BiF_3	−14.643	−2.820	
	$Mg_{41}Pr_5$	0.8913		$I4/m$	$Mg_{41}Ce_5$	−6.726	−1.293	
	$Mg_{12}Pr$	0.9231		Im/mmc	$Mg_{12}Ce$	−4.816	−0.868	
Mg–Si	Mg_2Si	0.6667	C1	$Fm\bar{3}m$	CaF_2	−21.745	−2.681	−17.70[b]
Mg–Sr	Mg_2Sr	0.6667	C14	$P6_3/mmc$	$MgZn_2$	−7.331	0.378	−10.62[d]
	$Mg_{23}Sr_6$	0.7931	$D8_a$	$Fm\bar{3}m$	$Mn_{23}Th_6$	−4.665	0.982	−7.74[d]
	$Mg_{38}Sr_9$	0.8085		$P6_3/mmc$	$Mg_{38}Sr_9$	−4.344	1.019	−6.27[d]
	$Mg_{17}Sr_2$	0.8947		$P6_3/mmc$	$Ni_{17}Th_2$	−2.315	1.308	−4.80[d]
Mg–Y	MgY	0.5000	B2	$Pm\bar{3}m$	CsCl	−15.600	−3.790	
	Mg_2Y	0.6667	C14	$P6_3/mmc$	$MgZn_2$	−12.593	−2.182	
	$Mg_{24}Y_5$	0.8276	A12	$I\bar{4}3m$	α-Mn	−7.837	−1.259	
Mg–Zn	Mg_2Zn_{11}	0.1538		$Pm\bar{3}$	Mg_2Zn_{11}	−5.823	−1.943	
	$MgZn_2$	0.3333	C14	$P6_3/mmc$	$MgZn_2$	−11.884	−3.835	
	Mg_2Zn_3	0.4000		$B2/m$		−11.015	−3.672	
	MgZn	0.4800				−9.590	−3.197	
	Mg_7Zn_3	0.7183		$Immm$		−4.814	−1.000	

Sources: [a]Zhong *et al.*, 2005; [b]Zhang *et al.*, 2009; [c]Zhou *et al.*, 2007; [d]Zhong *et al.*, 2006b.
Notes: The reference for each system is given in Table 3.3. First-principles (F-P) predicted enthalpies of formation at 0 K are also shown for comparison.

(Hafner, 2000). The Kohn–Sham one-electron equation for DFT is (Kohn and Sham, 1965):

$$\hat{H}_{\mathrm{KS}}\psi_i = \varepsilon_i \psi_i \qquad [3.27]$$

where ψ_i are wave functions of single particles and ε_i the single-particle energies. The Kohn–Sham Hamiltonian is:

$$\hat{H}_{\mathrm{KS}} = -\frac{\hbar^2}{2m_e}\vec{\nabla}_i^2 + \frac{e^2}{4\pi\varepsilon_0}\int \frac{\rho(\vec{r}\,')}{|\vec{r}-\vec{r}\,'|}\mathrm{d}\vec{r}\,' + V_{\mathrm{xc}} + V_{\mathrm{ext}} \qquad [3.28]$$

where the first term is the kinetic energy operator for the electrons with mass m_e at position $\vec{r}$, the second term describes the Coulomb interactions between electrons, V_{xc} is the exchange-correlation operator, and the system-specific information of atomic species and crystal structure is contained within the external potential V_{ext}, that is, the electron–ion interaction. Unlike the Hartree–Fock method, the Kohn–Sham Hamiltonian treats only the exchange exactly and does not include the correlation. In DFT, both exchange and correction are treated approximately. To solve Equation [3.27], we first need to express the wave function ψ_i by a given basis set, such as the plane wave basis set used in the pseudopotential method of the VASP code (Kresse and Furthmuller, 1996a, 1996b). Secondly, we need a method to account for the electron–ion interactions, such as the ultra-soft pseudopotentials (USPP) or the projector augmented wave method (PAW) (Kresse and Joubert, 1999; Blöchl, 1994) used in the VASP code. Finally, we need a method to describe the exchange-correlation functional. The two widely used approximations herein include (i) the local density approximation (LDA) (Perdew and Zunger, 1981) and (ii) the improved LDA by adding the gradient of the density, that is, the generalized gradient approximation (GGA) (Perdew and Wang, 1992; Perdew *et al.*, 1996a). By construction, LDA is expected to adequately describe a slowly varying density, but it works well for most cases, especially in materials with band gaps (e.g., insulators) and for surface properties. GGA is typically superior to LDA for metallic systems. For example the ground-state of bcc Fe can be described correctly by GGA but not by LDA (Jones and Gunnarsson, 1989). Generally speaking, GGA overestimates the lattice parameters while LDA underestimates them, and both GGA and LDA underestimate the band gap widths. In order to overcome the drawbacks of GGA and LDA, several improved exchange-correlation functionals have been developed, such as the PBEsol GGA proposed by Perdew *et al.* (2008) which uses a reduced gradient dependence for better description of surface and solid properties. Hybrid functionals, which mix the (partial) exact exchange energy with a DFT exchange functional, offer a better description of band gaps and other properties. Examples of hybrid functionals include PBE0 (Perdew *et al.*, 1996b), HSE06 (Krukau *et al.*, 2006), and HSEsol (Schimka *et al.*, 2011), etc.

Two parameters are extremely important to the practice of first-principles calculations: the k-point sampling mesh and the energy cutoff for the wave functions. 'Bigger is better' for both of these parameters and will yield more accurate results; however, an increase in accuracy requires more computational time. For more details of DFT-based first-principles calculations, see Cottenier (2002); Hafner (2000); Jones and Gunnarsson (1989); Martin (2004).

3.4.2 First-principles thermodynamics: the quasiharmonic approach

Finite temperature thermodynamics of a sold phase can be predicted by first-principles in terms of the quasiharmonic approach, wherein Helmholtz energy at volume V and temperature T is usually approximated by (Wang *et al.*, 2004b; Shang *et al.*, 2007b; Shang *et al.*, 2010d):

$$F(V,T) = E_c(V) + F_{\text{el}}(V,T) + F_{\text{vib}}(V,T). \qquad [3.29]$$

Here, E_c is the static energy at 0 K predicted directly by first-principles, F_{el} the thermal electronic contribution, and F_{vib} the vibrational contribution of the lattice. The second and the third terms are related to contributions at finite temperatures, depended on the number of configurations, that is, the electronic density of states (DOS) and phonon DOS as a function of energy and frequency, respectively.

Energy vs volume equation of state (EOS)

In order to get the static energy term in Equation [3.29], the energy vs volume (E-V) data calculated via first-principles can be fitted by an EOS. The commonly adopted E-V EOS uses four parameters (Shang *et al.*, 2010d):

$$E(V) = a + bV^{-n/3} + cV^{-2n/3} + dV^{-3n/3} \qquad [3.30]$$

where a, b, c, and d are fitting parameters. When $n = 2$, Equation [3.30] is the widely used Birch–Murnaghan (BM) EOS (Birch, 1947; Birch, 1978), and when $n = 1$, it becomes the modified Birch–Murnaghan (mBM) EOS proposed by Teter *et al.* (1995). Note that four equilibrium properties can be evaluated from the four-parameter EOS: the equilibrium volume (V_0), energy (E_0), bulk modulus (B_0) and the derivative of the bulk modulus with respect to pressure ($B_0^{'}$, $B^{'}(V) = \partial B / \partial P$), consequently, the fitting parameters can be expressed by the predicted equilibrium properties. More details about EOS, such as the Murnaghan, Vinet, and Morse EOS, can be found in (Shang *et al.*, 2010d).

Thermal electronic contribution

Thermal electronic contribution to the Helmholtz energy is important to consider for metals due to the finite number of electrons present at the Fermi level. F_{el} can be determined by Mermin statistics, $F_{el} = E_{el} - TS_{el}$, using the electronic DOS as the input (Wang *et al.*, 2004b; Shang *et al.*, 2007b; Shang *et al.*, 2010d). The internal energy due to electronic excitations at a given V and T is given by:

$$E_{\mathrm{el}}(V,T) = \int n(\varepsilon) f \varepsilon \mathrm{d}\varepsilon - \int^{E_F} n(\varepsilon)\varepsilon \mathrm{d}\varepsilon \qquad [3.31]$$

where $n(\varepsilon)$ is the electronic DOS, ε the energy eigenvalues, ε_F the energy at the Fermi level, and f the Fermi distribution function, $f(\varepsilon, T, V) = 1/\{\exp(\varepsilon - \mu(T,V)/k_B T) + 1\}$. Here, k_B represents the Boltzmann's constant, and μ is the electronic chemical potential, which is calculated by keeping the number of electrons at T the same as the number of electrons present below ε_F at 0 K. The electronic entropy due to electronic excitations is written as:

$$S_{\mathrm{el}}(V,T) = -k_B \int n(\varepsilon)[f \ln f + (1-f)\ln(1-f)]\mathrm{d}\varepsilon \qquad [3.32]$$

Vibrational contribution from phonon calculations

The vibrational contribution to the Helmholtz energy can be obtained by utilizing the partition function of the lattice vibration, Z_{vib}:

$$F_{\mathrm{vib}}(V,T) = -k_B T \ln Z_{\mathrm{vib}} = k_B T \int_0^{\infty} \ln\left[2\sinh\frac{\hbar\omega}{2k_B T}\right] g(\omega)\mathrm{d}\omega \qquad [3.33]$$

where $\hbar$ is the reduced Planck constant, ω the frequencies, and $g(\omega)$ the phonon DOS. When $T = 0$ K, the zero-point vibrational energy is obtained, and Equation [3.33] reduces to:

$$F_{\mathrm{zero-vib}}(V) = \frac{1}{2}\int_0^{\infty} \hbar\omega g(\omega)\mathrm{d}\omega \qquad [3.34]$$

Based on the phonon DOS, the nth moment of the Debye cutoff frequency ω_n and the corresponding n^{th} moment of the Debye temperature are obtained via (Hellwege and Olsen, 1981; Arroyave and Liu, 2006; Shang *et al.*, 2010d):

$$\omega_n = \left[\frac{n+3}{3}\int_0^{\omega_{\max}} \omega^n g(\omega)\mathrm{d}\omega\right]^{1/n}, \text{ with } n \neq 0, n > -3 \quad [3.35]$$

$$\omega_0 = \exp\left[\frac{1}{3}+\int_0^{\infty} \omega^n \ln(\omega)\mathrm{d}\omega\right], \text{ with } n = 0 \quad [3.36]$$

The nth moment Debye temperature is estimated by:

$$\Theta_D(n) = \frac{\hbar\omega_n}{k_B} \quad [3.37]$$

For different n in Equation [3.37], the obtained Debye temperature corresponds to different thermodynamic properties (Arroyave and Liu, 2006). The value obtained at $\Theta_D(2)$ is commonly used and is linked to the Debye temperature obtained from heat capacity data (Arroyave and Liu, 2006; Shang *et al.*, 2010d).

There are two leading models in the realm of first-principles phonon calculations: the linear response method and the supercell method (Baroni *et al.*, 2001; van de Walle and Ceder, 2002). Each has its own advantages and disadvantages. The linear response method evaluates directly the dynamical matrix for a set of q points in the Brillouin zone through the density functional perturbation theory without the approximation of the cutoff in neighboring interactions, where the LO–TO splitting (longitudinal and transverse optical phonon splitting) for polar materials due to dipole–dipole interactions can be predicted. The limitations of the linear response method arise from (i) the use of small number of q points, (ii) the time-consuming calculations for some q points (not the Γ point) in complex systems, and (iii) the difficulty of implementation in the existing first-principles codes. In comparison, the supercell method is conceptually simple and can be easily applied to quite complex systems and implemented in the existing first-principles codes, and the phonon frequencies at the exact wave vectors can be calculated accurately without further approximations (Wang *et al.*, 2010f), though the cutoff exists in neighboring interactions, and the LO–TO splitting cannot be estimated directly. Recently Wang *et al.* (2010f) proposed a parameter-free mixed-space approach to phonon calculations applied easily for any systems, where the accurate force constants are evaluated from the real-space supercell approach and the dipole-dipole interactions, which result in LO–TO splitting, are calculated from the linear response method at the Γ point of reciprocal space. More details of the mixed-space approach to phonon can be found in for example Wang *et al.* (2010c, 2010e, 2010f) and Shang *et al.* (2012).

Vibrational contribution from Debye model

As an alternative to the more computationally expensive phonon calculations, the empirical Debye model can be used to quickly evaluate the vibrational contribution to Helmholtz energy, even for unstable structures. The only input required by this model is the Debye temperature Θ_D (Shang *et al.*, 2010d):

$$F_{\mathrm{vib}}(V,T)=\frac{9}{8}k_B\Theta_D+k_BT\left\{3\ln\left[1-\exp\left(-\frac{\Theta_D}{T}\right)\right]-D\left(\frac{\Theta_D}{T}\right)\right\} \quad [3.38]$$

The first term is the zero-point vibrational energy, and the Debye function $D(x)=3/x^3\int_0^x t^3/[\exp(t)-1]\mathrm{d}t$. In order to estimate Equation [3.38], Θ_D can be calculated using the Debye–Grüneisen model (Moruzzi *et al.*, 1988; Shang *et al.*, 2010d) or the Debye–Wang model (Wang *et al.*, 2004a). For more details regarding the Debye model, see Shang *et al.* (2010d).

First-principles thermodynamics of hcp Mg

Pure hcp Mg is selected to demonstrate the capabilities of the first-principles quasiharmonic approach with respect to calculating thermodynamic properties. The VASP code (Kresse and Furthmuller, 1996a, 1996b) is used for first-principles calculations in combination with the PAW method (Kresse and Joubert, 1999; Blöchl, 1994) and the GGA–PBE potential (Perdew *et al.*, 1996a). The phonon calculations are performed using the supercell method as implemented in the ATAT code (van de Walle, 2009) and again using VASP as the computational engine. A 3 × 3 × 2 supercell with 36 atoms is used for this case. More details of first-principles and phonon calculations by VASP and ATAT codes are given in (Shang *et al.*, 2007b, 2007d, 2010b, 2010c, 2010d).

Figure 3.2 illustrates the first-principles calculated static energies of hcp Mg as a function of volume and the E–V fittings by the EOSs in Equation [3.30] and the ones in (Shang *et al.*, 2010d). Table 3.6 shows the predicted equilibrium properties of V_0, E_0, B_0, and B_0' from each of these EOSs, together with the fitting errors estimated by (Shang *et al.*, 2010d):

$$\sqrt{\frac{\sum[(E_{\mathrm{fit}}-E_{\mathrm{calc}})/E_{\mathrm{calc}}]^2}{n}} \quad [3.39]$$

Table 3.6 Predicted properties for hcp Mg by different EOSs together with experimental data, including equilibrium volume V_0 (Å³/unit cell), energy E_0 (eV), bulk modulus B_0 (GPa) and its derivative with respect to pressure B_0'. EOS fitting errors estimated by Equation [3.39] are also shown

Method	V_0	E_0	B_0	B_0'	Error (× 10^{-4})
mBM	45.724	−3.0837	36.31	4.07	0.019
BM	45.723	−3.0836	36.28	4.08	0.043
LOG	45.731	−3.0837	36.46	3.98	0.179
Murnaghan	45.714	−3.0836	36.05	4.19	0.304
Vinet	45.727	−3.0837	36.38	4.04	0.071
Morse	45.727	−3.0837	36.36	4.04	0.055
Expt.	46.41[a]		36.9[b]		

[a]Villars and Calvert (1991); [b]Slutsky and Garland (1957).

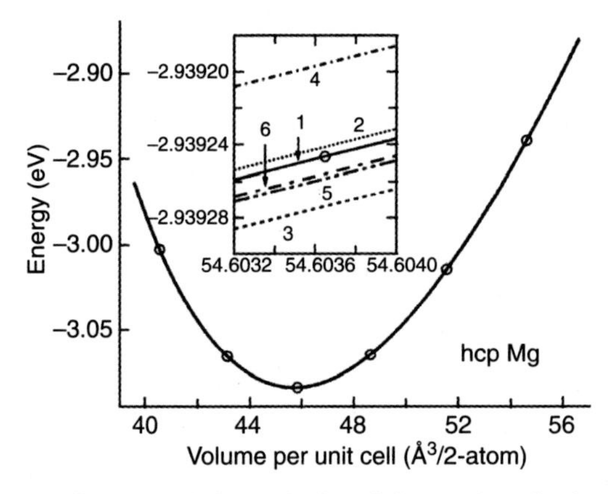

3.2 First-principles calculated data points for hcp Mg (symbols) fitted by different *E-V* EOS's (lines): 1-mBM, 2-BM, 3-LOG, 4-Murnaghan, 5-Vinet, and 6-Morse.

where E_{fit} and E_{calc} are fitted and first-principles calculated energies, respectively, and n is the total number of the data points. Figure 3.2 and Table 3.6 indicate the mBM EOS gives the best fitting quality, while the Murnaghan EOS is the worst. The BM EOS gives the second best fit. Additionally, the predicted equilibrium volumes agree well with the measurements (Villars and Calvert, 1991) with a difference of ~1.5%. The predicted bulk moduli are also in good agreement with experimental results (Simmons and Wang, 1971) with an error of only ~1.6%.

Figure 3.3 shows the predicted electronic DOS for hcp Mg. Below the Fermi level, the electrons in this simple metal behave like free electrons.

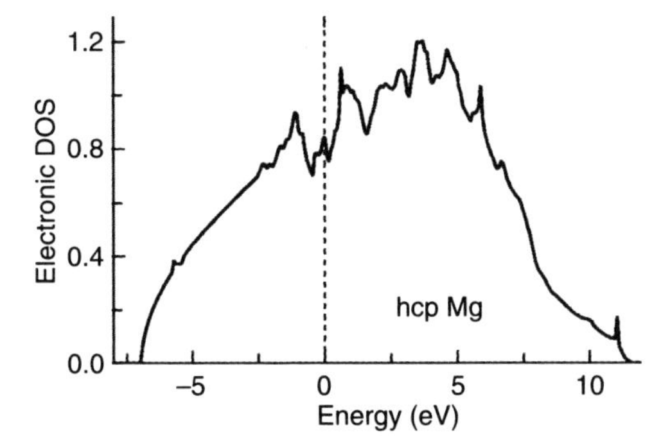

3.3 First-principles predicted electronic DOS for hcp Mg. The dashed line indicates the Fermi level.

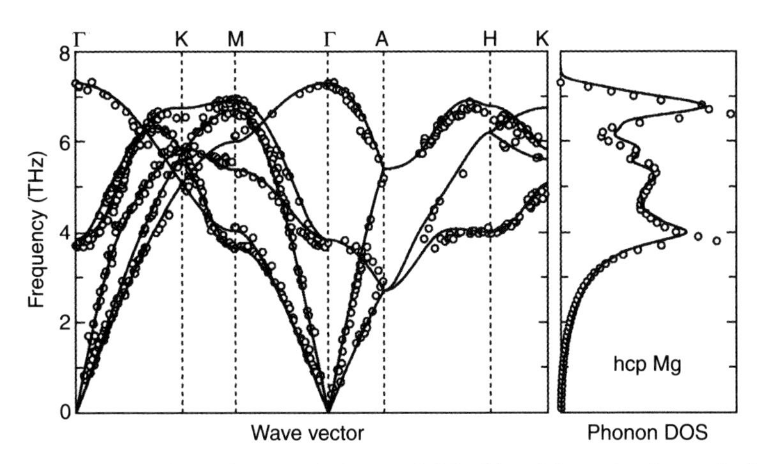

3.4 Phonon dispersions and phonon DOS of hcp Mg at the theoretical equilibrium volume (lines) predicted by first-principles together with experimental data (Hellwege and Olsen, 1981).

Because the electronic DOS has a finite value at the Fermi level, the thermal electronic contribution must be considered when calculating the Helmholtz energy (Shang *et al.*, 2010d). Figure 3.4 shows the phonon dispersions along the high-symmetry directions and the phonon DOS of hcp Mg at the theoretical equilibrium volume. Both the dispersion relations and the phonon DOS are in prefect agreement with experimental measurements (Hellwege and Olsen, 1981). The predicted phonon DOS will be used to calculate the vibrational contribution to the Helmholtz energy using Equation [3.33].

Figure 3.5 shows the predicted thermodynamic properties of hcp Mg after considering both the vibrational and thermal electronic contributions before

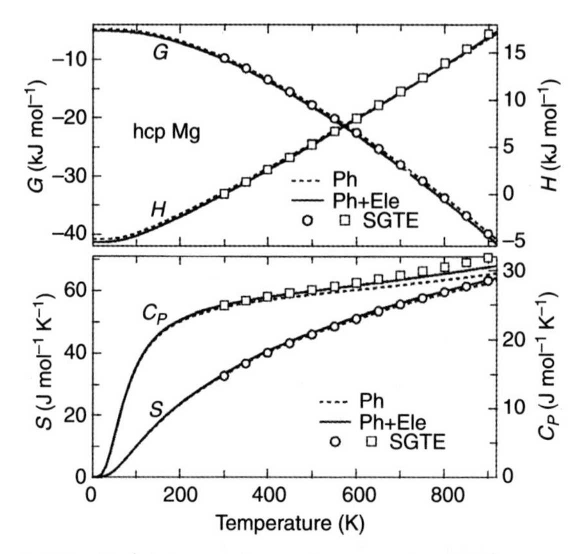

3.5 Predicted thermodynamic properties (Gibbs energy *G*, entropy *S*, enthalpy *H*, and heat capacity C_P at zero external pressure) of hcp Mg using first-principles quasiharmonic approach, together with SGTE data (Dinsdale, 1991).

finally using Equation [3.29] and the derivatives of *G* given in Section 3.2.1. The values from the SGTE database (Dinsdale, 1991) are also shown for comparison. The reference state of hcp Mg is the commonly used one in CALPHAD community, that is, the enthalpy at 298.15 K and ambient pressure as mentioned in Section 3.2.1. Figure 3.5 indicates that the properties predicted with the first-principles quasiharmonic approach agree well with SGTE data, especially when both vibrational and thermal electronic contributions are included.

Figure 3.6 shows the predicted coefficients of linear thermal expansion (LTE) and the predicted isothermal and isentropic bulk moduli derived from the predicted Helmholtz energy according to the thermodynamic relations shown in Section 3.2.1. The predicted LTE agrees with experimental data (Corporate-Author, 1972) with differences less than 6% at high temperatures. The predicted bulk moduli are lower than the measured isentropic data (Slutsky and Garland, 1957), but the trend relative to temperature agrees well with experimental measurements.

Besides the example of hcp Mg shown herein, finite temperature thermodynamic properties have been predicted widely using first-principles calculations, for example, the Mg-containing compounds of $Mg_{17}Al_{12}$ (Zhang *et al.*, 2010), Mg_2Si, and Mg_2Ge (Wang *et al.*, 2010a; Yu *et al.*, 2010).

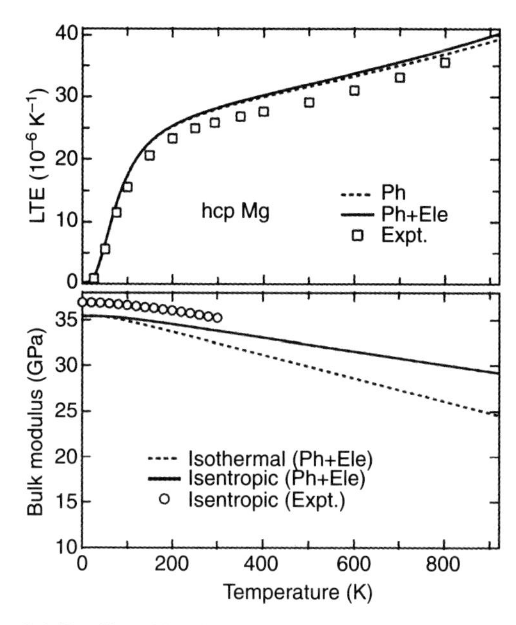

3.6 Predicted isothermal and isentropic bulk moduli and the average coefficient of linear thermal expansion (LTE) of hcp Mg using first-principles quasiharmonic approach. The symbols are experimental data of bulk modulus (Slutsky and Garland, 1957) and LTE (Corporate-Author, 1972).

3.4.3 Treatment of disordered phases for first-principles calculations

The methodology described in Section 3.4.2 is for ordered phases. For disordered or partially disordered phases, the degree of randomness must be considered. Currently, three approaches stand out among first-principles based models (Ghosh *et al.*, 2008; Shang *et al.*, 2011b): the coherent potential approximation (CPA), the cluster expansion method (CEM), and SQS. Each method has its own advantages and disadvantages. For the CPA, the average occupations of A and B atoms are assumed to be in a structureless, uniform average medium (Zunger *et al.*, 1990); therefore, local structural relaxations are excluded. This conflicts with experimental observations because distances between A–A, B–B, and A–B atoms are generally different (Weightman *et al.*, 1987; Renaud *et al.*, 1988). All CPA-type methods are excluded herein because they are incapable of capturing the intrinsic nature of disordered alloys (Shang *et al.*, 2011b). The CEM is derived from statistical lattice theory, which assumes a direct correspondence between a given structure and a set of correlation functions

(Shang *et al.*, 2010f, 2011b). Usually, the configuration-dependent properties of ordered and disordered alloys are determined by applying correlation functions to the phase of interest and by accounting for the effective cluster interactions (ECIs). Within the CEM, the truncated error associated with determining ECIs is unavoidable, and all local structural information of the disordered structure is also left out. The fundamental idea of the SQS approach proposed by Zunger *et al.* (1990) is the same as that of the CEM approach, namely, a given structure can be characterized by a set of correlation functions (Shang *et al.*, 2011b). An SQS is in fact an ordered supercell that mimics the most relevant pair and multisite correlation functions of the disordered phase. In contrast to the global and non-structural natures of CPA and CEM, an SQS is a local structural model, giving one of the down-selected microstates of a disordered phase (Shang *et al.*, 2011b). Besides the commonly used CPA, CEM and SQS, another approach for describing disordered phases has been recently proposed: the partition function method by considering the competing local configurations (Shang *et al.*, 2010a, 2010f; Wang *et al.*, 2008, 2009, 2010d).

Special quasirandom structures (SQS)

The key to generating a 'good' SQS is to encompass all possible structure configurations with a given number of atoms. The ATAT code (van de Walle, 2009) is the tool of choice for this task. The ordered structure obtained from this procedure (i.e., the SQS) possesses the most relevant pair and multisite correlation functions of the disordered phase. The essence of the SQS method is, therefore, to use an *ordered* structure to mimic the properties of the corresponding *disordered* structure (Shang *et al.*, 2011b). Currently, SQS structures have been generated for some of the symmetrical crystal structures. For the case of binary A_xB_{1-x} alloys, the generated SQSs include: (i) fcc-based 8-atom, 16-atom, and 32-atom SQSs with $x = 0.25$ and 0.5 (Shang *et al.*, 2011b); (ii) bcc-based 4-atom, 8-atom, and 16-atom SQSs with $x = 0.25$ and 0.5 (Jiang *et al.*, 2004); and (iii) hcp-based 8-atom and 16-atom SQSs with $x = 0.25$ and 0.5 (Shin *et al.*, 2006). For the case of ternary alloys, the generated SQSs include: (i) fcc-based 24-atom SQSs for $A_{1/3}B_{1/3}C_{1/3}$ and $A_{1/2}B_{1/4}C_{1/4}$ (Shin *et al.*, 2007) and (ii) bcc-based 32-atom SQSs for $A_{1/2}B_{1/4}C_{1/4}$, 36-atom SQSs for $A_{1/3}B_{1/3}C_{1/3}$, and 64-atom SQSs for $A_{2/8}B_{3/8}C_{3/8}$ and $A_{6/8}B_{1/8}C_{1/8}$ (Jiang, 2009). Applications of SQSs can be found in these sources: Zhong *et al.* (2005) for the Al–Mg system, Zhong *et al.* (2006a) for the Ca–Mg system, Shang *et al.* (2007a) for the Ba-Ni-Ti system, and the citations for the work of Zunger *et al.* (1990).

Partition function method

Applying knowledge of statistical physics, one can define the macroscopic disordered phase as a collection of numerous microscopic ordered configurations. The macroscopic disordered system can therefore be described by the canonical partition function under the *NVT* framework (a constant number of particles *N*, volume *V*, and temperature *T*) (Shang *et al.*, 2010a, 2010f; Wang *et al.*, 2008, 2009, 2010d):

$$Z = \sum_{\sigma} w^{\sigma} \sum_{i\in\delta,\rho\in\delta} \exp[-\beta\varepsilon_i(N,V,\rho)] = \sum_{\sigma} w^{\sigma} \exp(-\beta F^{\sigma}) = \sum_{\sigma} Z^{\sigma} \qquad [3.40]$$

where $\beta = 1/(k_B T)$, w^σ is the degeneracy factor of the distinguishable electronic state σ, i the vibrational state belonging to σ, ρ the electronic distributions associated with σ, and ε_i (N, V, ρ) the eigenvalue of the corresponding microscopic Hamiltonian. Summations over i and ρ yield the Helmholtz energy (F^σ) of state σ as a function of V and T. Here it is proposed that F^σ for each state σ can be depicted by the first-principles quasiharmonic approach according to Equation [3.29]. Based on the partition function, various thermodynamic functions can be obtained for a disordered phase. The total Helmholtz energy is given by:

$$F = \frac{-\log(Z)}{\beta} = \sum_{\sigma} x^{\sigma} F^{\sigma} - TS_{conf} \qquad [3.41]$$

where $x^\sigma = Z^\sigma/Z$ is the thermal population of σ as a function of T with the degeneracy factor w^σ included. The structural configurational entropy S_{conf} due to the competing configurations σ is introduced automatically by the partition function given in Equation [3.40]:

$$S_{\text{conf}} = -k_B \sum_{\sigma} \left[x^{\sigma} \log(x^{\sigma}) - x^{\sigma} \log(w^{\sigma}) \right] \qquad [3.42]$$

3.4.4 First-principles elasticity and thermal expansion

Theories of elasticity and thermal expansion

In terms of the strain energy defined by Equation [3.20] or Hooke's law, first-principles elastic constants c_{ij} can be determined based on the strain vs strain energy method or the strain vs stress method, respectively. With the current capabilities of first-principles calculations, the stresses, defined as the first derivatives of first-principles total energies with respect to a given set of strains, can be evaluated directly in most first-principles codes (e.g.,

VASP code), making the strain vs stress method simpler and more efficient than the conventional strain vs strain energy method (Shang *et al.*, 2010b). The strain vs stress method avoids the numerical difficulties often encountered with evaluations of the strain vs strain energy, and all c_{ij} values are computed simultaneously rather than as independent sums of c_{ij}. Only the strain vs stress method is presented herein.

Based on the methodology proposed by Shang *et al.* (2007c), a set of independent strains given by Equation [3.17] can be written as:

$$\boldsymbol{\varepsilon} = (\varepsilon_1\ \varepsilon_2\ \varepsilon_3\ \varepsilon_4\ \varepsilon_5\ \varepsilon_6\ ) \qquad [3.43]$$

where ε_1, ε_2, and ε_3 are the normal strains and the others are shear strains (see Equation [3.17]). Considering a strain ε applied to lattice vectors R in Cartesian coordinates:

$$\mathrm{R} = \begin{pmatrix} a_1 & a_2 & a_3 \\ b_1 & b_2 & b_3 \\ c_1 & c_2 & c_3 \end{pmatrix} \qquad [3.44]$$

where a_1 is the first (or x^{th}) component of the lattice vector **a**, and so on, a deformed crystal with lattice vectors $\bar{\mathrm{R}}$ results:

$$\bar{\mathbf{R}} = \mathbf{RD} = \mathbf{R}\begin{pmatrix} 1+\varepsilon_1 & \varepsilon_6/2 & \varepsilon_5/2 \\ \varepsilon_6/2 & 1+\varepsilon_2 & \varepsilon_4/2 \\ \varepsilon_5/2 & \varepsilon_4/2 & 1+\varepsilon_3 \end{pmatrix} \qquad [3.45]$$

Due to the applied strain $\boldsymbol{\varepsilon}$ (see Equations [3.17] and [3.43]), a set of relative stresses with respect to the stresses of the original crystal with lattice vectors R is generated:

$$\boldsymbol{\sigma} = (\sigma_1\ \sigma_2\ \sigma_3\ \sigma_4\ \sigma_5\ \sigma_6) \qquad [3.46]$$

When determining these stresses via first-principles, the atomic positions in the strained (and fixed) unit cell are allowed to relax. Based on n sets of strains and the resulting n sets of stresses, the elastic constant matrix **C** (see Equation [3.21]) is obtained based on Hooke's law:

$$\mathbf{C} = \Sigma^{-1}\Delta \qquad [3.47]$$

Here, $\mathbf{\Sigma}$ and $\mathbf{\Delta}$ are the $n \times 6$ strain and stress matrices, respectively. Each set of strains and stresses is given by the matrices of Equation [3.43] and Equation [3.46], respectively. The superscript '-1' on Σ represents the pseudo-inverse. For the sake of simplicity the following six linearly-independent sets of strains are chosen (Shang *et al.*, 2007c):

$$\Sigma = \begin{pmatrix} x & 0 & 0 & 0 & 0 & 0 \\ 0 & x & 0 & 0 & 0 & 0 \\ 0 & 0 & x & 0 & 0 & 0 \\ 0 & 0 & 0 & x & 0 & 0 \\ 0 & 0 & 0 & 0 & x & 0 \\ 0 & 0 & 0 & 0 & 0 & x \end{pmatrix} \qquad [3.48]$$

where each row is one set of strains. Commonly selected values x are ± 0.007, ± 0.01, and ± 0.013. According to Equation [3.47], the static elastic constants c_{ij} at 0 K (without the effect of zero-point vibrational energy) can be determined by first-principles calculations for a crystal with any symmetry at a specific volume of interest. With c_{ij} values determined for several volumes, one can develop an expression for $c_{ij}(V)$. Similarly, using the relaxed lattice vectors R at 0 K at a series of fixed cell volumes (the charge of cell shape is allowed at each fixed volume), the strains as a function of volume $\varepsilon_i(V)$ can also be obtained (Equation [3.45]).

It has been observed that the change of elastic properties at elevated temperatures is mainly due to volume change (Swenson, 1968; Gülseren and Cohen, 2002; Ledbetter, 2006). By ignoring the effects of anharmonicity, kinetic energy, and fluctuations of microscopic stress tensors, the isothermal elastic constants can be estimated using the quasistatic approach (Shang *et al.*, 2010g; Wang *et al.*, 2010g), outlined as follows:

i. predict the static elastic constants at 0 K as a function of volume $c_{ij}(V)$;
ii. predict the volume change as a function of temperature $V(T)$ or $T(V)$ according to the quasiharmonic approach (see Section 3.4.2);
iii. obtain the isothermal elastic constants as a functional of V, $C_{ij}(T) = C_{ij}(T(V))$.

Similar to the isothermal elastic constants from the quasistatic approach, the temperature-dependent strains are obtained: $\varepsilon_i(T) = \varepsilon_i(T(V))$, and, in turn, the direction-dependent thermal expansions can be determined based on Equation [3.18]. Note that the thermodynamic relations between the isentropic elastic constants and the isothermal elastic constants have been given previously, that is, Equation [3.7] and Equation [3.25].

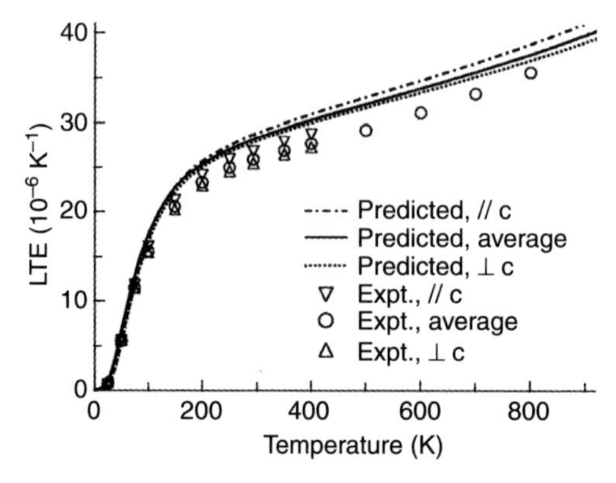

3.7 First-principles predicted direction-dependent coefficients of linear thermal expansion (LTE) of hcp Mg, together with experimental data (Corporate-Author, 1972).

Thermal expansion and elasticity for Mg-based alloys and compounds

Figure 3.7 shows the first-principles predicted coefficients of LTE of hcp Mg, including the average LTE and the direction-dependent LTEs: parallel (//) and perpendicular (⊥) to the *c*-axis (see also Equation [3.19]). In the calculations, the VASP code (Kresse and Furthmuller, 1996a, 1996b) has been used together with the PAW method (Kresse and Joubert, 1999; Blöchl, 1994) and the GGA–PW91 potentials (Perdew and Wang, 1992). Phonon calculations have been performed with the ATAT code (van de Walle, 2009). Further details can be found in Section 3.4.2. It is worth mentioning that the GGA–PBE potential (Perdew *et al.*, 1996a) is unable to make an accurate prediction of the direction-dependent LTEs of hcp Mg though the average LTEs predicted with GGA–PW91 and GGA–PBE are similar (see Figs 3.6 and 3.7). Hence, the GGA–PW91 has been selected to predict the direction-dependent LTE. Figure 3.7 shows that the predicted LTEs are only slightly higher than experimental data (Corporate-Author, 1972), and both the experiments and predictions indicate that the coefficient of LTE parallel to the *c*-axis is larger than that perpendicular to the *c*-axis.

Figure 3.8 illustrates the isothermal and isentropic c_{ij}'s for hcp Mg predicted by first-principles calculations, and the measured isentropic c_{ij} (Slutsky and Garland, 1957) are also shown for comparison. Note that the GGA–PBE has been used for this case, but the other settings are the same as for previous first-principles calculations. Figure 3.8 shows that the isentropic c_{ij} values are larger than isothermal values, the exception being c_{44} (isothermal) = c_{44} (isentropic) due to the zero thermal expansion in the hcp case,

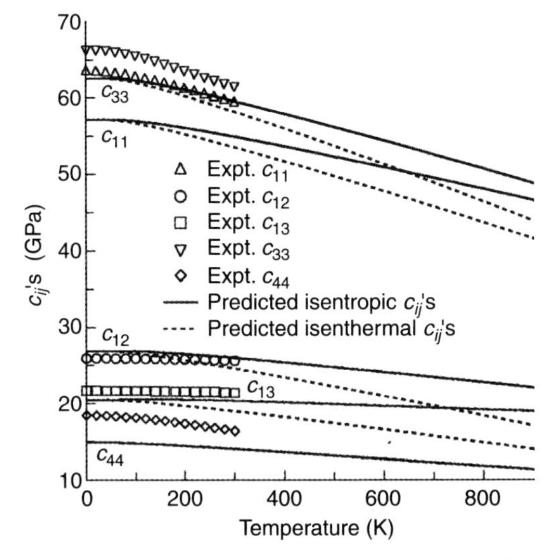

3.8 First-principles predicted isothermal and isentropic c_{ij}'s for hcp Mg, together with isentropic measurements (Slutsky and Garland, 1957).

that is, $\alpha_4 = \alpha_5 = \alpha_6 = 0$. The calculated isentropic c_{ij}'s in Fig. 3.8 show similar trends with respect to temperature as those of measured values, although the predicted c_{11}, c_{33}, and c_{44} are lower than the experimental measurements. Regarding c_{12} and c_{13}, the predictions agree well with measurements. Using the quasistatic approach, the predicted c_{ij} values as a function of temperature available in the literature include the following: $Mg_{17}Al_{12}$ (Zhang *et al.*, 2010), Mg_2X with X = Si, Ge, Sn, and Pb (Ganeshan *et al.*, 2010b), α- and θ-Al_2O_3 (Shang *et al.*, 2010g), the cubic elements Al, Cu, Ni, Mo, and Ta, and the compounds NiAl, and Ni_3Al (Wang *et al.*, 2010g).

For the binary compounds Mg-X (X = As, Ba, Ca, Ce, Cd, Cu, Ga, Ge, La, Ni, Pb, Si, Sn, and Y), the first-principles predicted c_{ij} at 0 K are listed in Table 3.7 (Ganeshan *et al.*, 2009b; Shang *et al.*, 2009). In addition, the bulk modulus B, shear modulus G, and B/G ratio have also been predicted based on these c_{ij} values. Table 3.8 lists the effects of the alloying elements X (X = Al, Ba, Ca, Cu, Ge, K, Li, Ni, Pb, Si, Y, and Zn) on the elastic constants of hcp Mg (Ganeshan *et al.*, 2009a). Details and discussions about these Mg-based alloys and compounds can be found in Ganeshan *et al.* (2009a, 2009b). In addition to the results given in Table 3.7, more first-principles c_{ij} at 0 K have been predicted for Mg-containing alloying and compounds, such as: $Mg_{17}Al_{12}$ and Mg_2Pb (Duan *et al.*, 2011), Mg_3As_2 and Mg_3Sb_2 (Tani *et al.*, 2010), Mg_3Gd (Tang *et al.*, 2010), Mg_7Gd (Gao *et al.*, 2011), Mg_2Si (Yu *et al.*,

Table 3.7 First-principles elastic properties of Mg-X compounds predicted by strain vs stress method (Ganeshan *et al.*, 2009b), where *B* and *G* are bulk modulus and shear modulus, respectively, estimated by Voigt approach

Compound	SG	c_{11}	c_{12}	c_{13}	c_{33}	c_{44}	c_{66}	*B*	*G*	*B*/*G*
Mg_3As_2	$Ia\bar{3}$	76.2	36.1			32.9		49.4	27.7	1.78
$MgAs_4$	$P4_12_12$	67.5	33.2	26.5	83.3	26	34.2	43.4	26	1.67
$Mg_{23}Ba_6$	$Fm\bar{3}m$	41.5	15.9			17.2		24.4	15.4	1.58
Mg_2Ba	$P6_3/mmc$	39.8	14.5	9.8	46.2	11.0		21.5	13.0	1.65
Mg_2Ca	$P6_3/mmc$	59.5	17.8	12.6	66.0	17.4		30.1	20.6	1.46
Mg_2Ca[a]		61.2	17.6	15.0	65.5	19.2		31.4	21.3	1.47
MgCe[b]	$Pm\bar{3}m$	37.4	23.9			46.7		28.4	22.3	1.27
MgCd[c]	*Pmma*	71.4	30.9	31.3	82.9	18.7	21.6	45.8	22.1	2.07
$MgCd_3$	$P6_3/mmc$	58.4	38.3	43.6	66.5	15.6		48.2	12.1	3.99
$MgCd_3$[a]		59.2	37.2	29.7	50	10.2		40.1	11	3.63
Mg_3Cd	$P6_3/mmc$	61.5	35.4	26.3	75.2	24.4		41.5	19.7	2.11
$MgCu_2$	$Fd\bar{3}m$	107.9	79			34.6		88.6	26.5	3.34
$MgCu_2$[a]		125	71.7			42.3		89.4	36	2.48
MgGa	$I4_1/a$	72.5	30.8	38.2	78.9	26.3	8.8	48.7	20	2.43
Mg_2Ga	$P\bar{6}2c$	81.4	22.3	42.2	36.7	19.5		45.8	19.9	2.31
Mg_2Ga_5	$I4/mmm$	90.5	44.8	23.9	105.4	25.0	33.7	52.4	29.6	1.77
$MgGa_2$[d]	*Pbam*	74.4	34.9	35.9	83.6	11.7	27.3	49.8	18.8	2.64
Mg_5Ga_2[e]	*Ibam*	79.7	26	30	71	20.3	13.6	43.9	22	2
Mg_2Ge	$Fm\bar{3}m$	107.3	21.1			41.8		49.8	42.3	1.18
Mg_2Ge[a]		117.9	23			46.5		54.6	46.8	1.17
MgLa	$Pm\bar{3}m$	46.7	27.8			36.2		34.1	25.5	1.34
Mg_2La	$Fd\bar{3}m$	58.4	24.9			21.8		36	19.7	1.82
Mg_3La	$Fm\bar{3}m$	59.2	26.2			36		37.2	28.2	1.32
Mg_2Ni	$P6_222$	113.4	45.1	34.9	123.7	19.4		64.4	30.3	2.13
$MgNi_2$	$P6_3/mmc$	214.1	87.5	70.1	225.3	47.3		123.2	59.9	2.05
Mg_2Pb	$Fm\bar{3}m$	55.2	23.3			24.3		33.9	20.9	1.62
Mg_2Pb[a]		71.7	22.1			30.9		38.6	28.4	1.36
Mg_2Si	$Fm\bar{3}m$	116.7	23.1			45.3		54.3	45.9	1.18
Mg_2Si[a]		121	22			46.4		55	47.6	1.15
Mg_2Sn	$Fm\bar{3}m$	69.8	25.9			31.1		40.5	27.4	1.48
Mg_2Sn[a]		82.4	20.8			36.6		41.3	34.2	1.21
MgY	$Pm\bar{3}m$	51.8	35.8			37.3		41.1	25.5	1.61
MgY[a]		53.3	36.4			39		42	26.8	1.57
$Mg_{24}Y_5$	$I\bar{4}3m$	73.9	22.1			18		39.3	21.1	1.86

[a]Experimental data, see details in Ganeshan *et al.* (2009b); [b]First-principles results for ferromagnetic phase (Shang *et al.*, 2009); [c]Three more c_{ij}s with c_{22} = 79.9, c_{23} = 27.1, and c_{66} = 21.6 GPa; [d]Three more c_{ij}s with c_{22} = 87.0, c_{23} = 31.0, and c_{66} = 27.3 GPa; [e]Three more c_{ij}s with c_{22} = 76.2, c_{23} = 28.2, and c_{66} = 13.6 Gpa.

2010), B2-MgRE (RE = Sc, Y, La-Lu) (Tao *et al.*, 2008), Mg–Li compounds (Phasha *et al.*, 2010), $AMgNi_4$ (A = Y, La, Ce, Pr, and Nd) (Wang *et al.*, 2011), $Mg(Cu_{1-x}Zn_x)_2$ alloys (Wu *et al.*, 2011), and $Zn_{1-x}Mg_xSe$ alloys (Surucu *et al.*, 2011).

Table 3.8 Calculated coefficients of linear regression, δ (GPa/at%), for elastic properties of hcp-based Mg-X alloys based on first-principles calculations and available experiments (Ganeshan *et al.*, 2009a). For hcp Mg, the first-principles and measured elastic properties are also shown as the basis of linear regression

X	δ-c_{11}	δ-c_{12}	δ-c_{13}	δ-c_{33}	δ-c_{44}	δ-B	δ-G
Al	0.74	0.36	−0.24	1.06	−0.60	0.27	−0.03
Ba	−3.16	−0.14	0.27	−2.42	0.25	−0.87	−0.61
Ca	−2.58	0.98	0.81	−2.04	−0.27	−0.21	−1.12
Cu	−0.69	−0.14	2.06	−1.19	0.22	0.61	−0.41
Ge	0.34	0.51	0.05	0.56	−0.41	0.28	−0.14
K	−3.18	−0.63	0.61	−3.47	−0.56	−0.95	−1.18
Li	−1.65	−0.13	1.16	−4.35	−0.11	−0.36	−0.85
Li[a]	−0.25	−0.12	−0.10	−0.25	−0.07	−0.15	−0.07
Ni	−0.02	0.39	1.37	1.30	1.41	0.84	0.40
Pb	−0.15	0.00	−0.12	0.45	−0.61	−0.03	−0.24
Si	0.52	0.88	−0.12	1.48	−0.29	0.43	−0.03
Y	−1.44	0.88	0.59	−0.54	1.33	0.09	−0.07
Zn	−0.43	0.23	1.10	0.05	−0.43	0.46	−0.46
Material	c_{11}	c_{12}	c_{13}	c_{33}	c_{44}	B	G
Mg	63.5	24.9	20.0	66.0	19.3	35.8	18.5
Mg[b]	63.48	25.94	21.70	66.45	18.42	36.89	

[a]Experimental data (Wazzan and Robinson, 1967). [b]Experimental data (Slutsky and Garland, 1957).

3.5 Future trends

Due to the advances in first-principles methodologies and computer resources, thermodynamic properties can be predicted efficiently and accurately. Apart from the thermochemical, thermal expansion, and elastic properties discussed herein, more properties can be predicted with first-principles. In particular, the ongoing and future research includes the following:

1. Helmholtz energies for metastable and unstable phases, especially disordered phases. These properties are crucial for CALPHAD-type modeling and particularly for the automation of thermodynamic database development (Shang *et al.*, 2010e). As an alternative to the quasiharmonic approach, *ab initio* modular dynamics (AIMD) is an effective way to predict the energetics of unstable phases (Ozolins, 2009).
2. Defect energies, defined as the energy difference before and after the formation of defects. These include vacancy formation energy, stacking fault energy (Wang *et al.*, 2010b) and the generalized stacking fault energy (Han *et al.*, 2011), APB energy (Manga *et al.*, 2010), surface and interfacial energies (Wang *et al.*, 2007).
3. Diffusion coefficients of solid phases. Predicted vacancy formation and migration energies, together with phonon calculations and transition

state theory, can be used to calculate diffusion coefficients of solid phases completely from first-principles (Mantina *et al.*, 2008, 2009; Ganeshan *et al.*, 2010a, 2011; Shang *et al.*, 2011a).

4. Creep properties. It is proposed that the creep rate $\dot{\varepsilon}$ of Mg and Mg alloys for the climb-controlled dislocation creep is given by (Somekawa *et al.*, 2005): $\varepsilon = 3.6 \times 10^{11} (\gamma_{SFE} / Gb)^3 (Gb / kT) (\sigma / G)^5 D_{eff}$, where γ_{SFE} is the stacking fault energy, G the shear modulus, b the magnitude of the Burgers vector, k the Boltzmann's constant, T the temperature, σ the flow stress, and D_{eff} the effective diffusion coefficient.
5. Tensile and shear strengths, which can be predicted from first-principles through the stresses resulting from crystal deformation (Ogata *et al.*, 2002; Clatterbuck *et al.*, 2003).
6. Solute strengthening by first-principles calculations. See details of the predictions for Al-based alloys (Leyson *et al.*, 2010) and Mg-based alloys (Yasi *et al.*, 2010) in the literature.

3.6 Acknowledgements

This work is funded by National Science Foundation through grant DMR-1006557 and Office of Naval Research under the contract number of N0014–07–1–0638. We greatly appreciate the program managers David Shifler and Alan J. Ardell for their support and encouragement, and Ms. Alyson Lieser for reading the manuscript.

3.7 References

Agarwal, R., Fries, S. G., Lukas, H. L., Petzow, G., Sommer, F., Chart, T. G. and Effenberg, G. (1992) 'Assessment of the Mg-Zn system', *Z. Metallkd.,* **83**, 216.

Allison, J., Liu, B., Boyle, K., Beals, R. and Hector, L. (2008) Integrated computational materials engineering (ICME) for magnesium: an international pilot project. In Pekguleryuz, M. O., Neelameggham, N. R., Beals, R. and Nyberg, E. A. (Eds.) *Magnesium Technology 2008.* New Orleans, Minerals, Metals and Materials Society/AIME, 184 Thorn Hill Road, Warrendale, PA.

Andersson, J. O., Helander, T., Hoglund, L. H., Shi, P. F. and Sundman, B. (2002) 'Thermo-Calc and Dictra, computational tools for materials science', *CALPHAD*, **26**, 273. doi: 10.1016/S0364–5916(02)00037–8.

Ansara, I., Dinsdale, A. and Rand, M. (1998) *COST 507: Thermochemical database for light metal alloys*, Luxembourg, European Commission.

Arroyave, R. and Liu, Z. K. (2006) 'Intermetallics in the Mg-Ca-Sn ternary system: Structural, vibrational, and thermodynamic properties from first principles', *Phys. Rev. B*, **74**, 174118. doi: 10.1103/PhysRevB.74.174118.

Arroyave, R., Shin, D. and Liu, Z. K. (2005) 'Modification of the thermodynamic model for the Mg-Zr system', *CALPHAD*, **29**, 230. doi: 10.1016/j.calphad.2005.07.004.

Bale, C. W., Belisle, E., Chartrand, P., Decterov, S. A., Eriksson, G., Hack, K., Jung, I. H., Kang, Y. B., Melancon, J., Pelton, A. D., Robelin, C. and Petersen, S. (2009) 'FactSage thermochemical software and databases – recent developments', *CALPHAD*, **33**, 295. doi: 10.1016/j.calphad.2008.09.009.

Baroni, S., De Gironcoli, S., Dal Corso, A. and Giannozzi, P. (2001) 'Phonons and related crystal properties from density-functional perturbation theory', *Rev. Mod. Phys.*, **73**, 515. doi: 10.1103/RevModPhys.73.515.

Birch, F. (1947) 'Finite elastic strain of cubic crystals', *Phys. Rev.*, **71**, 809. doi: 10.1103/PhysRev.71.809.

Birch, F. (1978) 'Finite strain isotherm and velocities for single-crystal and polycrystalline NaCl at high-pressures and 300 K', *J. Geophys. Res.*, **83**, 1257. doi: 10.1029/JB083iB03p01257.

Blöchl, P. E. (1994) 'Projector augmented-wave method', *Phys. Rev. B*, **50**, 17953. doi: 10.1103/PhysRevB.50.17953.

Born, M. and Huang, K. (1998) *Dynamical theory of crystal lattices*, Oxford, Clarendon Press.

Cao, W., Chen, S. L., Zhang, F., Wu, K., Yang, Y., Chang, Y. A., Schmid-Fetzer, R. and Oates, W. A. (2009) 'PANDAT software with PanEngine, PanOpti mizer and PanPrecipitation for multi-component phase diagram calculation and materials property simulation', *CALPHAD*, **33**, 328. doi: 10.1016/j.calphad.2008.08.004.

Clatterbuck, D. M., Chrzan, D. C. and Morris, J. W. (2003) 'The ideal strength of iron in tension and shear', *Acta Mater.*, **51**, 2271. doi: 10.1016/s1359–6454(03)00033–8.

Corporate-Author (1972) *American institute of physics handbook*, New York, McGraw-Hill.

Cottenier, S. (2002) *Density Functional Theory and the family of (L)APW-methods: a step-by-step introduction*, Belgium, KU Leuven.

Davies, G. F. (1974) 'Effective elastic-moduli under hydrostatic stress. 1. Quasi-harmonic theory', *J. Phys. Chem. Solids*, **35**, 1513. doi: 10.1016/S0022–3697(74)80279–9.

Dinsdale, A. T. (1991) 'SGTE data for pure elements', *CALPHAD*, **15**, 317. doi: 10.1016/0364–5916(91)90030-N.

Duan, Y. H., Sun, Y., Peng, M. J. and Guo, Z. Z. (2011) 'First principles investigation of the binary intermetallics in Pb-Mg-Al alloy: Stability, elastic properties and electronic structure', *Solid State Sci.*, **13**, 455. doi: 10.1016/j.solidstatesciences.2010.12.011.

Fabrichnaya, O. B., Lukas, H. L., Effenberg, G. and Aldinger, F. (2003) 'Thermodynamic optimization in the Mg-Y system', *Intermetallics*, **11**, 1183. doi: 10.1016/S0966–9795(03)00156–0.

Ganeshan, S., Hector, L. G. and Liu, Z. K. (2010a) 'First-principles study of self-diffusion in hcp Mg and Zn', *Comput. Mater. Sci.*, **50**, 301. doi: 10.1016/j.commatsci.2010.08.019.

Ganeshan, S., Hector, L. G. and Liu, Z. K. (2011) 'First-principles calculations of impurity diffusion coefficients in dilute Mg alloys using the 8-frequency model', *Acta Mater.*, **59**, 3214. doi: 10.1016/j.actamat.2011.01.062.

Ganeshan, S., Shang, S. L., Wang, Y. and Liu, Z. K. (2009a) 'Effect of alloying elements on the elastic properties of Mg from first-principles calculations', *Acta Mater.*, **57**, 3876. doi: 10.1016/j.actamat.2009.04.038.

Ganeshan, S., Shang, S. L., Wang, Y. and Liu, Z. K. (2010b) 'Temperature dependent elastic coefficients of Mg_2X (X = Si, Ge, Sn, Pb) compounds from first-principles calculations', *J. Alloy. Compd.*, **498**, 191. doi: 10.1016/j.jallcom.2010.03.153.

Ganeshan, S., Shang, S. L., Zhang, H., Wang, Y., Mantina, M. and Liu, Z. K. (2009b) 'Elastic constants of binary Mg compounds from first-principles calculations', *Intermetallics*, **17**, 313. doi: 10.1016/j.intermet.2008.11.005.

Gao, L., Zhou, J., Sun, Z. M., Chen, R. S. and Han, E. H. (2011) 'First-principles calculations of the β'-Mg_7Gd precipitate in Mg-Gd binary alloys', *Chin. Sci. Bull.*, **56**, 1142. doi: 10.1007/s11434–010–4061-z.

Ghosh, G., Van DeWalle, A. and Asta, M. (2008) 'First-principles calculations of the structural and thermodynamic properties of bcc, fcc and hcp solid solutions in the Al-TM (TM = Ti, Zr and Hf) systems: A comparison of cluster expansion and supercell methods', *Acta Mater.*, **56**, 3202. doi: 10.1016/j.actamat.2008.03.006.

Grobner, J., Mirkovic, D., Ohno, M. and Schmid-Fetzer, R. (2005) 'Experimental investigation and thermodynamic calculation of binary Mg-Mn phase equilibria', *J. Phase Equilib. Diff.*, **26**, 234. doi: 10.1361/15477030523544.

Gülseren, O. and Cohen, R. E. (2002) 'High-pressure thermoelasticity of body-centered-cubic tantalum', *Phys. Rev. B*, **65**, 064103. doi: 10.1103/PhysRevB.65.064103.

Guo, C. P. and Du, Z. M. (2004) 'Thermodynamic assessment of the La-Mg system', *J. Alloy. Compd.*, **385**, 109. doi: 10.1016/j.jallcom.2004.04.105.

Guo, C. P. and Du, Z. M. (2005) 'Thermodynamic assessment of the Mg-Pr system', *J. Alloy. Compd.*, **399**, 183. doi: 10.1016/j.jallcom.2005.03.033.

Hafner, J. (2000) 'Atomic-scale computational materials science', *Acta Mater.*, **48**, 71. doi: 10.1016/S1359–6454(99)00288–8.

Han, J., Su, X. M., Jin, Z. H. and Zhu, Y. T. (2011) 'Basal-plane stacking-fault energies of Mg: A first-principles study of Li- and Al-alloying effects', *Scr. Mater.*, **64**, 693. doi: 10.1016/j.scriptamat.2010.11.034.

Hellwege, K. H. and Olsen, J. L. (1981) *Phonon states of elements. Electron states and Fermi surfaces of alloys*, Berlin, Springer.

Hill, R. (1952) 'The elastic behaviour of a crystalline aggregate', *Proc. Phys. Soc. London Sec. A*, **65**, 349. doi: 10.1088/0370–1298/65/5/307.

Hillert, M. (2008) *Phase equilibria, phase diagrams, and phase transformations: their thermodynamic basis*, Cambridge, Cambridge University Press.

Hillert, M. and Jarl, M. (1978) 'Model for alloying effects in ferromagnetic metals', *CALPHAD*, **2**, 227. doi: 10.1016/0364–5916(78)90011–1.

Ho, C. Y. and Taylor, R. E. (1998) *Thermal expansion of solids*, ASM International, Materials Park, OH (United States).

Hohenberg, P. and Kohn, W. (1964) 'Inhomogeneous electron gas', *Phys. Rev.*, **136**, B864. doi: 10.1103/PhysRev.136.B864

Ipser, H., Mikula, A. and Katayama, I. (2010) 'Overview: The emf method as a source of experimental thermodynamic data', *CALPHAD*, **34**, 271. doi: 10.1016/j.calphad.2010.05.001.

Jiang, C. (2009) 'First-principles study of ternary bcc alloys using special quasi-random structures', *Acta Mater.*, **57**, 4716. doi: 10.1016/j.actamat.2009.06.026.

Jiang, C., Wolverton, C., Sofo, J., Chen, L. Q. and Liu, Z. K. (2004) 'First-principles study of binary bcc alloys using special quasirandom structures', *Phys. Rev. B*, **69**, 214202 doi: 10.1103/PhysRevB.69.214202.

Jones, R. O. and Gunnarsson, O. (1989) 'The density functional formalism, its applications and prospects', *Rev. Mod. Phys.*, **61**, 689. doi: 10.1103/RevModPhys.61.689.

Kohn, W. and Sham, L. J. (1965) 'Self-consistent equations including exchange and correlation effects', *Phys. Rev.*, **140**, A1133. doi: 10.1103/PhysRev.140.A1133.

Kresse, G. and Furthmuller, J. (1996a) 'Efficiency of ab-initio total energy calculations for metals and semiconductors using a plane-wave basis set', *Comput. Mater. Sci.*, **6**, 15. doi: 10.1016/0927–0256(96)00008–0.

Kresse, G. and Furthmuller, J. (1996b) 'Efficient iterative schemes for ab initio total-energy calculations using a plane-wave basis set', *Phys. Rev. B*, **54**, 11169. doi: 10.1103/PhysRevB.54.11169.

Kresse, G. and Joubert, D. (1999) 'From ultrasoft pseudopotentials to the projector augmented-wave method', *Phys. Rev. B*, **59**, 1758. doi: 10.1103/PhysRevB.59.1758.

Krukau, A. V., Vydrov, O. A., Izmaylov, A. F. and Scuseria, G. E. (2006) 'Influence of the exchange screening parameter on the performance of screened hybrid functionals', *J. Chem. Phys.*, **125**, 224106. doi: 10.1063/1.2404663.

Kubaschewski, O., Alcock, C. B. and Spencer, P. J. (1993) *Materials thermochemistry*, New York, Pergamon Press.

Ledbetter, H. (2006) 'Sound velocities, elastic constants: Temperature dependence', *Mater. Sci. Eng. A*, **442**, 31. doi: 10.1016/j.msea.2006.04.147.

Leyson, G. P. M., Curtin, W. A., Hector, L. G. and Woodward, C. F. (2010) 'Quantitative prediction of solute strengthening in aluminium alloys', *Nat. Mater.*, **9**, 750. doi: 10.1038/nmat2813.

Liu, Z. K. (2009) 'First-principles calculations and CALPHAD modeling of thermodynamics', *J. Phase Equilib. Diff.*, **30**, 517. doi: 10.1007/s11669–009–9570–6.

Liu, Z. K., Zhang, H., Ganeshan, S., Wang, Y. and Mathaudhu, S. N. (2010) 'Computational modeling of the effects of alloying elements on elastic coefficients', *Scripta Mater.*, **63**, 686. doi: 10.1016/j.scriptamat.2010.03.049.

Lukas, H. L., Fries, S. G. and Sundman, B. (2007) *Computational thermodynamics: The Calphad method,* Cambridge, Cambridge University Press.

Manga, V. R., Saal, J. E., Wang, Y., Crespi, V. H. and Liu, Z. K. (2010) 'Magnetic perturbation and associated energies of the antiphase boundaries in ordered Ni_3Al', *J. Appl. Phys.*, **108**, 103509. doi: 10.1063/1.3513988.

Mantina, M., Shang, S. L., Wang, Y., Chen, L. Q. and Liu, Z. K. (2009) '3d transition metal impurities in aluminum: A first-principles study', *Phys. Rev. B*, **80**, 184111. doi: 10.1103/PhysRevB.80.184111.

Mantina, M., Wang, Y., Arroyave, R., Chen, L. Q., Liu, Z. K. and Wolverton, C. (2008) 'First-principles calculation of self-diffusion coefficients', *Phys. Rev. Lett.*, **100**, 215901. doi: 10.1103/PhysRevLett.100.215901.

Martin, R. M. (2004) *Electronic structure: basic theory and practical methods*, Cambridge, Cambridge University Press.

Meng, F. G., Liu, H. S., Liu, L. B. and Jin, Z. P. (2007) 'Thermodynamic optimization of Mg-Nd system', *Trans. Nonferrous Met. Soc. China*, **17**, 77. doi: 10.1016/S1003–6326(07)60051-X.

Miedema, A. R., Dechatel, P. F. and Deboer, F. R. (1980) 'Cohesion in alloys – fundamentals of a semi-empirical model', *Physica B and C*, **100**, 1. doi: 10.1016/0378–4363(80)90054–6.

Moruzzi, V. L., Janak, J. F. and Schwarz, K. (1988) 'Calculated thermal-properties of metals', *Phys. Rev. B*, **37**, 790. doi: 10.1103/PhysRevB.37.790.

Nye, J. F. (1985) *Physical properties of crystals: their representation by tensors and matrices*, Oxford, Clarendon Press.

Ogata, S., Li, J. and Yip, S. (2002) 'Ideal pure shear strength of aluminum and copper', *Science*, **298**, 807. doi: 10.1126/science.1076652.

Olson, G. B. (1997) 'Computational design of hierarchically structured materials', *Science*, **277**, 1237. doi: 10.1126/science.277.5330.1237.

Ozolins, V. (2009) 'First-principles calculations of free energies of unstable phases: The case of fcc W', *Phys. Rev. Lett.*, **102**, 065702. doi: 10.1103/PhysRevLett.102.065702.

Perdew, J. P., Burke, K. and Ernzerhof, M. (1996a) 'Generalized gradient approximation made simple', *Phys. Rev. Lett.*, **77**, 3865. doi: 10.1103/PhysRevLett.77.3865.

Perdew, J. P., Emzerhof, M. and Burke, K. (1996b) 'Rationale for mixing exact exchange with density functional approximations', *J. Chem. Phys.*, **105**, 9982. doi: 10.1063/1.472933.

Perdew, J. P., Ruzsinszky, A., Csonka, G., Aacute, Bor, I., Vydrov, O. A., Scuseria, G. E., Constantin, L. A., Zhou, X. and Burke, K. (2008) 'Restoring the density-gradient expansion for exchange in solids and surfaces', *Phys. Rev. Lett.*, **100**, 136406. doi: 10.1103/PhysRevLett.100.136406.

Perdew, J. P. and Wang, Y. (1992) 'Accurate and simple analytic representation of the electron-gas correlation-energy', *Phys. Rev. B*, **45**, 13244. doi: 10.1103/PhysRevB.45.13244.

Perdew, J. P. and Zunger, A. (1981) 'Self-interaction correction to density-functional approximations for many-electron systems', *Phys. Rev. B*, **23**, 5048. doi: 10.1103/PhysRevB.23.5048.

Phasha, M. J., Ngoepe, P. E., Chauke, H. R., Pettifor, D. G. and Nguyen- Mann, D. (2010) 'Link between structural and mechanical stability of fcc- and bcc-based ordered Mg-Li alloys', *Intermetallics*, **18**, 2083. doi: 10.1016/j.intermet.2010.06.015.

Pugh, S. F. (1954) 'Relations between the elastic moduli and the plastic properties of polycrystalline pure metals', *Philos. Mag.*, **45**, 823. doi: 10.1080/14786440808520496

Ranganathan, S. I. and Ostoja- Starzewski, M. (2008) 'Universal elastic anisotropy index', *Phys. Rev. Lett.*, **101**, 055504. doi: 10.1103/PhysRevLett.101.055504.

Redlich, O. and Kister, A. T. (1948) 'Algebraic representation of thermodynamic properties and the classification of solutions', *Ind. Eng. Chem.*, **40**, 345. doi: 10.1021/ie50458a036.

Renaud, G., Motta, N., Lancon, F. and Belakhovsky, M. (1988) 'Topological short-range disorder in $Au_{1-x}Ni_x$ solid-solutions – An extended X-ray-absorption fine-structure spectroscopy and computer-simulation study', *Phys. Rev. B*, **38**, 5944. doi: 10.1103/PhysRevB.38.5944.

Saunders, N. and Miodownik, A. P. (1998) *CALPHAD (Calculation of Phase Diagrams): A Comprehensive Guide*, Oxford, Pergamon.

Schimka, L., Harl, J. and Kresse, G. (2011) 'Improved hybrid functional for solids: The HSEsol functional', *J. Chem. Phys.*, **134**, 024116. doi: 10.1063/1.3524336.

Shang, S., Zhang, H., Ganeshan, S. and Liu, Z. K. (2008) 'The development and application of a thermodynamic database for magnesium alloys', *JOM*, **60**(12), 45. doi: 10.1007/s11837–008–0165–1.

Shang, S. L., Hector, L. G., Jr., Wang, Y. and Liu, Z. K. (2011a) 'Anomalous energy pathway of vacancy migration and self-diffusion in hcp Ti', *Phys. Rev. B*, **83**, 224104. doi: 10.1103/PhysRevB.83.224104.

Shang, S. L., Hector, L. G., Wang, Y., Zhang, H. and Liu, Z. K. (2009) 'First-principles study of elastic and phonon properties of the heavy fermion compound CeMg', *J. Phys. Condens. Matter*, **21**, 246001. doi: 10.1088/0953–8984/21/24/246001.

Shang, S. L., Liu, Z. J. and Liu, Z. K. (2007a) 'Thermodynamic modeling of the Ba-Ni-Ti system', *J. Alloy. Compd.*, **430**, 188. doi: 10.1016/j.jallcom.2006.03.096.

Shang, S. L., Saal, J. E., Mei, Z. G., Wang, Y. and Liu, Z. K. (2010a) 'Magnetic thermodynamics of fcc Ni from first-principles partition function approach', *J. Appl. Phys.*, **108**, 123514. doi: 10.1063/1.3524480.

Shang, S. L., Saengdeejing, A., Mei, Z. G., Kim, D. E., Zhang, H., Ganeshan, S., Wang, Y. and Liu, Z. K. (2010b) 'First-principles calculations of pure elements: Equations of state and elastic stiffness constants', *Comput. Mater. Sci.*, **48**, 813. doi: 10.1016/j.commatsci.2010.03.041.

Shang, S. L., Wang, Y., Du, Y. and Liu, Z. K. (2010c) 'Entropy favored ordering: Phase stability of Ni_3Pt revisited by first-principles', *Intermetallics*, **18**, 961. doi: 10.1016/j.intermet.2010.01.011.

Shang, S. L., Wang, Y., Kim, D. E. and Liu, Z. K. (2010d) 'First-principles thermodynamics from phonon and Debye model: Application to Ni and Ni_3Al', *Comput. Mater. Sci.*, **47**, 1040. doi: 10.1016/j.commatsci.2009.12.006.

Shang, S. L., Wang, Y., Kim, D. E., Zacherl, C. L., Du, Y. and Liu, Z. K. (2011b) 'Structural, vibrational, and thermodynamic properties of ordered and disordered $Ni_{1-x}Pt_x$ alloys from first-principles calculations', *Phys. Rev. B*, **83**, 144204. doi: 10.1103/PhysRevB.83.144204.

Shang, S. L., Wang, Y. and Liu, Z. K. (2007b) 'First-principles calculations of phonon and thermodynamic properties in the boron-alkaline earth metal binary systems: B-Ca, B-Sr, and B-Ba', *Phys. Rev. B*, **75**, 024302. doi: 10.1103/PhysRevB.75.024302.

Shang, S. L., Wang, Y. and Liu, Z. K. (2007c) 'First-principles elastic constants of α- and θ-Al_2O_3', *Appl. Phys. Lett.*, **90**, 101909. doi: 10.1063/1.2711762.

Shang, S. L., Wang, Y. and Liu, Z. K. (2010e) ESPEI: extensible, self-optimizing phase equilibrium infrastructure for magnesium alloys. In Agnew, S. R., Neelameggham, N. R., Nyberg, E. A. and Sillekens, W. H. (Eds.) *Magnesium Technology 2010*. Warrendale, PA, The Minerals, Metals, and Materials Society. 617.

Shang, S. L., Wang, Y. and Liu, Z. K. (2010f) 'Thermodynamic fluctuations between magnetic states from first-principles phonon calculations: The case of bcc Fe', *Phys. Rev. B*, **82**, 014425. doi: 10.1103/PhysRevB.82.014425.

Shang, S. L., Wang, Y., Mei, Z. G., Hui, X. D. and Liu, Z. K. (2012) 'Lattice dynamics, thermodynamics, and bonding strength of lithium-ion battery materials $LiMPO_4$ (M = Mn, Fe, Co, and Ni): A comparative first-principles study', *J. Mater. Chem.*, **22**, 1142. doi: 10.1039/C1JM13547C.

Shang, S. L., Wang, Y., Zhang, H. and Liu, Z. K. (2007d) 'Lattice dynamics and anomalous bonding in rhombohedral As: First-principles supercell method', *Phys. Rev. B*, **76**, 052301. doi: 052301 10.1103/PhysRevB.76.052301.

Shang, S. L., Zhang, H., Wang, Y. and Liu, Z. K. (2010g) 'Temperature-dependent elastic stiffness constants of α- and θ-Al_2O_3 from first-principles calculations', *J. Phys. Condens. Matter*, **22**, 375403. doi: 10.1088/0953–8984/22/37/375403.

Shin, D., Arroyave, R., Liu, Z.-K. and Van De Walle, A. (2006) 'Thermodynamic properties of binary hcp solution phases from special quasirandom structures', *Phys. Rev. B*, **74**, 024204. doi: 10.1103/PhysRevB.74.024204.

Shin, D., Van De Walle, A., Wang, Y. and Liu, Z. -K. (2007) 'First-principles study of ternary fcc solution phases from special quasirandom structures', *Phys. Rev. B*, **76**, 144204. doi: 10.1103/PhysRevB.76.144204.

Simmons, G. and Wang, H. (1971) *Single crystal elastic constants and calculated aggregate properties,* Cambridge (Mass.), MIT Press.

Slutsky, L. J. and Garland, C. W. (1957) 'Elastic constants of magnesium from 4.2 K to 300 K', *Phys. Rev.*, **107**, 972. doi: 10.1103/PhysRev.107.972.

Somekawa, H., Hirai, K., Watanabe, H., Takigawa, Y. and Higashi, K. (2005) 'Dislocation creep behavior in Mg-Al-Zn alloys', *Mater. Sci. Eng. A*, **407**, 53. doi: 10.1016/j.msea.2005.06.059.

Surucu, G., Colakoglu, K., Deligoz, E., Ciftci, Y. and Korozlu, N. (2011) 'Electronic, elastic and optical properties on the $Zn_{1-x}Mg_xSe$ mixed alloys', *J. Mater. Sci.*, **46**, 1007. doi: 10.1007/s10853–010–4864-y.

Swenson, C. A. (1968) 'Equation of state of cubic solids – some generalizations', *J. Phys. Chem. Solids*, **29**, 1337. doi: 10.1016/0022–3697(68)90185–6.

Tang, B. Y., Chen, P., Li, D. L., Yi, J. X., Wen, L., Peng, L. M. and Ding, W. J. (2010) 'First-principles investigation of the structural and mechanical properties of β'' phase in Mg-Gd alloy system', *J. Alloy. Compd.*, **492**, 416. doi: 10.1016/j.jallcom.2009.11.127.

Tani, J., Takahashi, M. and Kido, H. (2010) 'Lattice dynamics and elastic properties of Mg_3As_2 and Mg_3Sb_2 compounds from first-principles calculations', *Physica B*, **405**, 4219. doi: 10.1016/j.physb.2010.07.014.

Tao, X. M., Ouyang, Y. F., Liu, H. S., Feng, Y. P., Du, Y. and Jin, Z. P. (2008) 'Elastic constants of B2-MgRE (RE = Sc, Y, La-Lu) calculated with first-principles', *Solid State Commu.*, **148**, 314. doi: 10.1016/j.ssc.2008.09.005.

Teter, D. M., Gibbs, G. V., Boisen, M. B., Allan, D. C. and Teter, M. P. (1995) 'First-principles study of several hypothetical silica framework structures', *Phys. Rev. B*, **52**, 8064. doi: 10.1103/PhysRevB.52.8064.

Van DeWalle, A. (2009) 'Multicomponent multisublattice alloys, nonconfigurational entropy and other additions to the Alloy Theoretic Automated Toolkit', *CALPHAD*, **33**, 266. doi: 10.1016/j.calphad.2008.12.005.

Van DeWalle, A. and Ceder, G. (2002) 'The effect of lattice vibrations on substitutional alloy thermodynamics', *Rev. Mod. Phys.*, **74**, 11. doi: 10.1103/RevModPhys.74.11.

Villars, P. and Calvert, L. D. (1991) *Pearson's handbook of crystallographic data for intermetallic phases*, Newbury, OH, ASTM International.

Wang, H. F., Jin, H., Chu, W. G. and Guo, Y. J. (2010a) 'Thermodynamic properties of Mg_2Si and Mg_2Ge investigated by first principles method', *J. Alloy. Compd.*, **499**, 68. doi: 10.1016/j.jallcom.2010.01.134.

Wang, J. W., Yang, F., Fan, T. W., Tang, B. Y., Peng, L. M. and Ding, W. J. (2011) 'Theoretical investigation of new type of ternary magnesium alloys $AMgNi_4$ (A=Y, La, Ce, Pr and Nd)', *Physica B*, **406**, 1330. doi: 10.1016/j.physb.2011.01.028.

Wang, Y., Ahuja, R. and Johansson, B. (2004a) 'Mean-field potential approach to the quasiharmonic theory of solids', *Int. J. Quantum Chem.*, **96**, 501. doi: 10.1002/qua.10769.

Wang, Y., Chen, L. Q., Liu, Z. K. and Mathaudhu, S. N. (2010b) 'First-principles calculations of twin-boundary and stacking-fault energies in magnesium', *Scr. Mater.*, **62**, 646. doi: 10.1016/j.scriptamat.2010.01.014.

Wang, Y., Hector, L. G., Zhang, H., Shang, S. L., Chen, L. Q. and Liu, Z. K. (2008) 'Thermodynamics of the Ce gamma-alpha transition: Density-functional study', *Phys. Rev. B*, **78**, 104113. doi: 10.1103/PhysRevB.78.104113.

Wang, Y., Hector, L. G., Zhang, H., Shang, S. L., Chen, L. Q. and Liu, Z. K. (2009) 'A thermodynamic framework for a system with itinerant-electron magnetism', *J. Phys. Condens. Matter*, **21**, 326003. doi: 10.1088/0953–8984/21/32/326003.

Wang, Y., Liu, Z. K. and Chen, L. Q. (2004b) 'Thermodynamic properties of Al, Ni, NiAl, and Ni_3Al from first-principles calculations', *Acta Mater.*, **52**, 2665. doi: 10.1016/j.actamat.2004.02.014.

Wang, Y., Liu, Z. K., Chen, L. Q. and Wolverton, C. (2007) 'First-principles calculations of β''-Mg_5Si_6/α-Al interfaces', *Acta Mater.*, **55**, 5934. doi: 10.1016/j.actamat.2007.06.045.

Wang, Y., Saal, J. E., Wang, J. J., Saengdeejing, A., Shang, S. L., Chen, L. Q. and Liu, Z. K. (2010c) 'Broken symmetry, strong correlation, and splitting between longitudinal and transverse optical phonons of MnO and NiO from first principles', *Phys. Rev. B*, **82**, 081104. doi: 10.1103/PhysRevB.82.081104.

Wang, Y., Shang, S. L., Zhang, H., Chen, L. Q. and Liu, Z. K. (2010d) 'Thermodynamic fluctuations in magnetic states: Fe_3Pt as a prototype', *Philos. Mag. Lett.*, **90**, 851. doi: 10.1080/09500839.2010.508446.

Wang, Y., Wang, J. J., Saal, J. E., Shang, S. L., Chen, L. Q. and Liu, Z. K. (2010e) 'Phonon dispersion in Sr_2RuO_4 studied by a first-principles cumulative force-constant approach', *Phys. Rev. B*, **82**, 172503. doi: 10.1103/PhysRevB.82.172503.

Wang, Y., Wang, J. J., Wang, W. Y., Mei, Z. G., Shang, S. L., Chen, L. Q. and Liu, Z. K. (2010f) 'A mixed-space approach to first-principles calculations of phonon frequencies for polar materials', *J. Phys. Condens. Matter*, **22**, 202201. doi: 10.1088/0953–8984/22/20/202201.

Wang, Y., Wang, J. J., Zhang, H., Manga, V. R., Shang, S. L., Chen, L. Q. and Liu, Z. K. (2010g) 'A first-principles approach to finite temperature elastic constants', *J. Phys. Condens. Matter*, **22**, 225404. doi: 10.1088/0953–8984/22/22/225404.

Wazzan, A. R. and Robinson, L. B. (1967) 'Elastic constants of magnesium-lithium alloys', *Phys. Rev.*, **155**, 586. doi: 10.1103/PhysRev.155.586.

Weightman, P., Wright, H., Waddington, S. D., Vandermarel, D., Sawatzky, G. A., Diakun, G. P. and Norman, D. (1987) 'Local lattice expansion around Pd impurities in Cu and its influence on the Pd density of states – An extended X-ray-absorption fine-structure and auger study', *Phys. Rev. B*, **36**, 9098. doi: 10.1103/PhysRevB.36.9098.

Wu, M. M., Jiang, Y., Wang, J. W., Wu, J. A., Tang, B. Y., Peng, L. M. and Ding, W. J. (2011) 'Structural, elastic and electronic properties of $Mg(Cu_{1-x}Zn_x)_2$ alloys calculated by first-principles', *J. Alloy. Compd.*, **509**, 2885. doi: 10.1016/j.jallcom.2010.11.148.

Yasi, J. A., Hector, L. G. and Trinkle, D. R. (2010) 'First-principles data for solid-solution strengthening of magnesium: From geometry and chemistry to properties', *Acta Mater.*, **58**, 5704. doi: 10.1016/j.actamat.2010.06.045.

Yu, B. H., Chen, D., Tang, Q. B., Wang, C. L. and Shi, D. H. (2010) 'Structural, electronic, elastic and thermal properties of Mg_2Si', *J. Phys. Chem. Solids*, **71**, 758. doi: 10.1016/j.jpcs.2010.01.017.

Zhang, H., Shang, S. L., Saal, J. E., Saengdeejing, A., Wang, Y., Chen, L. Q. and Liu, Z. K. (2009) 'Enthalpies of formation of magnesium compounds from first-principles calculations', *Intermetallics*, **17**, 878. doi: 10.1016/j.intermet.2009.03.017.

Zhang, H., Shang, S. L., Wang, Y., Saengdeejing, A., Chen, L. Q. and Liu, Z. K. (2010) 'First-principles calculations of the elastic, phonon and thermodynamic properties of $Al1_2Mg_{17}$', *Acta Mater.*, **58**, 4012. doi: 10.1016/j.actamat.2010.03.020.

Zhang, H., Wang, Y., Shang, S. L., Chen, L. Q. and Liu, Z. K. (2008) 'Thermodynamic modeling of Mg-Ca-Ce system by combining first-principles and CALPHAD method', *J. Alloy. Compd.*, **463**, 294. doi: 10.1016/j.jallcom.2007.09.020.

Zhang, S. J. (2006) *Ph.D. thesis*, Pennsylvania State University.

Zhao, J. C. (2007) *Methods for phase diagram determination*, Amsterdam, Elsevier Science.

Zhong, Y., Ozturk, K., Sofo, J. O. and Liu, Z. K. (2006a) 'Contribution of first-principles energetics to the Ca-Mg thermodynamic modeling', *J. Alloy. Compd.*, **420**, 98. doi: 10.1016/j.jallcom.2005.10.033.

Zhong, Y., Sofo, J. O., Luo, A. A. and Liu, Z. K. (2006b) 'Thermodynamics modeling of the Mg-Sr and Ca-Mg-Sr systems', *J. Alloy. Compd.*, **421**, 172. doi: 10.1016/j.jallcom.2006.09.076.

Zhong, Y., Yang, M. and Liu, Z. K. (2005) 'Contribution of first-principles energetics to Al-Mg thermodynamic modeling', *CALPHAD*, **29**, 303. doi: 10.1016/j.calphad.2005.08.004.

Zhou, D. W., Peng, P. and Liu, J. S. (2007) 'Electronic structure and stability of Mg-Ce intermetallic compounds from first-principles calculations', *J. Alloy. Compd.*, **428**, 316. doi: 10.1016/j.jallcom.2006.03.046.

Zunger, A., Wei, S. H., Ferreira, L. G. and Bernard, J. E. (1990) 'Special quasirandom structures', *Phys. Rev. Lett.*, **65**, 353. doi: 10.1103/PhysRevLett.65.353.

Zuo, Y. and Chang, Y. A. (1993) 'Thermodynamic calculation of Mg-Cu phase-diagram', *Z. Metallkd.*, **84**, 662.

4
Understanding precipitation processes in magnesium alloys

C. L. MENDIS, Helmholtz Zentrum Geesthacht, Germany (formerly of National Institute for Materials Science, Japan) and K. HONO, National Institute for Materials Science, Japan

DOI: 10.1533/9780857097293.125

Abstract: Recent interest in developing high strength wrought magnesium alloys has revived studies of precipitation processes in magnesium alloys, as they can be used to control recrystallized microstructure and to add yield strength by age hardening after processing of wrought magnesium. In this chapter, we describe the basic information necessary for controlling precipitate microstructures by aging. Thereafter, precipitation processes and microalloying effects in representative age-hardenable magnesium alloys are reviewed.

Key words: precipitation, solid state phase transformations, age hardening.

4.1 Introduction

Magnesium alloys have attracted renewed interest as light alloys, to substitute some conventional structural materials for weight reduction in vehicles such as cars, trucks, trains and aircrafts. Cast alloys, widely used in interior and power-train components, account for more than 99% of magnesium alloys used today, while only a small number of wrought products are utilized. This is because magnesium alloys lack formability for wrought applications, and their high cost discourages the use of magnesium alloys for automotive applications.

Rare earth (RE) elements, such as CeGd, Nd and Y, are often used as major alloying elements in cast magnesium alloys because of their relatively high solubility in Mg and their effectiveness in precipitation hardening and creep resistance. In fact, several Mg–RE alloys show a notable age hardening response, which can lead to substantial strengthening in Mg (Khoarashahi, 1997; Nie and Muddle, 2000; Lorimer *et al.*, 1999). Unfortunately, the recent price increase of RE elements is a significant hindrance to the widespread use of Mg–RE alloys. The mechanical properties, especially the yield strength required for many structural components, are not yet at adequate levels in most magnesium alloys without RE elements. Hence, existing wrought magnesium alloys do not have much merit over wrought aluminum alloys in terms of both cost and mechanical properties. Thus, more effort is needed to develop low cost and high strength magnesium alloys without using RE elements.

Existing commercial wrought magnesium alloys, such as ZK60 and AZ61, are based on the Mg–Zn and Mg–Zn–Al systems, both of which are age-hardenable. ZK and AZ alloys attain good strength by hot-extrusion or rolling but age hardening, such as T6 (solution heat treatment and aging) and T8 (cold work and subsequent aging), does not give additional strengthening due to softening by recrystallization at the temperature for artificial aging. However, recent investigations have demonstrated that microalloyed Mg–Zn alloys show pronounced age hardening response using T6 treatment. Precipitation processes in magnesium alloy as a method to enhance the strength of wrought products are therefore worth revisiting. This chapter reviews the basic concept of precipitation processes in magnesium alloys and then reviews specific precipitation processes in various magnesium alloys based on the recent literature. Microalloying effects to enhance age hardening are also discussed.

4.2 Precipitation from supersaturated solid solution

4.2.1 Continuous and discontinuous precipitation

The precipitation reaction is a diffusive phase transformation involving the formation of a second phase with a different chemical composition from the supersaturated solid solution. There are two types of precipitation reactions: one is continuous precipitation and the other is discontinuous precipitation (Aaronson and Clark, 1968; Porter and Easterling, 1992). Continuous precipitation occurs by relatively uniform precipitation of secondary phase particles from a supersaturated solid solution, either by the nucleation and growth mechanism or by the spinodal decomposition mechanism; the composition of the matrix phase continuously changes during this process, as shown in Fig. 4.1a. The growth rate of the precipitate particles is controlled by the volume diffusion of the solute in the matrix phase; the particles assumed to be near spherical in general and follow the power law, $r = kt^{1/2}$, during the growth period. Here, k is the temperature-dependent constant, $k \propto \exp(Q/RT)$, and Q is the activation energy for the volume diffusion.

On the other hand, discontinuous precipitation occurs by the migration of the interface between a two phase lamellar colony and the matrix, as shown in Fig. 4.1b. The diffusion occurs along the matrix and colony interface, but there is no change in the chemistry in the unreacted matrix region.

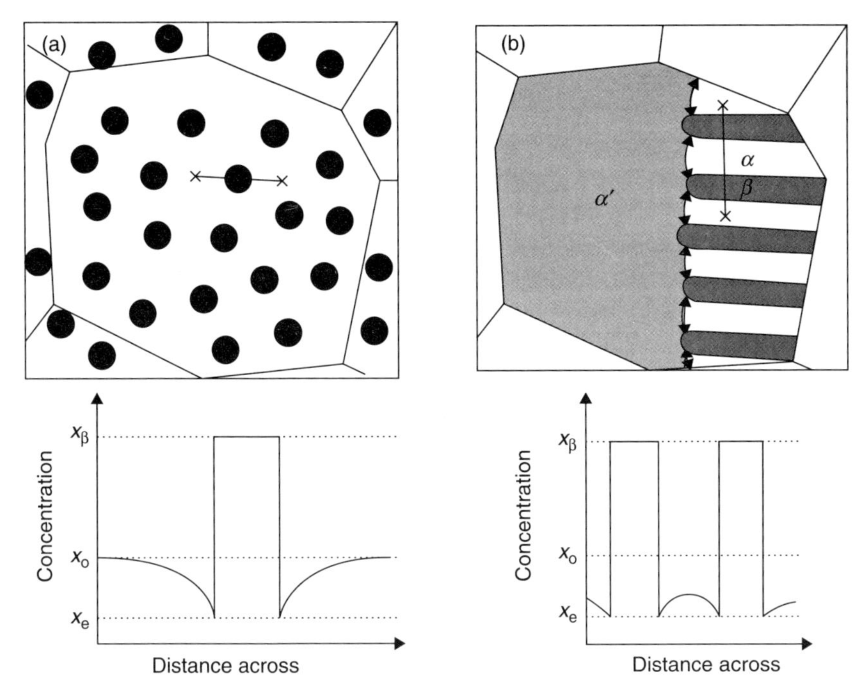

4.1 Schematic illustration of the concentration profiles surrounding the precipitation product during (a) continuous precipitation and (b) discontinuous precipitation from the supersaturated solid solution.

Discontinuous precipitation causes the formation of lamellar structures at grain boundaries (Porter and Easterling, 1992). Therefore, discontinuous precipitation must be avoided in aged alloys. Discontinuous precipitation often occurs at relatively low temperature, at which the kinetics for volume diffusion are too sluggish while there is a large driving force for the precipitation. Although Mg–Al–Zn alloys show a good age hardening response, a certain portion of precipitation occurs by discontinuous precipitation as shown in Fig. 4.2, so the alloy system is not suitable for age hardening.

4.2.2 Precipitation hardening

Given a constant volume fraction of a secondary phase, a higher degree of precipitation hardening is obtained with a finer particle size and a larger number density, as long as the Orowan mechanism governs (i.e., the precipitates are semi-coherent or non-coherent), and the increment of the shear stress should depend on the radius of the precipitates, $\Delta\tau_{\mathrm{Orowan}} \propto r^{-1}$. When the precipitate is coherent, precipitates are cut by dislocations provided that the size is not too small, and the increment of the shear stress changes

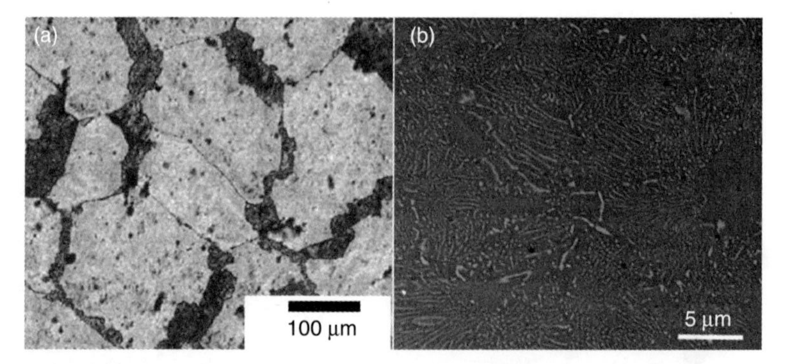

4.2 Discontinuous precipitation observed in a Mg–9Al–0.5Zn (at%) (Mg–9Al–1Zn (wt%)) (AZ91) alloy aged at 160°C for 100h following solution heat treatment at 400°C (a) optical microstructure showing many colonies of discontinuous precipitates and (b) scanning electron micrograph showing one such colony.

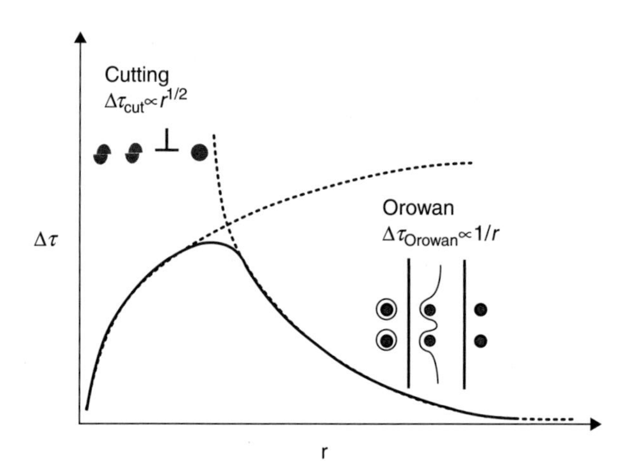

4.3 A schematic illustration of the strengthening increment observed for the increase in the particle size at a given volume fraction of precipitates with mechanisms of strengthening operational at a given particle size also represented.

in proportion to $r^{1/2}$ until an optimum size, that is, $\Delta\tau_{cut} \propto r^{1/2}$. Accordingly, the increment of the precipitation hardening will change as a function of the precipitate radius as shown in Fig. 4.3, for the same volume fraction of the precipitates. The critical radius for the peak increment varies depending on the type of precipitates: it is in the order of 7 nm for the $L1_2$ Ni_3Al particles (γ') in Ni–Al alloys (Munjal and Ardell, 1975) and ~3 nm for $MgZn_2$ (η') in Al–Zn–Mg alloys (Kovacs *et al.*, 1980).

Experimental work on the transition size from coherency strain hardening to cutting in magnesium alloy is not available, but it should not be much different from that for aluminum alloys. Thus, the precipitate size in heat-treatable magnesium alloys must be in the order of ~5 nm to achieve the maximum age hardening response. To achieve such a fine microstructure, the precipitation reaction must occur by the continuous precipitation mode with a large driving force, so that the nucleation rate can be fast. In the following section, the fundamental information on controlling microstructures by the precipitation reaction is described.

4.2.3 Nucleation of precipitates

The formation of a secondary phase with a different chemistry from the matrix phase involves the following three energy terms: (1) reduction of free energy associated with a given volume of precipitates $V\Delta G_v$, (2) increase in the free energy due to the formation of an interface between the precipitate and the matrix $A\gamma$ and (3) increase in the free energy due to the misfit between the matrix phase and the precipitate, $V\Delta G_s$. For a spherical particle with a radius r, the free energy change for the formation of a precipitate is

$$\Delta G = -\frac{4}{3}\pi r^3(\Delta Gv - \Delta G_S) + 4\pi r^3\gamma \qquad [4.1]$$

When the radius of the precipitation become larger than r^*, it will spontaneously grow without reverting back into the solution. The free energy barrier to be overcome for the nuclei to form is given by

$$\Delta G^* = \left(\frac{16\pi\gamma^3}{3(\Delta G_v - \Delta G_S)}\right) \qquad [4.2]$$

Thus, the nucleation rate for a homogeneous nucleation event can be written as

$$N_{\text{hom}} = \omega C_0 \exp\left(-\frac{\Delta G_m}{kT}\right).\exp\left(-\frac{\Delta G^*}{kT}\right) \qquad [4.3]$$

where, T is the temperature at which the event occurs, k_B is Boltzmann's constant, ω is the vibration frequency of atoms, C_o is the initial concentration of solute atoms per unit volume and ΔG_m is the free energy for atomic migration.

In real systems, however, homogeneous nucleation rarely occurs. In most cases, nucleation takes place heterogeneously at the sites that reduce the interfacial and strain energy terms, for example, vacancy clusters, dislocations,

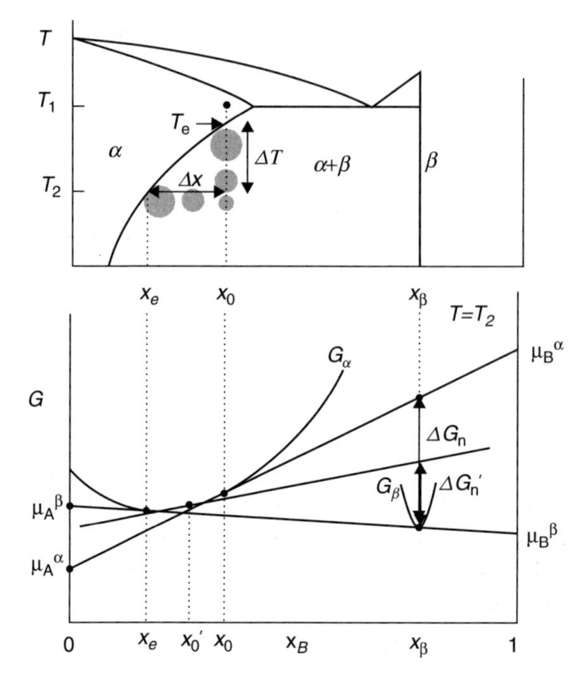

4.4 Schematic illustration of the free energy changes during precipitation. The increase in under cooling results in an increase in the supersaturation of solute and increase in the free energy reduction associated with the precipitation reaction resulting in smaller nuclei. Lower supersaturation at a given temperature results in formation of larger nuclei.

grain boundaries, solute clusters and metastable precipitates. The nucleation rate is strongly dependent on ΔG^* which largely changes with the degree of undercooling and supersaturation. For a solute concentration, x_0, the equilibrium solute concentration at temperature T is x_e, and the solvus temperature for x_0 is T_e as shown in Fig. 4.4. Then, the supersaturation Δx at T is $\Delta x = x_0 - x_e$ and the undercooling ΔT is $\Delta T = (T_e - T)$. The volume free energy ΔG_v in Equation [4.2] is defined as the volume free energy decrease associated with the formation of a unit volume of the β phase, $\Delta G_v = \Delta G_n / V_m$, known as driving force, where V_m is the molar volume of β. For dilute solutions, $\Delta G_v \propto \Delta x$. Since Δx increases as ΔT increases, the driving force, ΔG_n, for the precipitation increases as the supercooling increases, which reduces r^* and ΔG^*, making the nucleation rate faster. Accordingly, the size of the precipitate will be refined with a larger supercooling as shown in Fig. 4.4. However, at lower temperatures, atomic diffusion becomes too sluggish to allow the precipitation within a realistic period of time. Therefore, there will be an optimum temperature for achieving fine precipitate dispersion within an acceptable aging time.

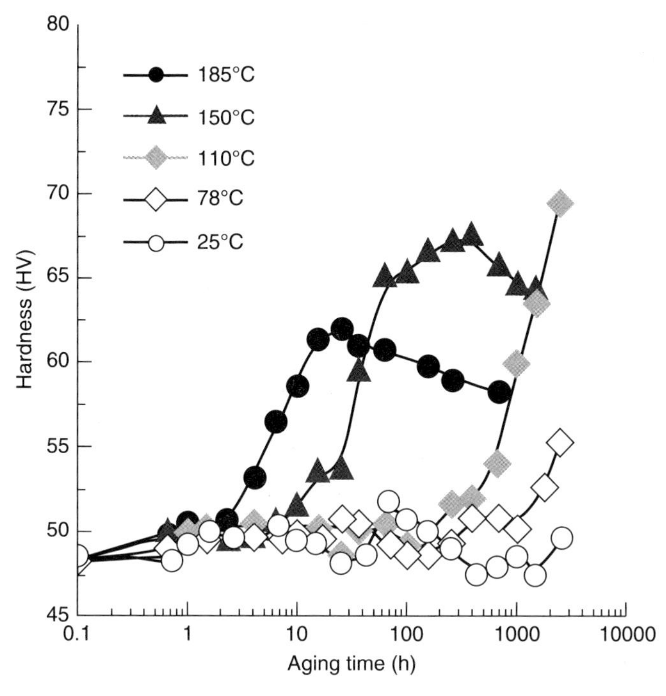

4.5 Age hardening response of Mg–1.8Zn (at%) alloy in the temperature range 25–185°C (After Mima and Tanaka, 1971a).

To refine the precipitate size and increase the number density for a larger age hardening, a larger supercooling is necessary. However, the precipitation kinetics becomes slow with larger undercooling. Thus, the shape of age hardening curves in general show a similar tendency regardless of the alloy systems. Figure 4.5 shows age hardening curves of an Mg–Zn alloy. At low temperature, the hardness increment is larger because of the smaller size and higher number density of precipitates, but the time to peak hardness is longer because of sluggish diffusion of Zn in Mg. At higher temperatures, the alloy reaches peak hardness after a short period of aging time, but the increment of the hardness is small because of coarse precipitate microstructure. Thus, the artificial aging temperature must be selected so that the alloy can reach peak hardness within a realistic aging time with sufficiently high age hardening increment.

4.2.4 Precipitate morphology

The growth direction of precipitates depends on the type of interface between the precipitate and the matrix phase. A coherent or semi-coherent interface results in slower rates of growth in the normal direction to the interface, as compared with an incoherent interface, which is highly mobile. This growth difference of the interfaces causes the formation of crossed platelets know as Widmanstatten microstructure, as typically observed in the γ' precipitates in Al–Ag alloys (Aaronson and Clark, 1968). The coherent or semi-coherent interfaces require the nucleation and growth of ledges, which is more sluggish than the growth of an incoherent interface. Thus, the plate-like precipitates are often observed in alloy systems. In magnesium alloys, precipitates tend to form coherent interface on the basal plane, as it is the most closely packed, so basal plates are a commonly observed precipitate morphology in magnesium alloys.

Since precipitation hardening occurs by the interaction of precipitates with dislocations, the morphology of precipitates largely influences the age hardening response. In aluminum alloys, a good example can be seen in the microalloyed Al–Cu based alloys. In Al-1.7at%Cu alloy, Guinier–Preston (GP) zones appear on the {001} planes initially, and they evolve to θ'' and θ' precipitates all on the {001} planes. However, by microalloying Mg and Ag, the habit plane of the platelets changes to {111} forming the Ω phase, whose structure and chemistry is close to the equilibrium θ (Al_2Cu). Without Ag and Mg additions, the θ phase precipitates incoherently without distinct habit planes with many orientation relationships. Microalloyed Ag and Mg segregate at α/Ω interfaces to keep the coherency at the {111} planes, which causes the growth of the precipitates on the {111} planes.

In magnesium alloys, laths, rods and plate-like precipitates are commonly observed. Many plate-like precipitates grow on the (0001) basal plane; however, some precipitates on prismatic planes, for example, $Mg_{12}Nd$ precipitates in an Mg–Nd–RE based WE54 alloy. Nie (2003) reported that the dislocation–precipitate interaction is more effective when plate-like precipitates form on prismatic planes rather than basal planes for the same volume fraction. Figure 4.6 shows how sphere, rod and plate-like precipitates with different habit planes contribute to the hardening of Mg alloys. Prismatic or pyramidal plates are most effective in strengthening, followed by rods with growth direction parallel to $[0001]_{Mg}$ and spherical precipitates. The least effective morphology is the basal plates, which are the most commonly observed precipitates in magnesium alloys. In magnesium alloys the majority of slip occurs through dislocation glide on basal planes, thus it is easier for the dislocations to bypass basal plane plates by cross-slipping. However, with prismatic plates it is difficult to cross-slip, due to the large surface area encountered by dislocations gliding in the basal plane, so the dislocation forms an Orowan loop or cut across the precipitate. The strengthening increment is inversely proportional to the inter-particle distance for a given volume fraction of precipitates. The inter-particle distances may be decreased by changing the number density of precipitates, as the increased number density increases the strengthening. For a given number density of

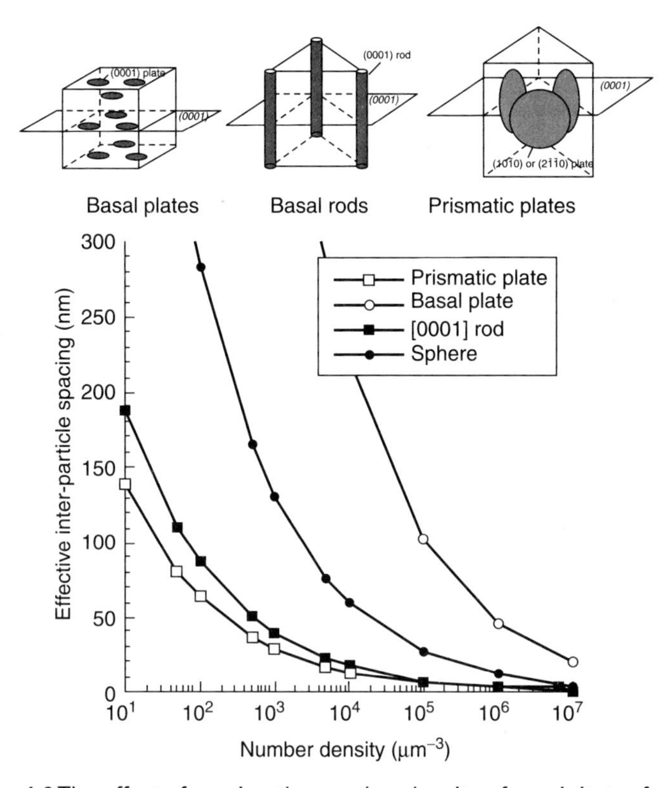

4.6 The effect of varying the number density of precipitates for a given volume fraction for precipitates with various orientations. (The figure is an adaptation of work by J.F. Nie, 2003.)

precipitates, prismatic plates have the least inter-particle distance in the slip plane, resulting in the highest increment in strength, while the largest inter-particle distance is for the basal precipitates.

4.2.5 Metastable precipitates

The precipitation hardening response cannot be enhanced by merely increasing the volume fraction of precipitates. To maximize the age hardening response, the refinement of the microstructure, that is, reduction in size and increase in the number density of precipitates, is essential as shown in Fig. 4.3. However, equilibrium phases tend to have large nucleation barriers, ΔG^* in Equation [4.2] and their direct precipitation often causes coarse microstructure that does not contribute to precipitation hardening. For example, Mg–Ca and Mg–Sn binary alloys show little age hardening response since equilibrium Mg_2Ca and Mg_2Sn phases precipitate coarsely. In order to increase the number density of precipitates, ΔG^* must be reduced to increase the nucleation rate. If the volume formation energy, ΔG_v, is increased, ΔG^* will decrease. Since $\Delta G_v = \Delta G_n / V_m$, the phase with the largest driving force, that is, an equilibrium phase, should have the largest nucleation rate. However, this is not the case for most precipitation systems, because the equilibrium phase usually has a structure totally different from the matrix phase and the nucleation of the equilibrium phase often causes large increase in the interfacial energy γ. Since $\Delta G^* \propto \gamma^3$, the contribution of γ is more influential to ΔG^* than ΔG_v, metastable phases with coherent interfaces precipitate prior to the precipitation of the equilibrium phases in many alloys.

GP zones in aluminum alloys are fully coherent solute enriched clusters with low interfacial energy, which are observed in many aluminum alloys such as Al–Cu, Al–Ag, Al–Zn, Al–Mg–Si and Al–Zn–Mg. Recently, GP zones were reported in many microalloyed magnesium alloys such as Mg–Ca–Al (Jayaraj *et al.*, 2010), Mg–Ca–Zn (Oh-ishi *et al.*, 2009) and Mg–RE–Zn Nie *et al.*, 2008). These findings adds new variations in the usage of precipitation reactions in magnesium alloys. Although coherent interfaces minimize the interfacial energy, they cause misfit strain, ΔG_s in Equation [4.1], associated with the formation of the precipitate. The atomic radius of Ca is 20% larger than that of Mg while the RE elements are on average ~12% larger, so if solute clusters form with these elements this alone will cause large misfit strain. Thus, GP zones do not form in the Mg–Ca and Mg–RE binary alloys. However, the addition of undersized atom such as Zn and Al with negative enthalpy of mixing with Ca and RE tend to occupy the site next to oversized atoms to reduce the strain energy, thereby reducing the strain energy term facilitating the precipitation of GP zones. This is why ordered GP zones form in the Mg–Ca–Zn and Mg–RE–Zn systems.

4.2.6 Heterogeneous nucleation

If precipitates are nucleated at defects such as dislocations, grain boundaries and phase interfaces, the interfacial energy, γ and strain energy, ΔG_s may be reduced. This is called heterogeneous nucleation, and it is often used to refine the precipitate microstructure in alloys. In aluminum alloys, there are three main processes to enhance the heterogeneous nucleation: (i) stretch aging, (ii) two step (double) aging and (iii) microalloying. In stretch aging, alloys are cold rolled after solution heat treatment to introduce dislocations as heterogeneous nucleation sites. This heat treatment is known as T8 temper. The additional sites for nucleation would result in an increased number density of precipitates and an enhancement in age hardening response. Similarly, Mg–Y–RE based WE54 (Hilditch *et al.*, 1998; Nie and Muddle, 2000) alloy is an example of an alloy that shows the refinement of precipitates by stretch aging. A rolling reduction of ~6% resulted in the enhancement of hardening response as observed in Fig. 4.7a, due to the refinement

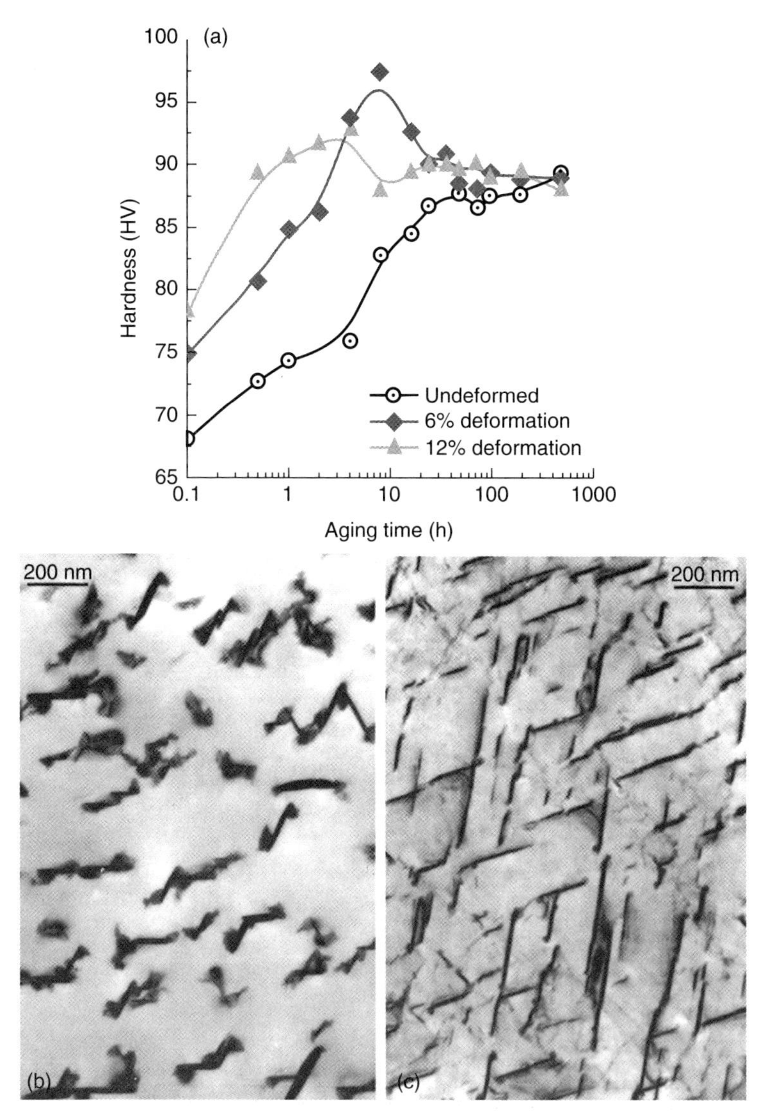

4.7 (a) The effect of plastic deformation prior to ageing on the age hardening of a Mg–2Y–1RE (at%) (WE54) alloy aged at 200°C. The transmission electron microscopy (TEM) micrographs typical of (b) undeformed alloy aged for 48 h at 200°C and (c) 6% deformed alloy aged for 4 h at 200°C (adapted from Hilditch *et al.*, 1998).

of the precipitate particles distribution, Fig. 4.7b and 4.7c. The samples that were not subjected to deformation prior to aging consisted of a coarse distribution of prismatic plate-like precipitates forming heterogeneously in the microstructure showing difficulties in nucleation associated with this phase,

Fig 4.7b. The 6% cold rolling deformation prior to aging, Fig. 4.7c showed a refined microstructure that contained a higher density of prismatic precipitates forming as plates. The refinement of the particle distribution alone resulted in the enhancement in the age hardening response.

Two step or double aging comprises low temperature aging and high temperature aging, which is used to refine the microstructure of Al–Zn–Mg and Al–Mg–Si alloys. In low temperature aging, solute clusters such as GP zones form due to homogeneous nucleation with a large number density because of large supercooling and supersaturation. However, the kinetics for precipitation is too slow, so it takes a long time to reach peak hardness at low temperature. Thus, the alloy aged at low temperature is subsequently aged at an elevated temperature so that it can reach peak hardness within a reasonable time (e.g., 24 h). The clusters that form during low temperature aging can act as heterogeneous nucleation sites for the precipitate that form by the artificial aging. To apply two step aging, the best combination of low temperature aging and high temperature aging conditions must be sought experimentally. If the clusters that form during the low temperature aging dissolve at the second step aging temperature before the nucleation of the high temperature precipitate, the pre-aging gives an adverse effect in age hardening in the second stage. Figure 4.8a compares the age hardening curves of direct aging and two step aging, and their peak-aged microstructures of an Mg–Zn–Mn alloy (Oh-ishi *et al.*, 2008). The age hardening response of the alloy was enhanced when pre-aged at 70°C for 48 h prior to aging at 150°C. The enhancement of the age hardening response was attributed to the presence of increased number density of the precipitates at peak hardness for the step aged alloy as compared to the alloy aged only at 150°C as shown in Fig. 4.8c. 3D atom probe tomography of the pre-aged alloy showed the presence of Zn-rich GP zones in the pre-aged stage, and these are considered to serve as heterogeneous nucleation sites for the rod-like precipitates ($MgZn_2$, β_1')(Oh-ishi *et al.*, 2008).

The addition of trace quantities of elements, microalloying, is a simple but effective method to enhance the age hardening responses, that can be used for both cast and wrought alloys. Microalloying has also been used commonly for the development of high strength aluminum alloys (Hardy, 1952; Silcock *et al.*, 1955; Polmear and Chester, 1989; Ringer *et al.*, 1995; Ringer and Hono, 2000; Bourgeois *et al.*, 2002). One type of additive is an element that has little solubility with the matrix phase, for example, Sn and In in Al–Cu alloys. Sn and In precipitate out earlier than the metastable phase, θ'', in an Al–Cu alloy and can act as heterogeneous nucleation site for θ', which causes uniform dispersion of the θ' precipitates even by high temperature aging (Hardy, 1952; Silcock *et al.*, 1955; Ringer *et al.*, 1995; Bourgeois *et al.*, 2002). A similar microalloying effect was reported in the Mg–Sn system with Na additions (Mendis *et al.*, 2006a). Another type of microalloying effect is the

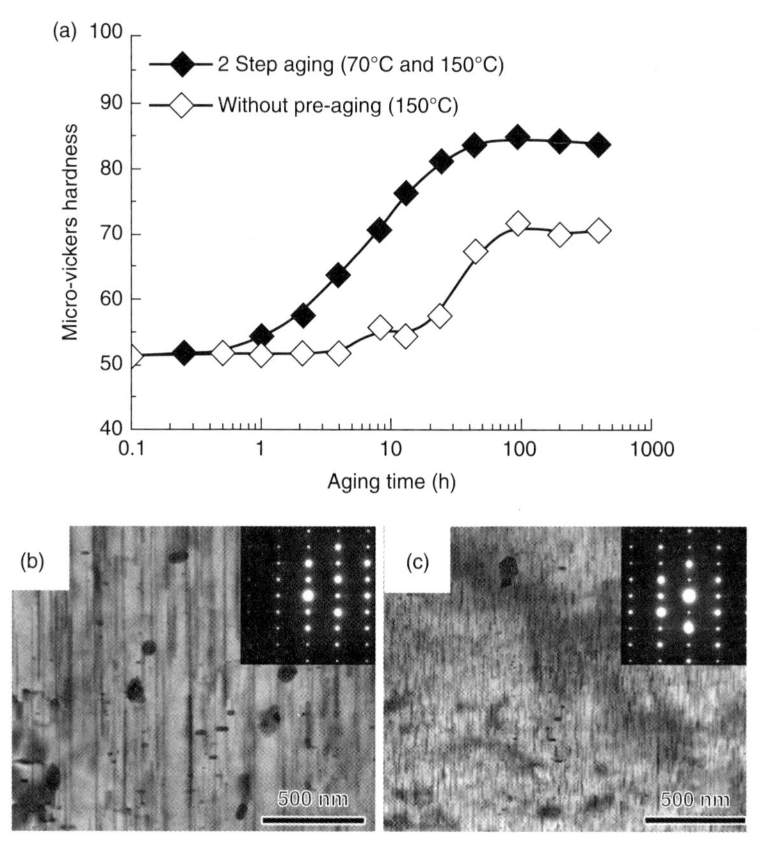

4.8 (a) The effect of step ageing on the age hardening response and the precipitate microstructures of Mg–2.4Zn–0.5Mn alloys (b) aged at 150°C for 96 h and (c) pre-aged at 70°C for 48 h prior to ageing at 150°C (from Oh-ishi *et al.*, 2008).

usage of clustering tendencies of microalloyed elements with sufficiently large solubility; a good example is the combined additions of Ag and Mg to an Al–Cu alloy (Polmear and Chester, 1989; Ringer *et al.*, 1995; Reich *et al.*, 1998; Ringer and Hono, 2000). In this case, Ag and Mg tend to form co-clusters, which later incorporate major solute element Cu to form metastable precipitates, Ω. Recently, a similar microalloying effect was found in the microalloyed Ag and Ca in Mg–Zn alloys (Mendis *et al.*, 2007). Elements such as Cu enhance the age hardening response of Mg–Al (Bettles, 1998) and Mg–Zn (Lorimer, 1987). However, Cu additions degrade the corrosion resistance of the magnesium alloys significantly (Lorimer, 1987), thus are not suitable as microalloying elements. Ni, Fe and Cr are also known to deteriorate the corrosion resistance, even at ppm level, and thus cannot be used as trace additives for magnesium alloys.

4.3 Precipitation hardening magnesium based alloy systems

Most commercial alloys of magnesium, such as AZ91 (Mg-9Al-1Zn (wt%)), WE54 (Mg-5Y-4RE), WE43 (Mg-4Y-3RE), QE22 (Mg-2Ag-2RE), ZE41(Mg-4Zn-1RE) and ZC63 (Mg-6Zn-3Cu) are age-hardenable. Other than WE43 and AZ91, the alloys listed here are aerospace alloys and are not widely used, especially in the automotive industry, due to the high cost associated with the RE additions, or due to high rates of corrosion observed due to the alloying additions such as Cu. In the following sections, some commercially viable alloys that exhibit precipitation hardening are described.

4.3.1 Mg–Al based alloys

The Mg–Al system is the basis of AZ91 commercial alloy which is one of the few alloys that is precipitation-hardened following sand casting. A large collection of work on the precipitation hardening of Mg–9Al and Mg–9Al–0.5Zn alloys has been conducted to understand the precipitation process (Celotto, 2000; Clark, 1968; Crawley and Miliken, 1974; Crawley and Lagowski, 1974; Fox and Lardner, 1943; Lagowski and Crawley, 1976; Lagowski, 1971). The phase diagram of the Mg–Al system shows a maximum solid solubility of approximately 12.7 at% at 437°C, which decreases to approximately 1 at% at 100°C (Nayeb-Hashemi and Clark, 1988). The expected equilibrium phase is $Mg_{17}Al_{12}$ with a bcc crystal structure (a = 1.057 nm and a space group I43 m (Villars and Calvert, 1991)) which generally forms as coarse laths on the basal plane of Mg as shown in Fig. 4.9. The

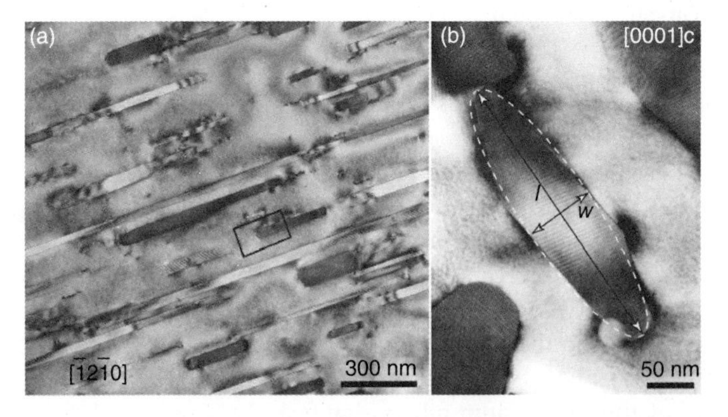

4.9 The precipitate microstructure of Mg–9Al–0.5Zn (at%) (AZ91) alloy showing basal precipitates of $Mg_{17}Al_{12}$ following isothermal ageing at 200°C 16 h (a) parallel to < $11\bar{2}0$> and (b) parallel to <0001> directions of magnesium (from Hutchinson Nie Grosse, 2005).

planner interface between the $Mg_{17}Al_{12}$ and Mg has been described as semi-coherent to incoherent, making the nucleation of such precipitates difficult. Due to the inherent difficulties of the nucleation of the phase, the precipitation of this alloy predominantly occurs by the discontinuous precipitation of $Mg_{17}Al_{12} + \alpha$ phases from at high angle grain boundaries as shown in Fig. 4.2 (Celotto, 2000; Clark, 1968; Crawley and Miliken, 1974; Crawley and Lagowski, 1974; Fox and Lardner, 1943; Lagowski and Crawley, 1976; Lagowski, 1971). For the continuous precipitation mode, $Mg_{17}Al_{12}$ phase form as laths on the basal plane, with the orientation commonly known as Burger's orientation (Celotto 2000; Clark 1968):

$$(0001)_{Mg}//\{110\}_{ppt} \text{ and } <11\bar{2}0>_{Mg}//<1\bar{1}1>_{ppt}.$$

Additionally a small number density of $Mg_{17}Al_{12}$ precipitates form perpendicular to the basal planes with two different orientation relationships with the matrix. The orientation relationships are:

i. $(0001)_{Mg}//\{1\bar{1}1\}_{ppt}$ and $<11\bar{2}0>_{Mg}//<1\bar{1}\bar{2}>_{ppt}$ (Crawley and Miliken, 1974; Crawley and Lagowski, 1974).
ii. $\{11\bar{2}0\}_{Mg}//\{1\bar{1}0\}_{ppt}$ and $<10\bar{1}0>_{Mg}//<110>_{ppt}$ (Porter, 1975; Duly *et al.*, 1993).

These orientations are more prevalent in alloys that have been subjected to lower temperature aging or duplex aging (Crawley and Miliken, 1974) with pre-aging done at a lower temperature or cold deformation (Duly *et al.*, 1993), but the density of these are too low to provide an enhancement of strength. Since the strengthening due to a given volume fraction of prismatic plate precipitates is more effective than the basal precipitation, stimulating the growth of the prismatic plate $Mg_{17}Al_{12}$ precipitates, instead of the basal variety, would lead to a larger precipitation hardening response. However, the stimulating the growth of prismatic plates in place of the basal $Mg_{17}Al_{12}$ has not been achieved to date.

The discontinuous precipitates cannot be suppressed by both cold work (Crawley and Lagowski, 1974; Duly *et al.*, 1993) and duplex aging (Crawley and Miliken, 1974). A significant hardening response has not been observed as a consequence. The addition of ternary elements such as Cu (Bettles, 1998) was reported to suppress the discontinuous precipitation with an increased number density of continuous precipitates, but the exact mechanism remains unclear.

4.3.2 Mg–Zn based alloys

Zn has a maximum equilibrium solubility of ~2.4 at% at 340°C which decreases to ~0.3 at% Zn at room temperature (Nayeb-Hashemi and Clark,

1988), so the alloy shows age hardening behavior as shown in Fig. 4.5. A hardening increment of 15 VH has been observed for Mg–Zn binary alloy at 150°C after aging for ~200 h. A higher hardening increment has been observed with the decrease in the aging temperature but the time to reach peak hardness increased significantly over 1000 h at temperatures below 100°C. The expected equilibrium intermetallic phase is MgZn phase with a rhombohedral crystal structure and a lattice parameters of a = 2.55 nm and c = 1.81 nm (Villars and Calvert, 1991). However, the precipitates observed during artificial aging and at peak hardness condition are metastable $MgZn_2$ phases, know as β_1' and β_2' phases. The precipitation sequence of the Mg–Zn binary allow below 80°C is

$$\text{SSSS} \rightarrow \text{G.P. Zones (below 80}\,^\circ\text{C)} \rightarrow \beta_1'([0001]_{\text{Mg}}\,\text{rods}) \rightarrow \beta_2'((0001)_{\text{Mg}}\,\text{plates})$$

The β_1' precipitates are either $MgZn_2$ (hexagonal phase with a = 0.521 nm and c = 0.86 nm (Villars and Calvert, 1991)) or Mg_4Zn_7 (monoclinic phase c = 2.596 nm, b = 1.428 nm, c = 0.524 nm, γ = 102.5° (Villars and Calvert, 1991)) depending on the aging temperature. The β_2' particles are $MgZn_2$ (hexagonal phase with a = 0.521 nm and c = 0.86 nm (Villars and Calvert, 1991)). Although β_2' precipitates subsequently to β_1', both precipitates coexist in the peak-aged and the over-aged conditions. The $MgZn_2$ phase forms coherent and semi-coherent interfaces with the matrix phase perpendicular to the $[0001]_{Mg}$ direction, thus relatively long rods form which are not of high density when aged at ~150 or 200°C (Clark, 1965; Mima and Tanaka, 1971a; Mima and Tanaka, 1971b), Fig. 4.10. Therefore, ZK60, the most commercially used wrought magnesium alloy, is rarely precipitation-hardened.

Pre-aging the Mg–Zn alloys below 80°C prior to aging at 150°C did not significantly enhance the age hardening response of Mg–Zn binary alloys (Mima and Tanaka, 1971a). However, recent investigation of the duplex aging of Mg–Zn based alloys indicated that these alloys show enhanced age hardening response by the two step aging (Oh-ishi *et al.*, 2008; Mendis *et al.*, 2010). This effect is attributed to the formation of GP zones at temperatures below 80°C. The trace addition of Au (Hall, 1968) and Ca (Bettles *et al.*, 2004) resulted in some enhancement in the age hardening but the combined addition of Ag and Ca (Mendis *et al.*, 2007) has been shown to enhance the hardening response substantially due to the refinement of the precipitate distribution as shown in Fig. 4.11. The age hardening response of the Mg–2.4Zn (at%) alloy containing trace amounts of Ag and Ca was ~3 times that of the binary magnesium alloy with a similar composition. This enhancement in hardening response is the result of the refinement of rod-

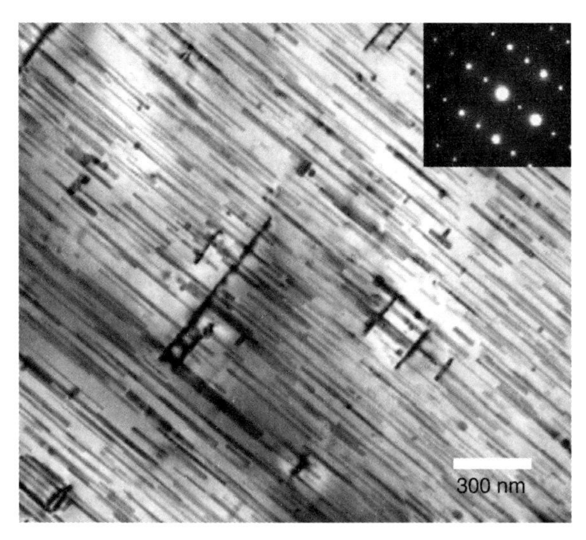

4.10 The precipitate microstructure of Mg–2.4Zn alloy following isothermal aging at 160°C for 120 h.

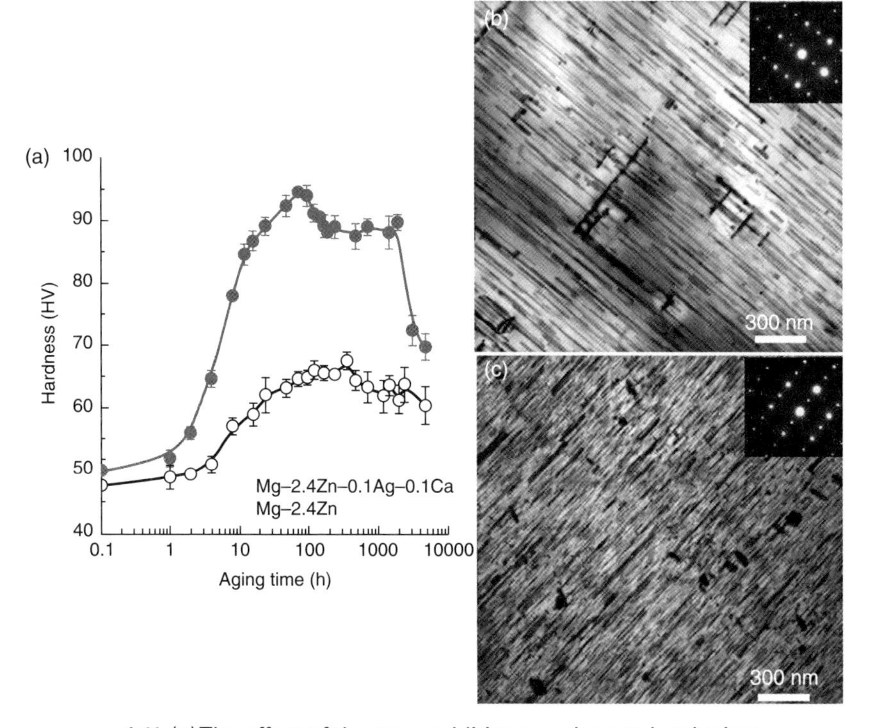

4.11 (a) The effect of the trace additives on the age hardening response of Mg–2.4Zn–0.1Ag–0.1Ca (at%) alloy aged at 160°C. TEM micrographs typical of (b) Mg–2.4Zn alloy aged at 160°C for 120 h and (c) Mg–2.4Zn–0.1Ag–0.1Ca alloy aged at 160°C for 72 h (adopted from Mendis *et al.*, 2007).

like precipitates of $MgZn_2$ with the addition of Ag and Ca to the alloy. As observed, the distribution of the rod-like precipitates was rather coarse in the binary Mg–2.4Zn alloy, Fig. 4.11b. This has been refined significantly in the presence of Ag and Ca, Fig. 4.11c which led to the enhancement of the age hardening response due to the reduction in inter-particle spacing making it difficult for the dislocations to bypass the precipitates.

4.3.3 Mg–Sn based alloys

The Mg–Sn system shows a high solid solubility of ~3.5 at%Sn at the eutectic temperature of 561°C with the solid solubility decreasing to ~0 at% at temperatures below 200°C (Nayeb-Hashemi and Clark, 1988). Additionally the precipitate phase Mg_2Sn (cubic phase with a = 0.677 nm (Villars and Calvert, 1991)) has a high melting temperature of 770°C, thus the alloy has received some attention for its potential as a creep-resistant alloy. However, high hardening response has not been observed for the binary alloy (Van Der Planken, 1969; Sasaki *et al.*, 2006; Mendis *et al.*, 2006a). This is because the Mg_2Sn phase directly precipitates from the supersaturated solid solution which results in a coarse distribution of basal laths as shown in Fig. 4.12b,c. The Mg–Sn system is not the basis of any commercial alloys at present but shows potential because of its excellent extrudability (Sasaki *et al.*, 2008). Several orientation relationships were observed including:

$$(0001)_{Mg}//\{111\}_{ppt} \text{ and } <11\bar{2}0>_{Mg}//<1\bar{1}0>_{ppt}$$ (Derge *et al.*, 1937).

$$(0001)_{Mg}//\{110\}_{ppt} \text{ and } <11\bar{2}0>_{Mg}//<1\bar{1}1>_{ppt}$$ (Derge *et al.*, 1937).

The dispersion of the precipitates was found to be substantially refined by the minor addition of Zn as shown in Fig. 4.12c; accordingly, the age hardening response is also enhanced as shown in Fig. 4.12. When solution treated alloy is extruded at elevated temperature, the precipitation of Mg_2Sn occurs dynamically, which pin the grain growth during recrystallization. This results in substantially refined recrystallized microstructure with the uniform dispersion of Mg_2Sn, giving rise to substantial increase in yield strength of over 300 MPa. Further addition of Al, Ag, Ca enhances the age hardening response (Sasaki *et al.*, 2011).

4.3.4 Mg–RE based alloys

The Mg–RE based alloys show the most significant age hardening response of any magnesium alloy. Table 4.1 shows the maximum solubility, the

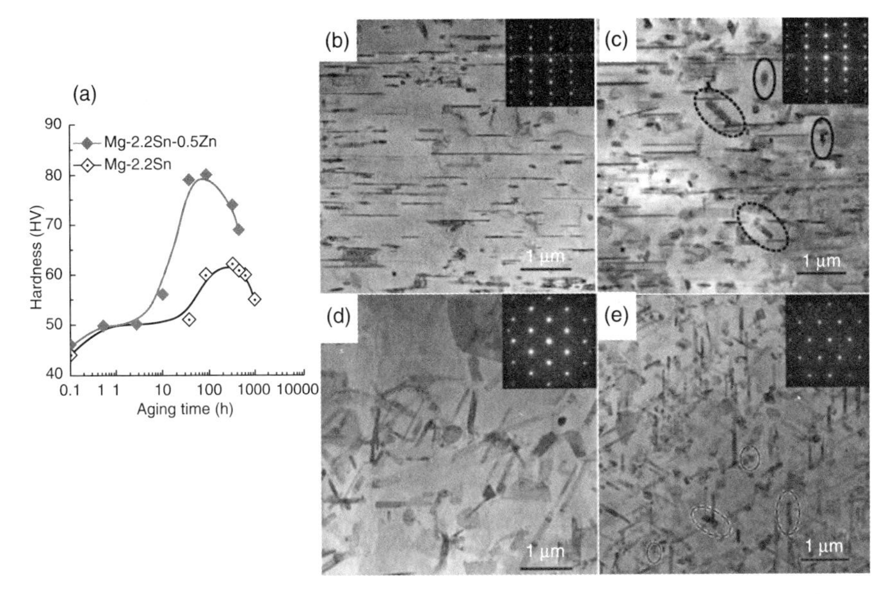

4.12 (a) Effect of Zn additions on the age hardening response of Mg–2.2Sn alloy at 200°C. The precipitate microstructures of (b,d) Mg-2.2Sn and (c,e) Mg–2.2Sn–0.5Zn alloy aged to peak hardness. The electron beam is parallel to (b,c) <11$\bar{2}$0 >M_g and (d,e) $[0001]_{Mg}$ (from Sasaki *et al.*, 2006.)

Table 4.1 The solid solubilities of the Mg–RE system with maximum and minimum solid solubility in magnesium

Alloy system	Maximum solubility (at%)	Temperature T_e °C	Minimum solubility (at%)
Mg–Ce	0.09	592	0
Mg–Dy	4.83	561	2
Mg–Er	6.9	584	2
Mg–Gd	4.53	548	~0
Mg–Ho	5.44	565	~2
Mg–La	0.54	613	0
Mg–Lu	8.8	616	~2
Mg–Nd	0.1	548	0
Mg–Pr	0.09	575	0
Mg–Sm	0.97	542	0
Mg–Tb	4.6	559	~1
Mg–Th	0.5	520	0
Mg–Tm	6.3	571	2
Mg–Y	3.75	566	0.5

Source: Data from Nayeb-Hashemi and Clark (1988).

eutectic temperatures and the solubility at room temperature of various Mg–RE systems. Majority of the RE containing alloys studied form prismatic plates as the main strengthening phase. The main motivation of RE addition in magnesium was to pin grain boundaries by the formation of high temperature intermetallic particles at the triple points; thus, the age hardening responses of Mg–RE alloys have not been thoroughly investigated. One of the promising age hardening alloys in the Mg–RE system is the commercial Mg-5Y–4RE (Nd rich) (wt%) alloy that forms the basis of high strength magnesium alloys used in the high temperature applications. The binary alloys such as Mg–Ce and Mg–Nd do not show a high age hardening response due to the limited solubility (Hisa, 1995). The addition of Y to Mg–Nd system resulted in the development of high strength WE54 alloy that shows a relatively high hardening response due the high density of prismatic plates (Nie and Muddle, 2000; Hilditch *et al.*, 1998; Vostry *et al.*, 1988). As shown in Fig. 4.7, the WE54 alloy shows enhanced precipitation due to stretch aging attesting to the difficulties of nucleation of precipitates. The $Mg_{12}Nd$ precipitate on the prismatic planes.

The addition of Zn to Mg–RE alloys, for example, Mg–Gd and Mg–Y, results in significant enhancement in the age hardening response. The enhancement has been attributed to the refinement of precipitate distribution by the replacement of coarse prismatic plates being replaced by either basal plates or long period ordered (LPO) structures. In the Mg–2.5Gd–1Zn (at%) alloy, following extrusion and heat treatment, the microstructure consisted of fine basal plates with a 14 H stacking forming parallel to the basal planes of magnesium (Yamasaki *et al.*, 2005). The addition of Zn to the Mg–1Gd (at%) binary alloy enhanced the age hardening response of the alloy by forming two atomic layer thick plate-like GP zones forming on the basal planes of magnesium, which had a different structure from the LPO structures described above (Nie Gao and Zhu., 2005; Nie *et al.*, 2008). Similar observations have been reported for the addition of Zn to Mg–Y system (Zhu *et al.*, 2010) and Mg–Nd–Zn system (Wilson *et al.*, 2003). The potential for developing high strength magnesium alloys through extrusion and precipitation hardening using Mg–RE alloys is illustrated by Honma *et al.* (2007) in the Mg–1.8Gd–1.2Y–0.07Zn–0.2Zr alloy; the alloy shows peak hardness exceeding 120 HV, which is usually not attainable without RE. This alloy showed substantially high tensile and compressive yield strengths of 473 and 524 MPa respectively (Honma *et al.*, 2008).

4.3.5 Mg–Ca based alloys

The addition of Ca significantly enhances the ignition resistance of magnesium, thus it is a preferred alloying element. However, the maximum solid solubility is limited to only 0.5 at% Ca at 516°C with the equilibrium Mg_2Ca

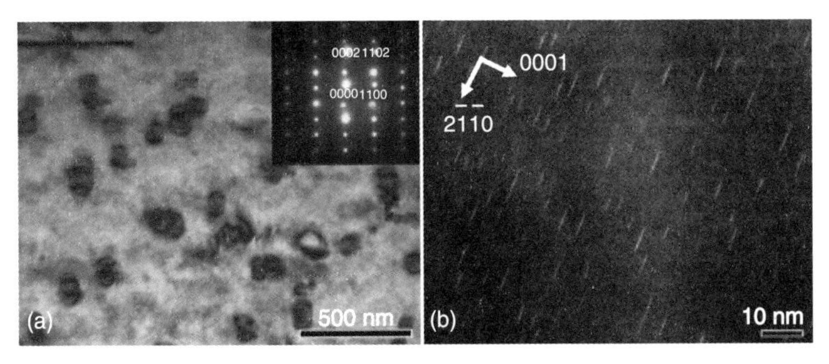

4.13 Effect of Zn additions on the microstructure of Mg–Ca alloy aged at 200°C (a) Mg–0.3Ca and (b) Mg–0.3Ca–0.6Zn (from Oh-ishi *et al.,* 2009).

phase stable to temperatures up to 713°C (Nayeb-Hashemi and Clark, 1988). However, due to the low supersaturation and the incoherent nature of the Mg_2Ca phase (hexagonal phase with $a = 0.605$ nm, $c = 981$ nm (Villars and Calvert, 1991)), the number density of Mg_2Ca precipitates is small and only a negligible hardening response is observed in binary alloys (Nie and Muddle, 1997; Oh *et al.*, 2005; Oh-ishi *et al.*, 2009). The atomic radius of Ca is 20% larger than that of Mg, so if solute clusters form with these elements only, this will cause large misfit strain. However, the addition of undersized atoms such as Zn and Al with negative enthalpy of mixing with Ca tends to occupy the site next to oversized atoms to reduce the strain energy, thereby, facilitating the precipitation of Ca and Zn(Al) containing clusters on the basal plane as shown in Fig. 4.13. While only a low number density of coarse Mg_2Ca precipitates are observed in the Mg–0.3Ca alloy, a large number density of fine precipitates of <10 nm are uniformly dispersed on the basal plane in the Mg–0.3Ca–0.6Zn alloy (at%). The habit plane of these platelets was found to be altered to the prismatic planes by the addition of In (Mendis *et al.*, 2011a), which gave rise to a higher degree of precipitation hardening.

4.4 Role of precipitation hardening in the development of high strength magnesium alloys

Precipitation hardening has been used to enhance the strength of cast alloys. However, it has not been used to give additional strength to wrought alloys. This is because the artificial aging often causes softening after the wrought process, due to the recrystallization that occurs at the artificial aging temperature. In order to use the age hardening in wrought alloys, the age hardening response must surpass the softening at the artificial aging temperature. Figure 4.14a shows the hardness of wrought Mg–2.4Zn–0.1Ag–0.1Ca–0.2Zr alloy and its change after the T6 and T5 (artificial aging after wrought process) treatments. The extruded alloy shows only a minor hardness increase by the T5 treatment, because of the low supersaturation after the hot-extrusion process at 350°C. On the other hand, the T6 treated alloy shows a more pronounced age hardening at the peak hardness condition. In T6, the hardness drops initially because of the recovery and recrystallization of the extruded Mg–2.4Zn–0.1Ag–0.1Ca–0.2Zr alloy by the solution heat treatment; thereafter, the hardness increases more than the as-wrought condition. If sufficient age hardening response can be obtained by the T5 condition, wrought products can be more readily strengthened by post-wrought annealing. To develop such heat-treatable wrought alloys, it is necessary to enhance the relatively low age hardening responses in magnesium binary alloys. In this section, one example of successful development of a precipitation-hardened magnesium alloy based on the Mg–2.4Zn system is described.

The extruded Mg–2.4Zn–0.1Ag–0.1Ca alloy with 0.16 at%Zr additions alloy has a fine grain size and a fine-scale distribution of solid state precipitates formed during extrusion, Fig. 4.14b–d. The mechanical properties of the as-extruded and precipitation-hardened Mg–2.4Zn–0.1Ag–0.1Ca–0.16Zr alloy show a yield strength of approximately 290 MPa with an ultimate tensile strength (UTS) of 350 MPa and an elongation to failure of 17%, Fig. 4.14d. Following a T6 treatment, both the yield strength and the UTS increased to 325 MPa and 360 MPa, respectively, with the elongation to failure of ~14%. The increased yield strength observed for the alloy aged to maximum hardness was attributed to the formation of $[0001]_{Mg}$ β_1' rods with a long aspect-ratio compared with the shorter rods observed in the as-extruded alloy, Fig. 4.13c. The extruded and T6 treated alloy showed a yield anisotropy of ~0.9, which is much higher than that observed for the commercial wrought alloys (Mendis *et al.*, 2009). Recent studies by Robson *et al.* (2011) proposed that the dispersion of fine precipitates is effective in suppressing the twin deformation in compression, thus is considered to improve the yield anisotropy of magnesium alloys. In this regard, the usage of precipitation is being considered to be important for the development of wrought magnesium alloys. The age hardening was also shown to be useful to increase the yield strength of twin roll and hot rolled sheets (Mendis *et al.*, 2011b).

4.5 Conclusions and future trends

While most of the wrought aluminum alloys are strengthened by precipitation hardening, the same approach has not been taken in magnesium alloys. This is because the precipitation response in magnesium alloys is too small to gain strength by heat treatment compared to the softening at the temperature for artificial aging due to grain growth. Microalloying has been demonstrated to be an effective means of enhancing the precipitation hardening.

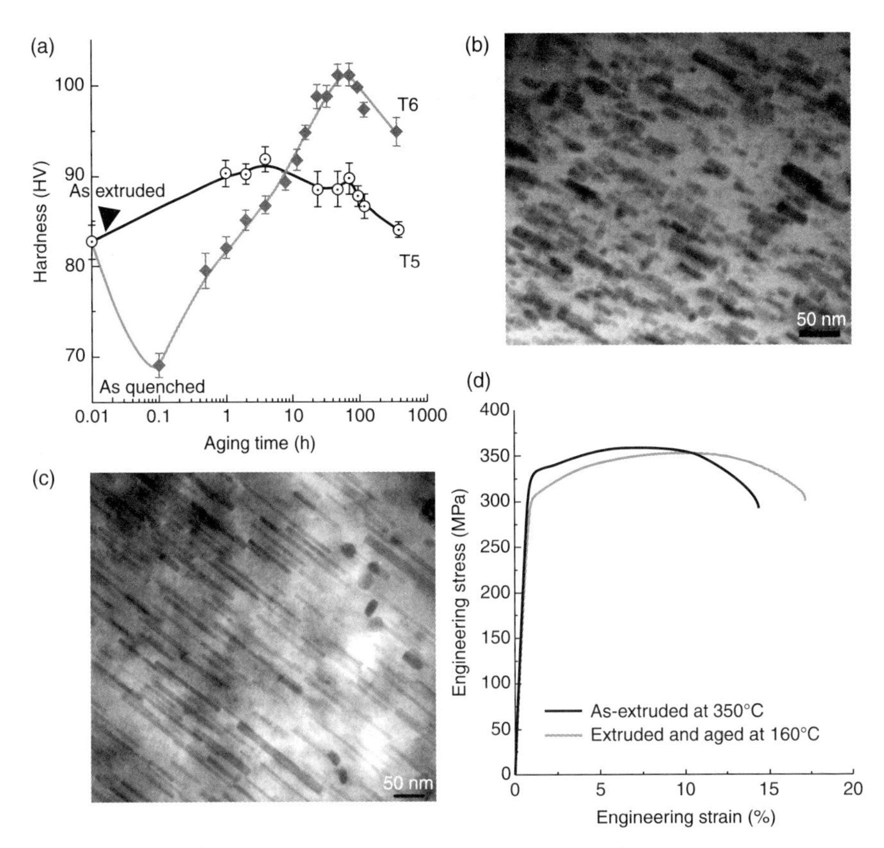

4.14 (a) The age hardening response and the microstructure of Mg–2.4Zn–0.1Ag–0.1Ca–0.16Zr alloy (b) in the as-extruded condition and (c) following isothermal ageing at 160°C 48 h. (d) The tensile properties of the extruded and the heat treated Mg–2.4Zn–0.1Ag–0.1Ca–0.16Zr alloy (adapted from Mendis *et al.*, 2009).

The choice of the right trace addition is important and requires many trial-and-error type experiments. However, the framework developed recently (Mendis *et al.*, 2006a) helps to determine which elements are likely to be more effective at providing heterogeneous sites for nucleation. The use of thermodynamic calculations based on Calphad framework (Schmid-Fetzer and Grobner, 2001) and the incorporation of such calculations in the prediction of favorable elemental interactions will play a very important role in increasing the accuracy of the selection of trace additions. Additionally, there are first principle calculations on the interactions between various elements and vacancies and between two different elements in a magnesium matrix (Shin and Wolverton, 2010a, 2010b). Such thermodynamic information is necessary for the prediction of suitable trace additions.

4.6 Sources of further information and advice

Phase transformations

R.D. Doherty Chapter 15: 'Diffusive phase transformations in the solid state' in *Physical Metallurgy* 4th edition (Ed. R.W. Cahn and P. Haasen) Elsevier Science (1996) Amsterdam.
D.A. Porter and K.E. *Easterling Phase Transformations in Metals and Alloys*, 2nd Edition, Chapman and Hall (1992) London.
R. W. Balluffi, S.M. Allen and W.C. *Carter Kinetics of Materials* Wiley (2005) New Jersey.
R. Dehoff, *Thermodynamics of Materials Science* 2nd Edition, Taylor and Francis (2006) Boca Raton.

Precipitation hardening

A. Kelly and R.B. Nicholson 'Precipitation hardening' *Progress in Materials Science*, **10** (1963) 149.
A.J. Ardell 'Precipitation hardening' *Metallurgical Transactions A*, **16A** (1985) 2131.

Magnesium

G.V. Raynor, *Physical Metallurgy of Magnesium and Its Alloys*, Pergamon Press (1959) London.
E.F. Emley, *Principles of Magnesium Technology*, Pergamon Press (1966) London.

4.7 References

Aaronson H.I. and Clark J.B. (1968) 'Influence of continuous precipitation upon the growth kinetics of the cellular reaction in an Al-Ag alloy', *Acta Metall*, **16**, 845–855.
Bettles C.J. (1998). in Kainer K.U and Mordike B.L. (Eds) *Magnesium Alloys and Their Applications* Wolfsburg, Werkstoff-Informationsfesellschaf, 265–270.
Bettles C.J., Gibson M.A. and Venkatesan K. (2004) 'Enhanced age-hardening behaviour in Mg-4wt%Zn micro-alloyed with Ca', *Scripta Mater*, **51**, 193–197.
Bourgeois L., Nie J.-F. and Muddle B.C. (2002) 'On the role of Tin in promoting nucleation of the θ' Phase in Al-Cu-Sn', *Mater Sci Forum*, **396–402**, 789–794.
Celotto S. (2000) 'TEM study of continuous precipitation in Mg-9wt%Al-1wt%Zn alloy', *Acta Mater*, **48**, 1775–1787.
Clark J.B. (1965) 'Transmission electron microscopy study of the age hardening in a Mg-5wt%Zn alloy', *Acta Metall*, **13**, 1281–1289.
Clark J.B. (1968) 'Age hardening in a Mg-9wt%Al alloy', *Acta Metall*, **10**, 141–152.

Crawley A.F. and Miliken K.S. (1974) 'Precipitate morphology and orientation relationships in an aged Mg-9%Al-1%Zn-0.3%Mn alloy', *Acta Metall*, **22**, 557–562.

Crawley A.F. and Lagowski B. (1974) 'Effect of two step aging on the precipitate structure in magnesium alloy AZ91', *Metall Trans*, **5**, 949–951.

Derge G., Kommel A.R. and Mehl R.F. (1937) 'Studies upon the widmanstatten structure IX- the Mg-Mg_2Sn and Pb-Sb systems', *Trans AIME*, **124**, 367–378.

Duly D., Audier M. and Brechet Y. (1993) 'In the influence of plastic deformation on discontinuous precipitation in Mg-Al', *Scripta Metall Mater*, **29**, 1593–1596.

Fox F.A. and Lardner E. (1943) 'An investigation of the effects of precipitation treatment of binary magnesium-aluminium alloys', *J Inst Metall*, **69**, 373–396.

Hardy H.K. (1952) 'The aging characteristics of ternary aluminum-copper alloys with cadmium, indium, or tin', *J Inst Metals*, **80**, 483–493.

Hall E.O. (1968) 'The age-hardening characteristics of two magnesium-zinc alloys' *J Inst Metall*, **96**, 21–27.

Hisa M. (1995) Ph D thesis, University of Queensland, Australia.

Hilditch T., Nie J.F. and Muddle B.C. (1998) 'The effect of cold work on precipitation in alloy WE54', in Kainer K.U. and Mordike B.L. (Eds.) *Magnesium Alloys and Their Applications,* Wolfsburg, Werkstoff-Informationsfesellschaf, 339–344.

Honma T., Kunito N. and Kamado S. (2009) 'Fabrication of extraordinary high-strength magnesium alloy by hot extrusion', *Scripta Mater*, **61**, 644–647.

Honma T., Ohkubo T., Kamado S. and Hono K. (2007) 'Effect of Zn additions on the age-hardening of Mg–2.0Gd–1.2Y–0.2Zr alloys', *Acta Mater*, **55,** 4137–4150.

Hutchinson C.R., Nie J.F. and Grosse S. (2005) 'Modeling the precipitation processes and strengthening mechanisms in a Mg-Al-(Zn) AZ91 alloy', *Metall Mater Trans A*, **36A**, 2093–2105.

Jayaraj J., Mendis C.L., Oh-ishi K., Ohkubo T. and Hono K. (2010) 'Enhanced precipitation hardening of Mg–Ca alloy by Al addition', *Scripta Mater*, **63**, 831–834.

Khoaroshahi R.A. (1997) Ph.D Thesis. Manchester University, United Kingdom.

Kovacs I., Lendval J., Ungar T., Groma G. and Lakner J. (1980) 'Mechanical properties of AlZnMg alloys', *Acta Metall*, 1621–1630.

Lagowski B. and Crawley A.F. (1976) 'The effect of prior cold work on the precipitation and mechanical properties of a heat treated Mg-9Al-2Zn (AZ92) alloy', *Metall Trans A*, **7A** 773–775.

Lagowski B. (1971) 'Aging of Mg-9Al-Zn alloys', *AFS Trans*, **79**, 115–120.

Lorimer G.W., Khoaroshahi R.A. and Ahmed A. (1999) 'Precipitation reactions in two magnesium alloys containing rare earths' in *Proc. Int. Conf. on Solid-Solid Phase Transformations '99 (JIMIC-3)*, Eds. Koiwa M., Otsuka K. and Miyazaki T., *Jpn Institute of Met.*, **12**(I), 185–192.

Lorimer G.W. (1987) 'Structure property relationships in cast magnesium alloys' in *Proc Magnesium Technology* Eds. Lorimer G.W. and Unsworth W. London Institute of Metals, 47–53.

Mendis C.L., Bettles C.J., Gibson M.A., Gorsse S. and Hutchinson C.R. (2006a) 'The refinement of precipitate distributions in an age hardenable Mg-Sn alloy through microalloying', *Phil Mag Lett*, **86**, 443–456.

Mendis C.L., Oh-ishi K. and Hono K. (2007) 'Enhanced age hardening in a Mg–2.4 at.% Zn alloy by trace additions of Ag and Ca', *Scripta Mater*, **57**, 485–488.

Mendis C.L., Oh-ishi K., Kawamura Y., Honma T., Kamado S. and Hono K. (2009) 'Precipitation-hardenable Mg-2.4Zn-0.1Ag-0.1Ca-0.16Zr (at%) wrought magnesium alloy', *Acta Mater*, **57**,749–760.

Mendis C.L., Oh-ishi K. and Hono K. (2010) 'Effect of Al additions on the age hardening response of the Mg–2.4Zn–0.1Ag–0.1Ca (at.%) alloy – TEM and 3DAP study', *Mater Sci Eng A*, **527,** 973–980.

Mendis C.L., Oh-ishi K., Ohkubo T. and Hono K. (2011a) 'Precipitation of prismatic plates in Mg–0.3Ca alloys with In additions', *Scripta Mater*, **64**, 137–140.

Mendis C.L., Bae J.H., Kim N.J. and Hono K. (2011b) 'Microstructures and tensile properties of twin roll cast and heat treated Mg-2.4Zn-0.1Ag-0.1Ca-0.1Zr alloy', *Scripta Mater*, **64**, 335–338.

Mima G. and Tanaka Y. (1971a) 'The aging characteristics of magnesium-4wt%zinc alloy', *Trans Jap Inst Met*, **12**, 71–75.

Mima G. and Tanaka Y. (1971b) 'Mechanism of precipitation hardening of magnesium-zinc alloys', *Trans Jap Inst Met*, **12**, 323–328.

Munjal V. and Ardell A.J. (1975) 'Precipitation hardening of Ni-12.19 at.% Al alloy single crystals', *Acta Metall*, **23,** 513–520.

Nayeb-Hashemi A.A. and Clark J.B. (1988) (eds) *Phase Diagrams of Binary Magnesium Alloys,* ASM International, Metals Park.

Nie J.F. and Muddle B.C. (1997) 'Precipitation hardening of Mg-Ca(-Zn) alloys', *Scripta Mater*, **37**, 1475–1481.

Nie J.F. and Muddle B.C. (2000) 'Characterisation of strengthening precipitate phases in a Mg–Y–Nd alloy', *Acta Mater*, **48**, 1691–1703.

Nie J.F. (2003) 'Effect of precipitate shape and orientation on dispersion strengthening in Mg alloys', *Scripta Mater*, **48,** 1009–1015.

Nie J.F., Gao X. and Zhu S.M. (2005) 'Enhanced age hardening response and creep resistance of Mg–Gd alloys containing Zn', *Scripta Mater*, **53,** 1049–1053.

Nie J.F., Oh-ishi K., Gao X. and Hono K. (2008) 'Solute segregation and precipitation in a creep-resistant Mg–Gd–Zn alloy', *Acta Mater*, **68,** 6061–6076.

Oh J.C., Ohkubo T., Mukai T. and Hono K. (2005) 'TEM 3DAP characterization of an age-hardened Mg-Ca-Zn alloy', *Scripta Mater*, **53**, 675–679.

Oh-ishi K., Hono K. and Shin K.S. (2008) 'Effect of pre-aging and Al addition on age-hardening and microstructure in Mg-6 wt% Zn alloys', *Mater Sci Eng A*, **496**, 425–433.

Oh-ishi K., Watanabe R., Mendis C.L. and Hono K. (2009) 'Age-hardening response of Mg–0.3 at.%Ca alloys with different Zn contents', *Mater Sci Eng A*, **526,** 177–184.

Polmear I. J. and Chester R.J. (1989) 'Abnormal age hardening in an Al-Cu-Mg alloy containing silver and lithium', *Scripta Metall*, **23,** 1213–1218.

Porter D.A. (1975) Ph.D. Thesis. University of Cambridge.

Porter D.A. and Easterling K.E. (1992) *Phase Transformations in Metals and Alloys,* 2nd Edition. Chapman and Hall, London.

Reich L., Murayama M. and Hono K. (1998) 'Evolution of Ω phase in a Al-Cu-Mg-Ag alloy- a three dimensional atom probe study', *Acta Mater*, **46**, 6053–6062.

Ringer S.P., Hono K. and Sakurai T (1995) 'The effect of trace additions of Sn on precipitation in Al-Cu alloys: an atom probe field ion microscopy study', *Metall Mater Trans*, **26A**, 2207–2217.

Ringer S.P. and Hono K. (2000) 'Microstructural evolution and age hardening in aluminium alloys: atom probe field-ion microscopy and transmission electron microscopy studies', *Mater Charact*, **44**, 101–131.

Robson J.D., Stanford N. and Barnett M.R. (2011) 'Effect of precipitate shape on slip and twinning in magnesium alloys', *Acta Mater*, **59,** 1945–1956.

Sasaki T.T., Oh-ishi K., Ohkubo T. and Hono K. (2006) 'Enhanced age hardening response by the addition of Zn in Mg-Sn alloys', *Scripta Mater*, **55**, 251–254.

Sasaki T.T., Yamamoto K., Honma T., Kamado S. and Hono K. (2008) 'A high-strength Mg–Sn–Zn–Al alloy extruded at low temperature', *Scripta Mater*, **59**, 1111–1114.

Sasaki T.T., Oh-ishi K., Ohkubo T. and Hono K. (2011) 'Effect of double aging and microalloying on the age hardening behavior of a Mg-Sn-Zn alloy' in press. *Mater Sci Eng A*, **530**, 1–8.

Schmid-Fetzer R. and Grobner J. (2001) 'Focused development of magnesium alloys using the Calphad approach', *Adv Eng Mater*, **3**, 947–961.

Shin D. and Wolverton C. (2010a) 'First-principles study of solute–vacancy binding in magnesium', *Acta Mater*, **58**, 531–540.

Shin D. and Wolverton C. (2010b) 'First-principles density functional calculations for Mg alloys: A tool to aid in alloy development', *Scripta Mater*, **63**, 680–685.

Silcock J.M., Heal T.J. and Hardy H.K. (1955) 'The structural ageing characteristics of ternary aluminium-copper alloys with cadmium, indium or tin', *J Inst Metals*, **84**, 23–31.

Van Der Planken J. (1969) 'Precipitation hardening in magnesium-tin alloys', *J Mater Sci* **4**, 927–929.

Villars P. and Calvert L.D. (1991) *Pearson's Handbook of Crystallographic Data for Intermetallic Phases. 2nd ed*, ASM International, Materials Park Ohio.

Vostry P., Stulikova I., Smola B., Cieslar M. and Mordike B.L. (1988) 'A study of the decomposition of supersaturated Mg-Y-Nd, Mg-Y and Mg-Nd alloys', *Z Metallkde*, **79**, 340–344.

Wilson R., Bettles C.J., Muddle B.C. and Nie J.F. (2003) 'Precipitation hardening in Mg-3wt%Nd(-Zn) casting alloys', *Mater Sci Forum*, **419–422**, 267–272.

Yamasaki M., Anan T., Yoshimoto S. and Kawamura Y. (2005) 'Mechanical properties of warm extruded Mg-Zn-Gd alloys with coherent 14H long periodic stacking ordered structure precipitate', *Scripta Mater*, **53,** 799–803.

Zhu Y.M., Morton A.J. and Nie J.F. (2010) 'Improvement in the age-hardening response of Mg–Y–Zn alloys by Ag additions', *Acta Mater*, **58,** 2936–2947.

5

Alloying behavior of magnesium and alloy design

M. PEKGULERYUZ, McGill University, Canada

DOI: 10.1533/9780857097293.152

Abstract: Magnesium exhibits significant property improvement via alloying. The addition of thorium to magnesium has to be on the record as leading to one of the highest strengthening responses in a metal. Magnesium alloy research in the past two decades has further shown that magnesium can be designed to improve creep and corrosion resistance and formability. This chapter presents an updated perspective on magnesium alloy design. The solute and second-phase effects of the alloying additions are discussed separately. The spotlight is placed on alloy properties rather than alloying elements. Surface-active elements in magnesium and recent findings on alloy design with respect to wrought texture are highlighted.

Key words: magnesium, alloy design, intermetallics, creep, corrosion, recrystallization, texture, surface-active elements.

5.1 Introduction

The development of magnesium alloys had a golden period in the 1940s and 1950s. Military applications of magnesium led to the development of various magnesium alloys, and the post-war period later led to the use of some of these alloys in civilian applications. Notably, the 1970s VW Beetle and the Super Beetle, the prime user of Mg, featured an air-cooled Mg engine block. In the 1980s cost sensitive industries such as the automotive sector shifted their interest to other materials, mainly steel and cast iron, while the aerospace industry started looking into polymer composites. The 1980s and 1990s may be considered as a dormant period in terms of interest in magnesium.

Since the 1990s, there has been renewed interest in using Mg in weight-sensitive applications. The transport industries have been facing weight reduction objectives to attain fuel economy and high performance in their vehicles since the beginning of the twenty-first century[1–3] and have started to look for materials that are lightweight. At the beginning of that period, there were really three or four major commercial Mg alloy systems, such as Mg–Al–Zn, Mg–Al, and Mg–Zn–Rare earth (RE) and Mg–RE. In 1990s, most magnesium usage for structural applications was in die-casting (36 kt), and 90% of this was in one alloy, AZ91D (Mg–9wt%Al–1wt%Zn). Later, Mg–Al (AM) casting alloys without Zn were used due to their higher ductility. The RE containing alloys, though used in the aerospace industry, were not readily adopted by car makers due to their high cost.

The beginning of 2000s brought the development of casting alloys with high creep resistance and low-to-moderate cost, for use in powertrain applications. Mg–Al alloys with RE and alkaline earth elements (RE, Ca, Sr) made their debut.[1–3] Magnesium–aluminum–strontium alloy (AJ62), developed by Noranda, was used in BMW engine blocks from 2004 to 2009.[4–6] These activities were not sufficient to develop a steady Mg supplier base, either in North America or in Europe, and, despite the low cost production of Mg in China in the latter part of the decade (2000–2010), wide ranging application of magnesium in the transport industries was not achieved. The new decade from 2010 onward is seeing interest in Mg wrought alloys for use in the automotive body (closures, front-end structure, roof pillars). Extrusion and sheet alloys are currently being developed to improve formability in wrought magnesium. Interestingly, these alloys, like creep-resistant casting alloys, are also based on RE and Sr additions to magnesium.

A lot more research is needed to develop magnesium alloys with improved properties. There is also interest in the use of magnesium in civilian aircraft and in biomedical applications. Mg alloys need to meet the requirements of high ignition resistance or corrosion resistance. This chapter gives an overview of the alloying behavior of magnesium and provides an insight into how certain materials properties can be obtained in magnesium via alloying.

5.2 Alloy design: solid solution alloying of magnesium

Pure metals are rarely used in structural applications. Alloys, which are strengthened via mechanisms such as solid solution strengthening, precipitation hardening and dispersion strengthening (that occur when a metal is combined with one or more elements), form the basis of structural metallic materials. Also, properties of an alloy can be altered and improved via microstructural design and modification.

A solid solution alloy is characterized by the complete dissolution of the alloying element (solute) in the lattice of the base metal (solvent), either substitutionally (substituting the lattice sites of the host) or in the interstitial sites. Mg does not have any major interstitial alloys. According to the

Hume–Rothery rules, extensive solid solutions cannot be formed if the atomic sizes of the solvent and solute differ by more than 15%. In the case of Mg, which has an atomic diameter of 3.2 Å, the elements that fall in this favorable size range are Li, Al, Ti, Cr, Zn, Ge, Yt, Zr, Nb, Mo, Pd, Ag, Cd, In, Sn, Sb, Te, Nd, Hf, W, Re, Os, Pt, Au, Hg, Tl, Pb and Bi.[7–9] Additionally, a metal forms extensive solid solutions with metals of similar electronegativity and crystal structure. Valency also plays a role; a metal of low valency is more likely to dissolve one of higher valency than the reverse (relative valency effect) because the addition of extra electrons to a metal increases its bond-forming capacity and hence the stability of the metal structure. According to this rule, among the elements that have favorable size factors, divalent Mg would dissolve trivalent and higher valency elements in addition to divalent ones. However, as the solute valency increases (groups IV–VII) so does the difference between the electrochemical characteristics of the solute and Mg, which is highly electropositive. In these cases, Mg forms second phases or stable compounds (intermetallic compounds) rather than solid solutions. If these criteria are not satisfied, there are still possibilities for limited solid solutions.

Figure 5.1[7] shows the alloying behavior of Mg with respect to favorable size, relative valency and electronegative valency effects. The elements inside the curve present a group where there is at least a minimum of 0.5 wt% maximum solid solubility in Mg (percentages indicated in parentheses under the element). *In this chapter the alloy compositions are given in weight percent unless otherwise indicated.* The highest solid solubility exists when size factors and relative valency are very favorable (up to 15% relative size, and valency of 2 or 3). In this region, solid solubility can be as high as 53% for In. However, the strengthening effect is not appreciable for these cases, due to extreme similarity of atom sizes. At around 12–15% size factor, where border line sizes are involved, alloy systems of Mg offer considerable solid solution strengthening (Al, Zn, REs, etc.).

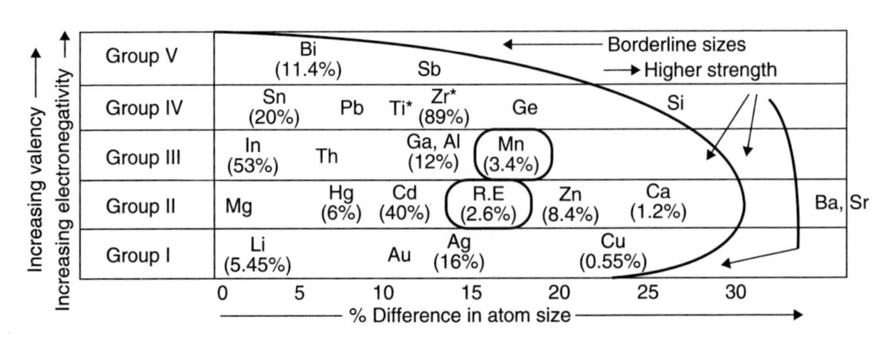

5.1 Alloying behavior of Mg.[7]

5.2.1 Solute hardening

The solute atoms which substitute solvent atoms at lattice sites provide strengthening to the base metal by (i) atom misfit creating elastic lattice strain and (ii) differences in elastic moduli of solute and solvent atoms. Yield strength of the solid solution, σ_y, would increase in proportion to the one-half power of the solute concentration, C. The strengthening is also proportional to the three-halves power of the lattice strain e_s created by the solute; the lattice strain is proportional to the atom radii difference between Mg and the solute:

$$\sigma_y \; \alpha \; \varepsilon_S^{\frac{3}{2}} C^{\frac{1}{2}}$$

where

$$\varepsilon_s \; \alpha \; \Delta r, \quad \Delta r = (r_{\mathrm{Mg}} - r_{\mathrm{solute}}).$$

An additional effect of the solute is its segregation to dislocations, grain and sub-grain boundaries. Solutes can pin dislocations producing solute drag stress, σ_{drag}, which increases the yield strength especially at room and moderate temperatures. Temperature, however, causes unpinning of the solute atmosphere, due to atom diffusion. The effectiveness of the solute pinning at elevated temperatures (>1/3 T_m) depends on solute diffusivity (and hence melting point, T_m). High melting point solutes such as Mn will maintain the effect up to high temperatures, while low melting point Al, Zn, Sn atmospheres around dislocations can be lost at moderate temperatures, leading to the loss of the drag stress.

5.2.2 Other effects of solutes

Solute effects on recrystallization, grain growth and preferred orientation in wrought alloys

Solute pinning can affect the mobility of sub-grain boundaries and grain boundaries, thereby slowing the recrystallization and grain-growth kinetics. Solute pinning of basal dislocations in magnesium can increase the critical resolved shear stress (CRSS) for basal slip, activate non-basal slip at moderate temperatures, and cause texture alteration. A solute effect of Ce has been postulated for Mg–Mn–Ce alloys, which has led to weakened textures after rolling and annealing.[10] This was attributed to a possible change in stacking fault energy (SFE) in another study.[11]

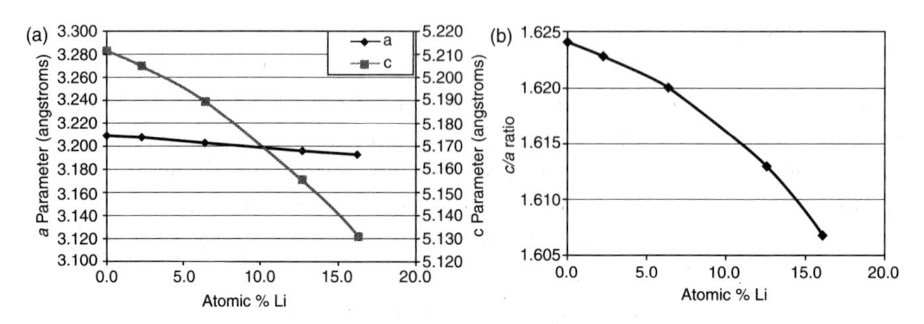

5.2 The effect of Li on the (a) lattice parameters and (b) the axial ratio of Mg.[13]

Axial ratio, deformation, twinning and edge cracking in wrought alloys

Lithium additions above 5.5 wt% change the hexagonal close-packed (HCP) structure of Mg to body-centered cubic (BCC).[8] Small amounts of lithium alloying are also known to improve the room temperature formability of magnesium, which is associated with the activation of non-basal slip with the decrease in the axial ratio (c/a) of magnesium as a function of lithium.[12,13] As lithium is added to magnesium, the c value of the hcp crystal falls faster than the a value (Fig. 5.2), which leads to a decrease in the c/a ratio.[13,14] As a result, the Peierls stress, which is proportional to $e^{-(d/b)}$ where d is the interplanar spacing and b is the Burgers vector (and consequently the CRSS), for basal slip increases relative to that for prismatic slip, and activates slip on non-basal planes. Texture modeling and transmission electron microscopy (TEM) studies indicate that lithium additions also increase the glide of $<c + a>$ dislocations on {1 1–2 2} pyramidal planes.[11,15] There is evidence that $<c + a>$ dislocations dissociate into partial dislocations,[16] and lithium may lower the energy of the stacking fault, and thereby increase the stability of the glissile dislocation configuration.[15] Lithium has been observed to alter the crystallographic texture by altering the balance of deformation mechanisms, which in turn influences the texture.[11]

The effect of low levels of Li on the c/a ratio and grain size of Mg was investigated by Becerra *et al.*[13,17] in Mg–Zn–Li and Mg–Zn–In–Li solid solution alloys. The mono-valent Li decreased the axial ratio (c/a) of magnesium from 1.624 to 1.6068 within the 0–16 at%Li range (Fig. 5.2). This was attributed to the decrease in e/a (number of electrons per atom) causing electron overlap from the second Brillouin to the first Brillouin zone; this causes a contraction of the c-spacing in real space and a decrease in c/a. The divalent zinc showed no effect on c/a in the 0.2–0.7 at% range, since, as explained by Vegard's law, the atom size caused a similar change in both a- and c-parameters. The trivalent indium increased the c/a of magnesium

Table 5.1 Axial ratio, grain structure and size, edge cracking index and texture intensity[18]

Alloy, wt%	a	c/a	Effective cast grain size D′, mm	Recry-stallized grain size, d, μm	Edge cracking index	Maximum Intensity	
						Roll	Roll + anneal
Mg	3.2088	1.6240	6.1	46	High/4	12.2	17.9
Mg–1.8Li–1Zn	3.2009	1.6190	1.8	48	Minimum/1	4.2	2.6
Mg–6Li–0.4Zn–0.2In	3.1991	1.6190	1.0	49	Minimum/1	6.2	3.0

*D_c = average diameter of columnar grains = $(D_{minor} + D_{major})/2$; D = average grain diameter of equiaxed region.

to 1.6261 as In increased towards 3.3 at%. An important effect of *c*/*a* was seen (Table 5.1) in the twinning behavior of the alloys and in turn on edge cracking during 150°C rolling.[18]

The changes in axial ratio and lattice parameters influence slip systems and twinning modes in HCP metals. The change in the deformation mechanism with reduced axial ratio can be attributed to the change in interplanar spacing, *d*, since the shear stress required to move dislocations is given by the Peierls stress,

$$\tau_\pi = \mathrm{P.}\, e^{\{-2s\, d/[b(1-v)]\}}, \quad [5.1]$$

where P is a factor depending on the shear modulus and Poisson's ratio, v; b is the magnitude of Burgers vector of the dislocation; and d is the interplanar spacing. The CRSS, τ_{CRSS}, is related to the Peierls stresses, τ_p. In a recent study, Uesugi *et al.*[19] have calculated τ_p from first principles as

$$\tau_p = \frac{Kb}{a'} \exp\left(-\frac{Kb}{2a'\tau\max}\right), \quad [5.2]$$

where K is the energy factor (depending on elastic constants and the type of dislocation); a' is interplanar spacing in the direction of dislocation sliding on the slip plane ($a' = a/2$ for basal slip where a is the lattice parameter and $a' = a$ for prismatic plane); b = Burgers vector; $b = (a/3)^{1/2}$ for Shockley partial dislocations of the basal plane and $b = a$ for edge dislocations of the prismatic plane; τ_{max} is the maximum restoring force (= maximum slope of the generalized SFE in the analysis). Equation [5.2] interestingly indicates that changing the a-spacing would not influence the Peierls stresses and CRSS for prismatic slip (because $b = 2a$ for edge dislocations), but it would alter the basal slip of partial dislocations because their $b = (a/3)^{1/2}$.

Effects of cast grain size were seen on the texture evolution of Mg alloys.[18] The 150°C rolling produced deformation structures and texture, and static recrystallization occurred during post-annealing. Maximum texture intensity during 150°C rolling and subsequent annealing correlated strongly with the cast grain size (Table 5.1). The texture weakened with decreasing grain size, which was attributed to the activation of non-basal slip by plastic compatibility stresses at grain boundaries. The *a* lattice parameter exerted a weaker influence on texture, possibly by facilitating the cross-slip of partial dislocations at grain boundaries.

The other important effect of *c*/*a* is on the twinning behavior of Mg. The twinning shear is related to the axial ratio in HCP crystals[20–22] as presented in Fig. 5.3, where the twinning systems with positive slope indicate compression (contraction) twins and those with negative slopes indicate tension (extension) twins along the *c*-axis. Figure 5.3 also presents the axial ratios of the HCP metals and the most common twinning modes activated in them. Within the range of the axial ratios of the HCP metals (1.568–1.886), the {10–12}<–1011> shear direction reverses at a *c*/*a* of $\sqrt{3}$, where this twinning mode becomes compressive for Zn and Cd, but is tensile for all other HCP metals (Be, Ti, Zr, Re, Mg). HCP metals that exhibit high ductility (Re, Zr, Ti) twin profusely in both tensile and compressive modes and those that are very brittle (Be, Zn) twin only by the most common type {10–12} tension twin.[20] In Mg, the {10–12} tension twins are followed by {10–11} compression twins. Generally, the lower the value of twinning shear of a given twin mode, the higher is the frequency of its occurrence. This correlation has been generally observed as indicated by Fig. 5.3 and also by those experiments in which the axial ratio was changed by alloying;[20] the additional factors are the ease with which atomic shuffling can occur to nucleate the twin and the ease of gliding parameters. The absence of {10–11}[10–12] compression twins in Be, despite the low twinning shear, is related to the complex atom shuffling, while the occurrence of {11–21} 1/3 [–1–126] twinning in Ti, Zr and Re in spite of its relatively large twinning shear is attributed to the relatively simple atomic shuffling.[20] It has been observed that subsequent slip-twin interactions largely determine the ductility of the HCP metals that undergo twinning.[20] If the twin boundaries become sinks for gliding dislocations the metal is ductile; if however, the dislocations are repelled by the twin boundary they pile up at the interface leading to eventual crack nucleation. In Mg, the basal dislocations are repelled by {10–12} tension twins[20] leading to edge cracking. It is known that the twinning modes in HCP metals are altered via alloying[20] that changes the axial ratio; for example, twinning is stopped in Mg–Cd alloy with $c/a = \sqrt{3}$, leading to extreme brittleness. It is, therefore, possible to influence the twinning behavior of Mg alloys via changes in *c*/*a*.

In the study by Pekguleryuz *et al.*,[18] the differences in the twinning modes of Mg–3Al–1Zn (AZ31) and Mg–2Li–1Zn and Mg–6Li–0.4Zn–0.2In,

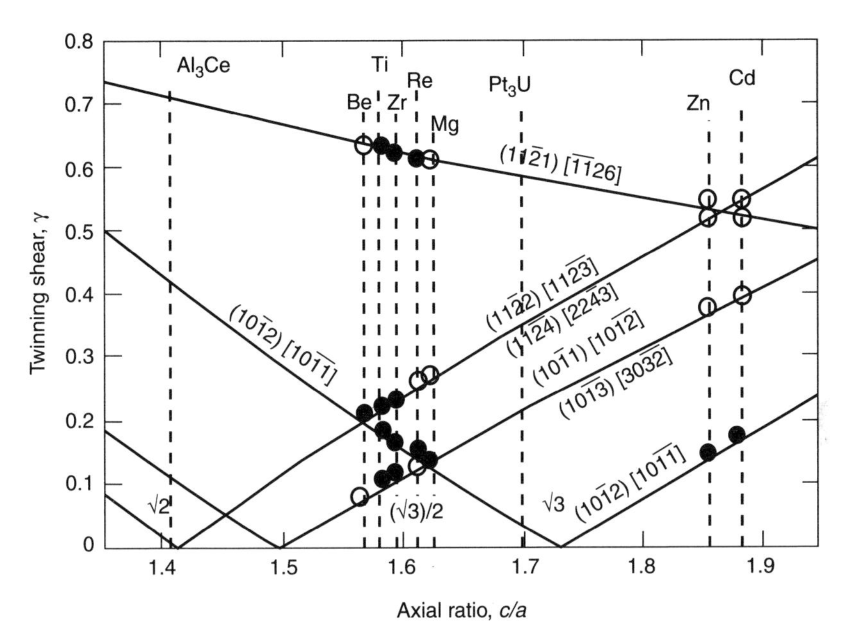

5.3 Twinning shear versus axial ratio in HCP metals.

which governed the edge cracking index (Ic) during 150°C rolling, have been related to the axial ratio. AZ31 alloy significantly edge-cracked during rolling, twinned in the tensile mode, and had a *c*/*a* (axial ratio) of 1.6247. At the *c*/*a* of 1.6247, the shears of {10–12} tension twins and {10–11} compression twins start to differ, making the atomic reshuffling more difficult, impeding double twinning in AZ31, and leaving a higher propensity of tension twins which are potent sites for crack nucleation. The Li-containing alloys (Mg–2Li–1Zn and Mg–6Li–0.4Zn–0.2In) had lower *c*/*a* (Table 5.1), double-twinned during rolling, exhibited a lower degree of twin-related brittleness and did not edge-crack during rolling.

Solute effects on creep

Creep resistance can be improved via solid solution if the solute can (i) decrease the homologous temperature, (ii) increase the elastic modulus or (iii) decrease diffusivity as proposed by Sherby[23] via

$$\varepsilon = S\left(\frac{\sigma}{E}\right)^{\eta} D \qquad [5.3]$$

where ε is steady-state creep rate, S is a structure term (grain morphology, dislocation density, distribution, etc.), σ is stress, E is elastic modulus, D is self-diffusivity and η is the stress exponent.

While most creep-resistant alloy-development work in the 1990s and 2000s focused on ternary additions to Mg–Al based alloys that can suppress creep-induced $Mg_{17}Al_{12}$ precipitation, one alloy family, Mg–RE–Zn, has re-explored the effects of REs on Mg–Zn alloys.[24,25] Zhu *et al.*[25] studied the addition of different RE elements (La-rich and Nd-rich) in Mg–2.5RE-0.6Zn alloys. They found that the primary creep (which is a competition between work hardening and work softening) was more extensive in the La-containing alloy, but there was no discernible difference in the steady-state creep rates of the two alloys. The authors related the difference in the primary creep of the two alloys to the higher solid solubility of Nd than that of La in Mg. The higher solubility can indeed provide a richer solute atmosphere around the dislocations, increasing work hardening in the Nd-alloy.

Mg alloys with Ce-based REs have been further developed into ternary compositions with low level additions of Mn. Mn was found to improve the creep resistance of Mg–Ce alloys by increasing the solid solubility of Ce in Mg at high temperatures. The work carried out in Germany before World War II showed that Mg–6RE–2Mn alloy has good creep resistance (300–315°C, ~ 14 MPa). Mg–2RE–1Mn alloy was observed to be much stronger at elevated temperatures than any of the cast Mn-free alloys.[8] Mg–2.25RE–0.5Mn and Mg–2RE–1.25Mn–0.2Ni alloys were also creep-resistant compositions.[26]

Diffusional creep can occur at relatively low temperatures (~0.4 T_m) when the stresses are high through diffusion along dislocation cores, which is called *pipe diffusion*. Solutes which segregate to dislocations can impede *pipe diffusion*. These should conceivably be solutes with significant difference in metallic radii so that enhanced segregation to dislocations can occur. However, at higher temperatures solutes in the dislocation atmosphere become mobile and dissociate from the dislocations, therefore, for high temperature service, high melting point solutes with lower diffusivity need to be selected.

Pipe diffusion has not been considered in many of the creep studies carried out on magnesium alloys. AZ91D alloy at stresses above 50MPa in the 125–175°C range gave an n value of 5 and Q value of 94 kJ/mole, which can be related to pipe diffusion.[27] AJ63 alloy also gave an activation energy, Q, of 92 kJ/mol at 50 MPa in the 150–200°C range, which can also be related to pipe diffusion.[6] In AJ63 the grain boundaries are stronger than AJ62, due to the higher amount of intermetallics, and the dislocation cores, in that case, would become more open to diffusion than the grain boundaries. Recently, Mg–Mn–Sr was found to be susceptible to *pipe diffusion* at moderate temperatures,[28] basically due to the low solubility of Sr in Mg. Age-hardened Mg–Gd–Nd–Zr at 250°C/50 MPa exhibited Q of 109 kJ/mol, where pipe diffusion can also be a possible mechanism.[29] These alloys would benefit from the addition of high temperature solutes with adequate solubility for impeding creep deformation related to pipe diffusion.

5.3 Alloy design: compound formation in magnesium alloys

When the Hume–Rothery rules for solid solution formation do not hold, compound formation occurs. Second phases are important in strength, creep resistance, and recrystallization and texture. Important intermetallic compounds found in Mg alloys are listed in Table 5.2,[30] some of which are described in some detail below.

5.3.1 $Mg_{17}Al_{12}$

The $Mg_{17}Al_{12}$ intermetallic phase has a cubic A12 crystal structure isomorphous with α-Mn (cI58) with the space group of $I\bar{4}3m$ (Table 5.2, Fig. 5.4) and lattice parameter a = 1.06 nm. The unit cell contains 34 Mg atoms and 24 Al atoms. The unit cell has the highest symmetry T_d^3 and its atomic coordinates are shown in Fig. 5.4.[31]

It has been explained[8] that Al–Al bonds are homopolar, each Al atom accepting an electron from the structure as a whole from the 24 Mg nearest neighbors, while the other ten Mg atoms retain their valency electrons. The structure is designated as twelve (12) Al pairs of two negative charges, twenty-four (24) positively charged Mg atoms and ten (10) Mg atoms in their normal state: $(Al_2)_{12}^{-2}Mg_{24}^{+}Mg_{10}$. The structure is interesting in that in one intermetallic compound both homopolar, heteropolar and normal metallic characteristics are present.[8] The planes (033)p and (411)p are both identified as close-packed planes, but there is also (877)p, which exhibits lower diffraction intensity but its d spacing is one-third of the d spacing of (033)p or (411)p, that is, 0.8289 Å.[32] The $Mg_{17}Al_{12}$ phase has also been reported with 44.5 at%Mg with a tetragonal structure of a = 10.50 nm, c = 10.19 nm.

The γ-$Mg_{17}Al_{12}$ is the main second phase[33] which forms eutectically in Mg–Al alloys and also in Mg–Al–Zn alloys with high Al:Zn ratio (Fig. 5.5a).

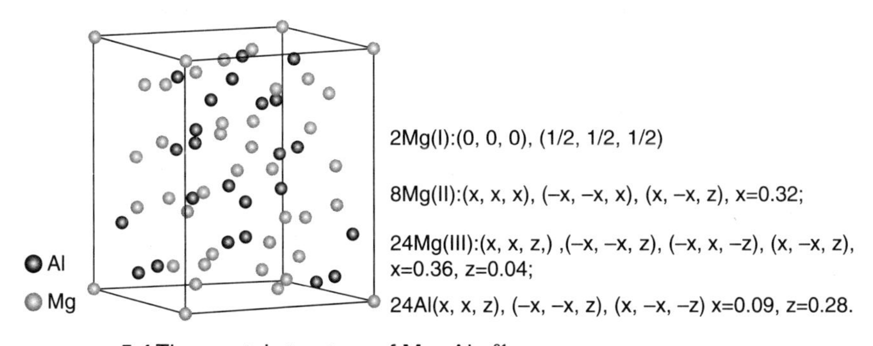

5.4 The crystal structure of $Mg_{17}Al_{12}$.[31]

Table 5.2 The crystal structure data of important equilibrium intermetallic phases in Mg alloys

Phase	Melting point (°C)	Crystal structure	Pearson symbol	Space group	Lattice parameters		
					a (nm)	*c* (nm)	*c/a* or *b**
$Mg_{17}Al_{12}$	450	Cubic A12	cI58	$I\bar{4}3m$	1.06	–	–
		Tetragonal	–	–	1.05	1.02	0.97
Mg_2Ca	714	Hexagonal C14	hP12	*P63/mmc*	0.62	1.02	
Al_2Ca	1079	Cubic C15	cF24	*Fd3m*	0.80	–	
$Al_2(Mg,Ca)$		Hexagonal C14	hP12	*P63/mmc*	0.56	0.90	1.6
		Hexagonal C36	hP24	*P63/mmc*	0.55	1.75	3.2
Mg_2Si	1081	Cubic C1	cF12	$Fm\bar{3}m$	0.63	–	–
$Mg_{12}Ce$	616	Tetragonal	tI26	*I4/mmm*	1.03	0.60	0.58
$Mg_{12}Nd$	548	Tetragonal	tI26	*I4/mmm*	1.03	0.59	0.57
B-$Mg_{14}Nd_2Y$		Cubic A12	cI58	$F\bar{4}3m$	2.2		
B_1-$Mg_{12}NdY$		Cubic C1	cF12	$Fm\bar{3}m$	0.74		
$Al_{11}RE_3$	1020†	Tetragonal	tI26	*I4/mmm*	0.44	1.01	2.30
		Orthorhombic	oI28	*Immm*			1.30*
Al_2RE	1455,1480	Cubic	cF24	*Fd3mS*	0.81		
$Mg_{17}Sr_2$	606	Hexagonal	hP38	*P63/mmc*	1.05	1.03	0.98
Al_4Sr	1040	Tetragonal	tI10	*I4/mmm*	0.44	1.11	2.52
$Mg_{24}Y_5$	567	Cubic A12	cI58	$I\bar{4}3m$	1.13	–	

*The value is for parameter *b* rather than *c/a*.
†Transformation *T*.

In non-equilibrium solidification, the α-Mg matrix becomes supersaturated in Al, especially in the interdendritic regions which give rise to the discontinuous precipitation of $Mg_{17}Al_{12}$ (γ_d in Fig. 5.5a) when exposed to elevated temperatures. The alloys can be heat-treated by dissolving the γ-phase through long (24 h) solution treatment and re-precipitating it as a discontinuous fine plate-like phase (Fig. 5.5b and 5.5c).

Six different orientation relationships (OR) are found in the Mg/γ-$Mg_{17}Al_{12}$ precipitation system.[127,128] The initial lattice correspondence between HCP and BCC structures (Fig. 5.6a and 5.6b) is described by the Pitsch–Schrader OR,[34] as shown in Fig. 5.6c: The Pitsch–Schrader OR requires that the close-packed planes in the BCC lattice should be parallel to the close-packed planes or near-close-packed planes in the HCP lattice, that is, (011)p // (0001)m and (011)p // (0110)m. Multiple morphologies, growth directions and habit planes are described as twining versions of the initial Pitsch–Schrader OR.

γ-$Mg_{17}Al_{12}$ is an intermetallic that has a composition range (Table 5.2). It strengthens Mg–Al alloys and the AZ91; however, it is also prone to deformation by slip. The γ phase increases the strength by creating obstacles to slip and accumulating dislocation pile-up at its interface up to a certain level of strain (2.1%); however, further strain (5.4%) does not change the

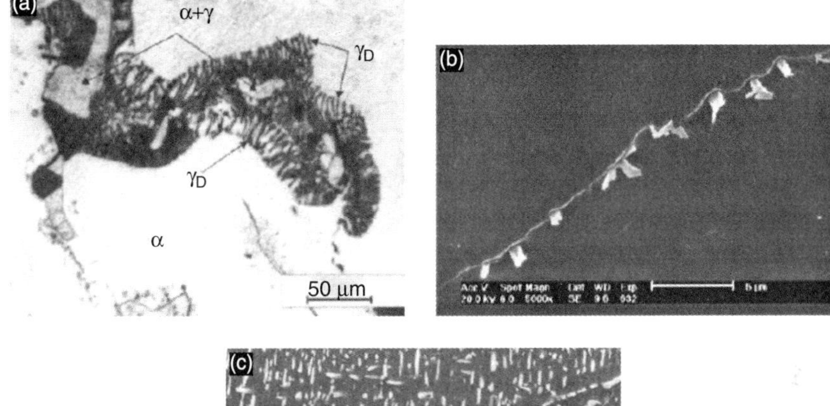

5.5 (a) Eutectic γ-$Mg_{17}Al_{12}$ precipitates and discontinuous γ_d-$Mg_{17}Al_{12}$ precipitates in AZ91. (b) Grain boundary γ-$Mg_{17}Al_{12}$ precipitates and (c) discontinuous γ_d-$Mg_{17}Al_{12}$ precipitates in AZ91 alloy after aging at 623 K; SEM.[33]

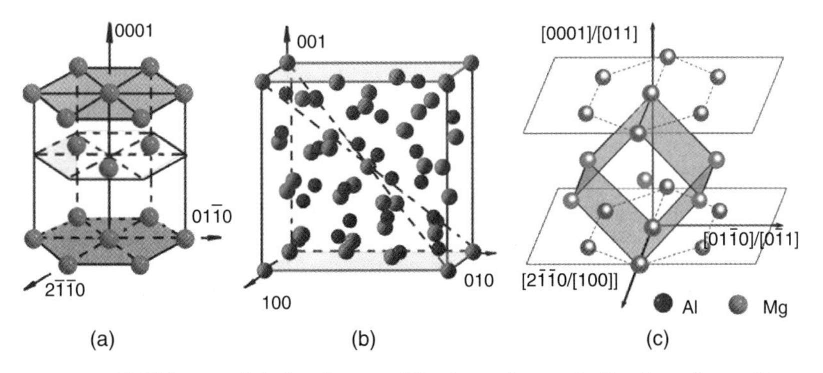

5.6 The crystal structures of the two phases in the transformation system Mg/γ-$Mg_{17}Al_{12}$: (a) HCP Mg; (b) BCC γ-$Mg_{17}Al_{12}$. (c) Schematic diagram showing the initial lattice correspondence between HCP and BCC structures, described as the Pitsch–Schrader OR.[33]

dislocation pile-up, but instead leads to dislocation accumulation inside the γ-phase (Fig. 5.7).[35]

The γ-phase softens at elevated temperatures, due to its partial metallic bonding, which adversely affects elevated temperature strength and creep

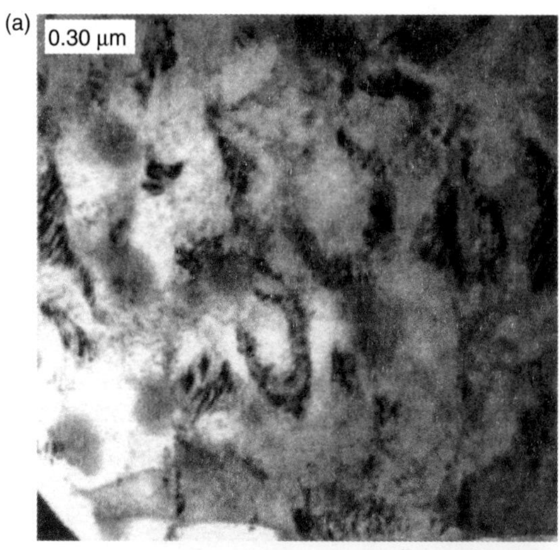

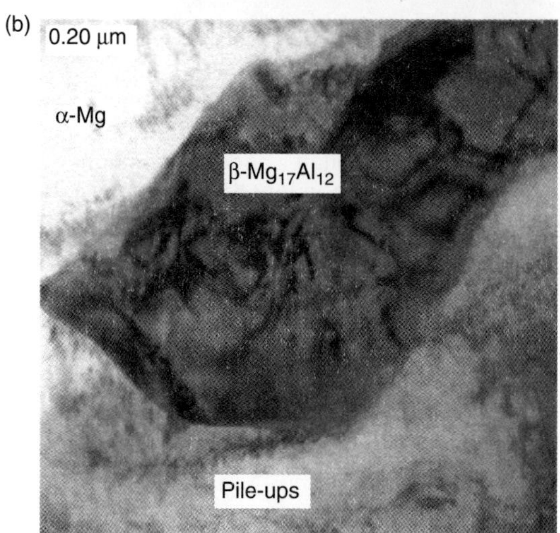

5.7 Dislocation arrangements in AZ91D alloy. (a) Dislocation pile-up after 2.1% deformation, (b) dislocations in the γ-$Mg_{17}Al_{12}$ phase.[35]

resistance. Its discontinuous precipitation from the supersaturated α-Mg matrix during creep leads to grain boundary deformation and migration. Many alloys have been developed to improve the creep resistance by adding Ca, Sr and REs to Mg–Al alloys to eliminate the discontinuous precipitation of $Mg_{17}Al_{12}$.

Zhou *et al.*[36] have conducted a first principles study to determine the structural stability of Ca alloying of the $Mg_{17}Al_{12}$ phase, and have concluded that the cohesive energy of $(Mg_{17-x}Ca_x)Al_{12}$ (x = 0, 1, 4, 12) phases

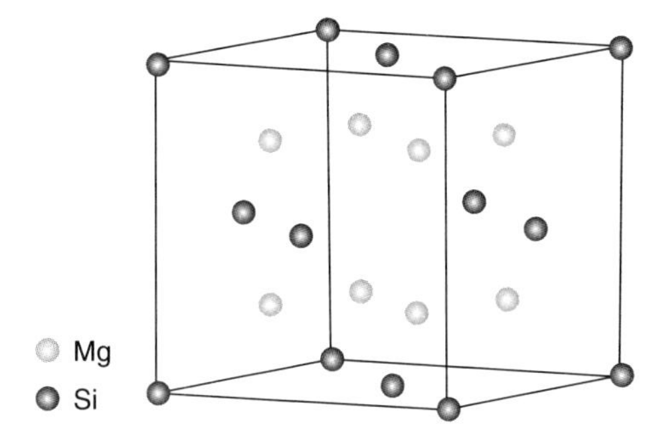

5.8 Mg_2Si unit cell.[37]

gradually increase when Mg atoms in I, II, III positions of $Mg_{17}Al_{12}$ phase are substituted by Ca. Their results have indicated that the $(Mg_5Ca_{12})Al_{12}$ with Ca substitution of the Mg(III) atoms has the highest alloying ability and the highest structural stability.

5.3.2 Mg_2Si

Mg_2Si crystallizes in an FCC (cubic C1) lattice with lattice parameter a = 6.339 Å (Fig. 5.8 and Table 5.2) and with 12 atoms per unit cell. Si atoms are located on the corners and surface centers of the face-centered cubic and Mg atoms are in the eight tetrahedral interstices of the crystal cell. The Mg_2Si has high thermal and mechanical stability.[38] It occurs as a eutectic precipitate in Mg–Al–Si alloys (AS41, AS31, AS21), in addition to the $Mg_{17}Al_{12}$ phase (Fig. 5.9), and renders slight improvement in creep resistance.[39] The morphology is coarse Chinese script when the alloy is cast at slow to moderate solidification rates. High solidification rates, as in high-pressure die-casting, refine the Mg_2Si into polygonal dispersed particles, leading to improved mechanical properties.[40] Trace levels of Ca, Sb, Sr[41–43] also modify and refine to different extent the Mg_2Si morphology (Fig. 5.10).

5.3.3 Intermetallics with Ce

$Mg_{12}Ce$

Two crystal structures have been previously reported for $Mg_{12}Ce$: a tetragonal structure ($ThMn_{12}$-type) below the eutectic temperature, and an ordered structure above the eutectic.[44,45] Figure 5.11a shows the unit cell of $Mg_{12}Ce$ with the $ThMn_{12}$ tetragonal (Pearson symbol tI26) structure,[46] which has two (2) cerium and twenty-four (24) magnesium atoms with all the cerium atoms at the 2a position, and magnesium atoms occupying the 8f, 8i and 8j

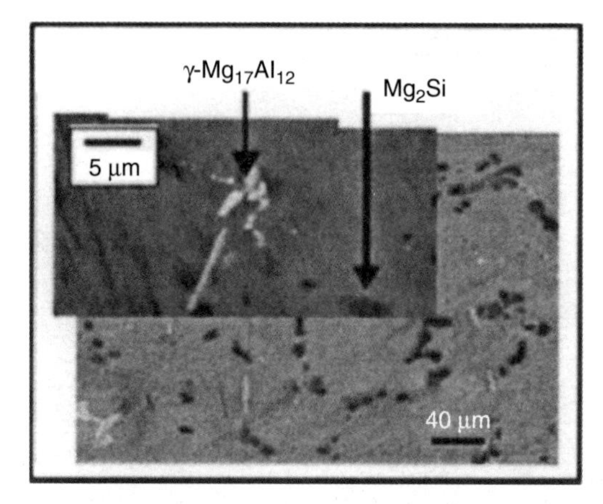

5.9 SEM image showing Mg_2Si and γ-$Mg_{17}Al_{12}$ precipitates in AS alloys.[39]

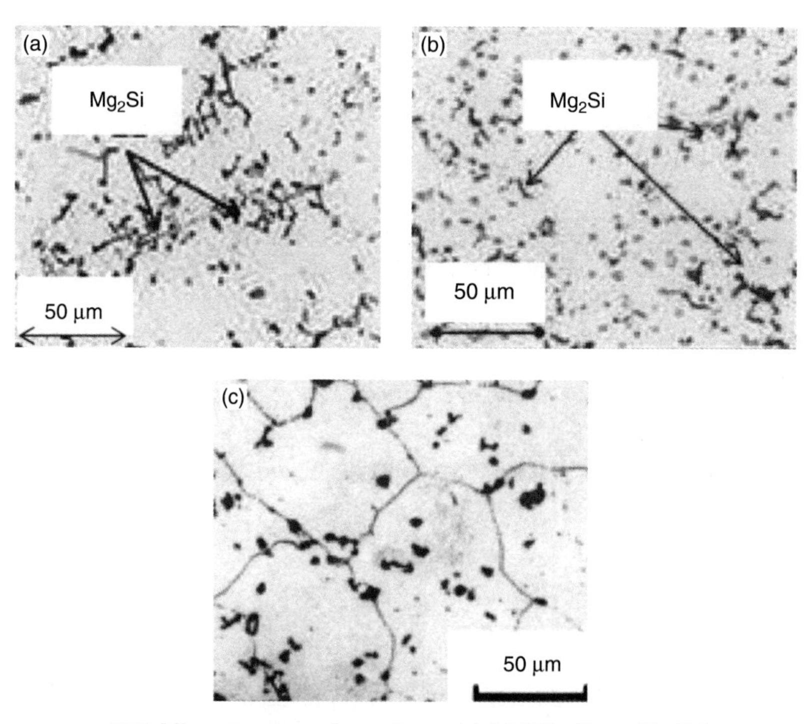

5.10 Microstructure of gravity cast (a) AS41 alloy with Chinese script Mg_2Si phase; (b) same alloy with 0.1 wt% Ca with refined Mg_2Si phase.[41] (c) Heat-treated AZS511 with 0.5 wt%Sb with fine Mg_2Si.[42]

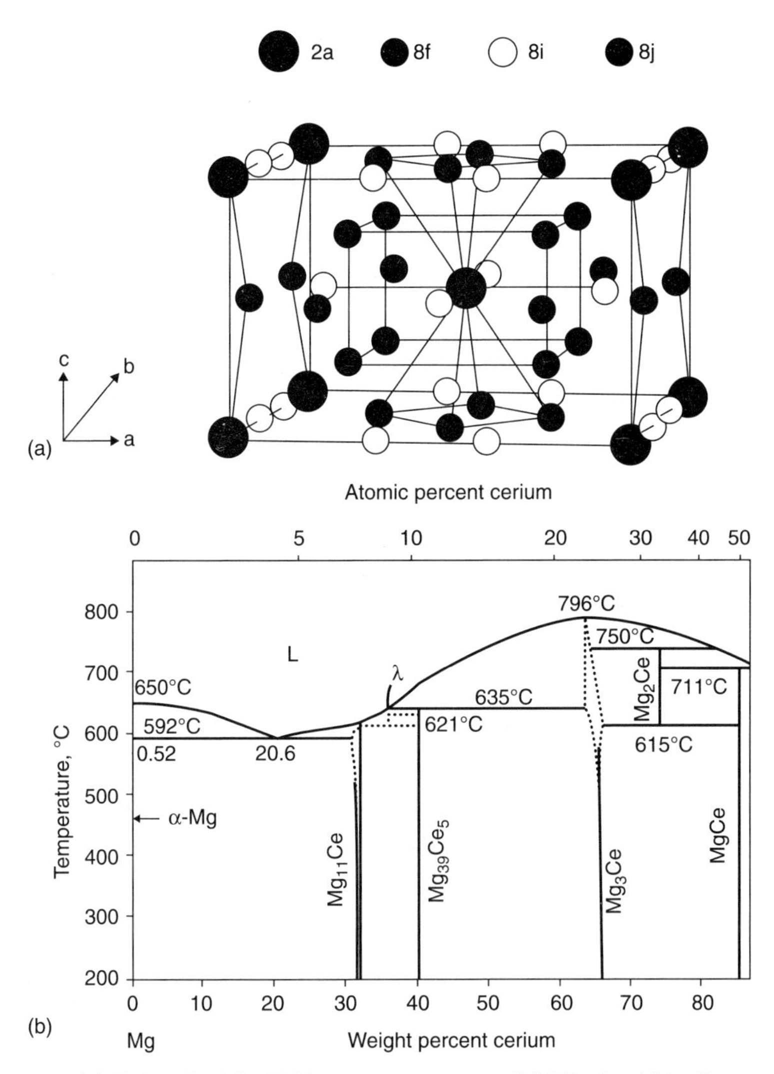

5.11 (a) Unit cell of the ThMn12 type structure.[46] (b) Revised Mg–Ce phase diagram up to 50 at% Ce.[48]

positions. The coordination number is 12. In the phase diagram (Fig. 5.11b) of the Mg–Ce binary system,[4] a dotted line to the left of $Mg_{12}Ce$, shows a small compositional range, which increases when the temperature increases above ~450°C.[4]

Xin *et al.*[46–48] have made an in depth study of the $Mg_{12}Ce$ along with the other Mg–Ce intermetallics in the binary phase diagram up to 50 at%Ce and have concluded the intermetallic in the region of 8–10 at%Ce to be a vacancy defect structure of $Mg_{12}Ce$ with the real composition of $Mg_{11}Ce$ and suggested the phase diagram shown in Fig. 5.11b.

Magnesium alloys containing Ce, which are commercially available, usually contain Zn, and/or Mn as well; in these alloys the β-$Mg_{12}Ce$ exists as a second-phase precipitate but exhibits solubility for Mn or Zn. Xin *et al.*[49] have made EPMA studies of the $(Mg, Mn)_{12}Ce$ in Mg–Ce–Mn alloys and have determined ~0.8 at% Mn in the intermetallic. Moreno *et al.*,[50] have observed 0.07 at% Mn and 1.2 at%Zn in the $(Mg,Zn)_{12}Ce$ of the MEZ alloy. These observations fit well with the vacancy defect structure of the $Mg_{12}Ce$. The Mn solubility of $Mg_{12}Ce$ has an important influence on the creep behavior of Mg–Ce–Mn alloys.[51] The dynamic precipitation of nano-particles of Mn from the intergranular $(Mg, Mn)_{12}Ce$ phases provides effective dislocation pinning in the grain boundary regions and prevents the local recovery phenomenon in the vicinity of the intermetallics. This local recovery involves the formation of high dislocation density (HDD) zones next to dislocation-free zones at the intermetallic/matrix interface, which lead to stress accumulation at the HDD zones and subsequently cracking of the intermetallic. The dynamic precipitation of nano-sized $Mg_{12}Ce$ dispersed inside the α-Mg grains is also observed during the creep of Mg–Ce–Mn alloys, a phenomenon due to the Ce supersaturation of the α-Mg. These particles improve the creep resistance by effective dislocation pinning.

$Al_{11}RE_3$ and Al_2RE

The $Al_{11}Re_3$ intermetallic has the tetragonal structure with the space group I4/mmm and lattice parameters $a = 4.365$, $b = 4.365$ and $c = 10.10$. Its melting point is 1235°C. The Al_2RE has cubic with the Fd3mS space group; its melting point is 1200°C. These intermetallics exist in the interdendritic regions of Mg–Al–RE alloys (e.g., AE42, AE44). In the AE42 alloy, the $Al_{11}RE_3$ that is present in the as-cast alloy is found to decompose when heated above 175°C into Al_2RE and $Mg_{17}Al_{12}$.[52] In the AE44 both intermetallics are present in the as-solidified alloy (Fig. 5.12).[53]

Pettersen *et al.*[54] suggested that when RE/Al weight ratio is above 1.4, all the Al is tied up as $Al_{11}RE_3$, in which case further precipitation of other phases such as another type of Al–RE phase or $Mg_{12}RE$ becomes possible. In samples produced by cold chamber die-casting, $Al_{11}RE_3$ and $Al_{10}RE_2Mn_7$ were observed. The TEM investigation has revealed that the major phase was $Al_{11}RE_3$ (a body-centered orthorhombic cell with lattice parameters a = 4.5 Å, b = 13.2 Å and c = 9.9 Å). The relative amounts of REs were reported to be (at%) $La_{27\pm2}Ce_{51\pm2}Pr_{6\pm1}Nd_{16\pm1}$. This phase co-existed with a minor phase $Al_{10}RE_2Mn_7$, the diffraction patterns of which were indexed to a hexagonal system with lattice parameters a = 9.0 Å and c = 13.1 Å. The relative compositions of the RE components was $La_{10\pm1}Ce_{50\pm3}Pr_{9\pm1}Nd_{31\pm2}$. La was said to go into the $Al_{11}RE_3$ phase, while Nd went into the $Al_{10}RE_2Mn_7$ phase. It must be noted

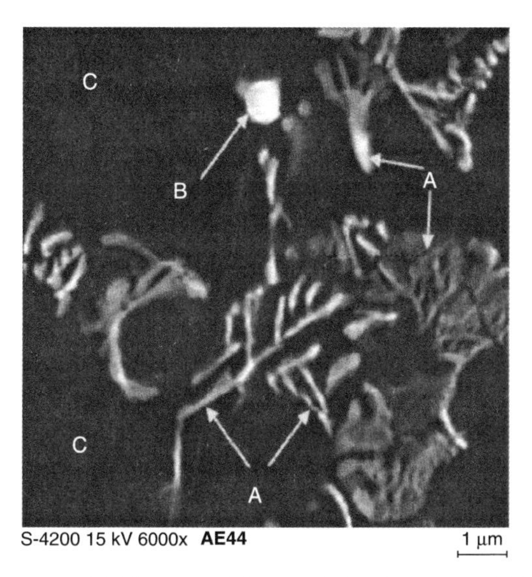

5.12 SEM image of die-cast AE44. Labels A: $Al_{11}RE_3$; B: Al_2RE; C: α-Mg.[53]

that the fast solidification rate of the die-casting process may have influenced the compositions and the crystal structures observed in these studies.

5.3.4 Intermetallics of Ca

In Mg–Al–Ca based alloys, Mg_2Ca and/or Al_2Ca phases are known to form in the interdendritic region in preference to the $Mg_{17}Al_{12}$ phase observed in the binary system. Mg_2Ca and Al_2Ca are Laves phases with crystal structures of C14 (hexagonal) and C15 (cubic), respectively. Al_2Ca and Mg_2Ca, have melting points of 1079°C and 714°C, respectively.[55–57] Mg–Al–Ca alloys also exhibit ternary intermetallics, which has been the topic of much research in order to understand if these are stable or metastable structures.[55,58–62]

Mg_2Ca

Mg_2Ca has a hexagonal C14 type structure with 12 atoms in a unit cell (Fig. 5.13). The lattice parameters are a = 6.2386 Å and c = 10.146 Å and atomic coordinates in the unit cell are shown in Fig. 5.13.[63]

Most of the atoms in the Mg_2Ca unit cell can be classified into two groups. The first group is composed of Ca and Mg(I) atoms, and the second is laid by Mg(II)atoms parallel to each other. The connection between the Ca–Ca–Mg(I) atom group and Mg(II) atom layers depends on three kinds of relatively weak bonds – B, D and F. This indicates that

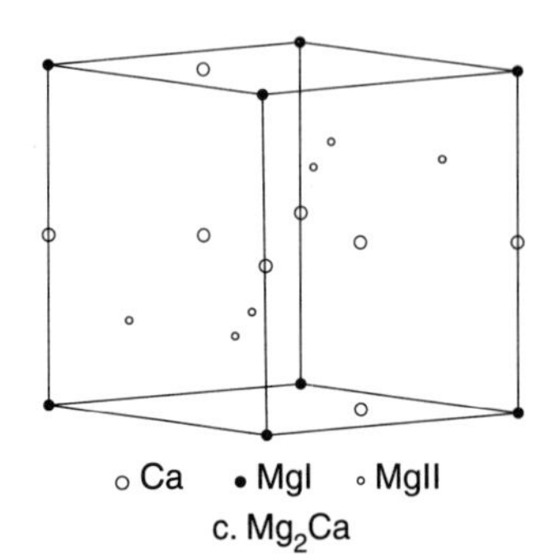

5.13 The crystal structure of Mg_2Ca.

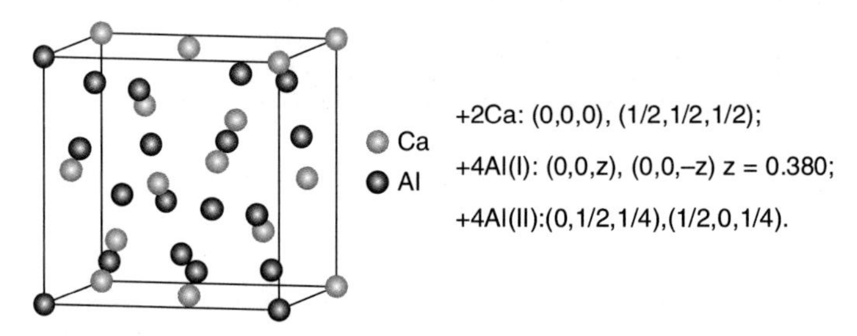

5.14 The crystal structure of Al_2Ca.

the stability of the Mg_2Ca compound is controlled by these weak bonds. When the temperature increases, the expansion of the lattice of Ma_2Ca will produce stress on the bonds connecting atoms. Because the networks of Ca–Ca–Mg(I) and Mg(II)–Mg(II) are so solid and difficult to elongate that the lattice will break on the weak bonds. The existence of the weak bond zones in the Mg_2Ca intermetallic structure is the main reason for its low stability.

Al_2Ca

Al_2Ca has an ordered cubic C15 structure with lattice constants $a = b = c =$ 8.020 Å. As shown in Fig. 5.13; its unit cell has the highest symmetry O_h^T, space group Fd3m. In a unit cell of the Al_2Ca phase, there are 24 atoms and their atomic coordinates are as shown in Fig. 5.14.

Mg–Al–Ca intermetallics

The possible existence of a Laves phase with C36 structure in the pseudo-binary system Mg_2Ca–Al_2Ca was shown by Ameroun *et al.*[56] who carried out first principles calculations and neutron diffraction (ND) analysis on 700°C-annealed samples. The authors showed that the structural sequence in the $CaAl_{2-x}Mg_x$

system is C15–$MgZn_{2(cubic)}$ → C36–$MgNi_{2(hexagonal)}$ → C14–$MgCu_{2(hexagonal)}$. The electron/atom ratio (*e*/*a*) was seen to govern the structural sequence: C15–$CaAl_2$ (cubic) structure is stable above *e*/*a* of 7.75 and until the composition of $CaAl_{1.76}Mg_{0.24}$. The hexagonal C14 is stable below the *e*/*a* value of 6.5 and above the composition of $CaAl_{0.49}Mg_{1.51}$. The C36–$CaAl_{2-x}Mg_x$, structure is stable in the composition range of $0.66 < x < 1.073$. Zhong *et al.*[57] made first principles calculations of three Laves phase structures at Al_2Ca and Al_2Mg compositions and have predicted the existence of various Laves phases in the Al_2Mg and Al_2Ca compositions. Their diffusion couple studies found the existence of two ternary Al_2(Mg,Ca) with C14 (a = 5.56 Å, c = 9.02 Å) and C36 (a = 5.45 Å, c = 17.51 Å) structures via TEM and Orientation Imaging Microscopy (OIM), respectively. The existence of two Laves phases was attributed to the very small energy difference between the Laves phase structures. Based on the explanations of Ameroun,[56] their existence may also be due to a mere change in *e*/*a* ratio with change in composition.

5.3.5 Intermetallics of Sr

Mg–Sr alloys are a new class of advanced Mg alloys; Mg–Al–Sr compositions have been developed as creep-resistant cast alloys[64–68] and Mg–Mn–Sr alloys have been recently investigated for creep resistance.[69] Mg–Sr based wrought alloys are also being currently being investigated.[70–77]

Al_4Sr and $Mg_{13}Al_3Sr$ or Mg_9Al_3Sr

Sr added to an Mg–Al system leads to the precipitation of Al_4Sr phase (Fig. 5.15). Mg–Al–Sr ternary compound formation is also observed in certain Mg–Al–Sr alloys (e.g., AJ52). The Sr/Al ratio seems to determine the phase selection: if the ratio is below 0.3 only Al_4Sr is seen, otherwise a ternary phase ($Mg_{13}Al_3Sr$ or Mg_9Al_3Sr) precipitates.

$Mg_{17}Sr_2$

$Mg_{17}Sr_2$ is the most Mg-rich stable compound in the Mg–Sr system. It has the $Ni_{17}Th_2$-type hP38 hexagonal structure. It forms in Mg–Sr alloys not containing Al,[69–72] but also in Mg alloys containing Al (e.g., AJ52) when the Sr/Al ratio is 0.318–0.361.

5.3.6 Intermetallics of Y and/or Nd

Mg–Nd/Y intermetallics (Fig. 5.16a) $Mg_{41}Nd_5$, $Mg_{24}Y_5$, β ($Mg_{14}Nd_2Y$), are found in alloys such as WE43, WE54. β_1 is seen in WE43 and β is in WE54.[80,81] Crystal structures of these phases are given in Table 5.2 and the TEM diffraction data of β is shown in Fig. 5.16b.

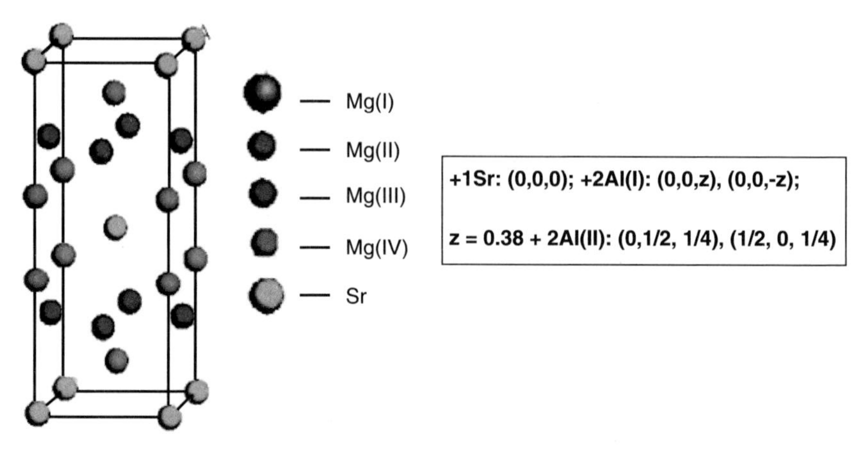

5.15 Model crystal unit cell of Al_4Sr.[79]

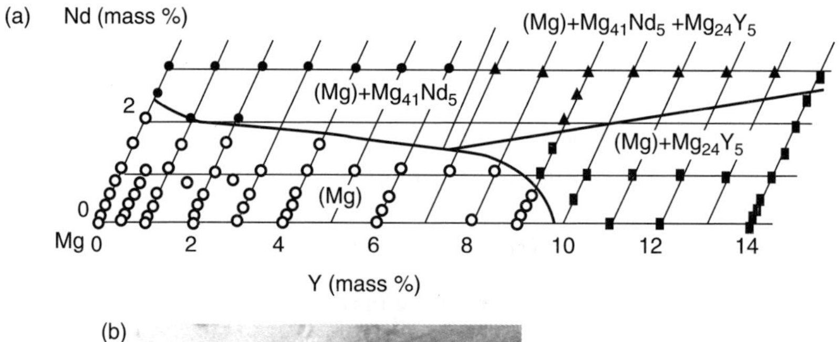

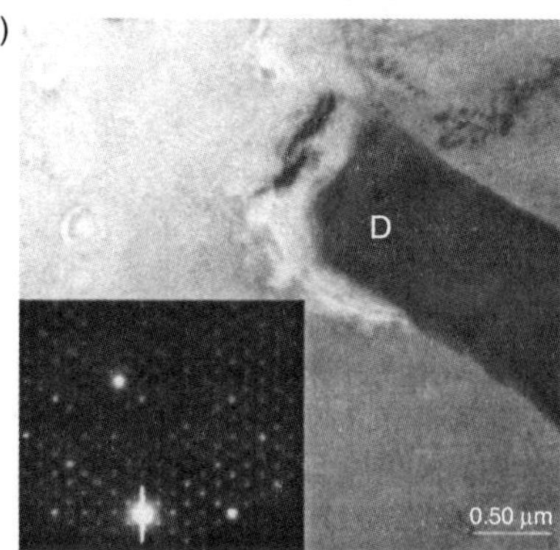

5.16 (a) Isothermal section of the Mg-Nd-Y phase diagram at 500°C.[80] (b) TEM image and SAED diffraction pattern of the β particle.[81] B// [-1–11].

5.4 The effects of second phases on the mechanical behavior of magnesium

Second phases that form in Mg alloys are usually intermetallic compounds, except for α-Mn which forms in Mg–Mn based systems and α-Li which forms in Mg–Li based alloys. Depending on their type and distribution, the second phases can contribute to strengthening and creep resistance. They can also play an important role in nucleating recrystallization and altering preferred orientation (texture) in wrought alloys.

5.4.1 Second-phase strengthening

Dispersion hardening (Orowan strengthening)

Non-coherent precipitates can provide strengthening by pinning dislocations if their size is small and they are closely spaced. The mechanism that causes the strengthening is the interaction of glide dislocations with the precipitates to increase the yield stress as

$$\tau_{\text{yield}} = \frac{(2\text{aGb})}{1 + \tau_a}$$

where τ_{yield} = the yield strength of the alloy with second phases, τ_a = the yield strength of the matrix without precipitates,

(2aGb/l) = stress required to bow out dislocations between precipitates

In Mg casting alloys the second phases are non-coherent but they are coarse and are located at the grain boundaries since quite a number of the alloy systems (Mg–Al, Mg–Zn, Mg–Al–Zn, Mg–Al–RE, Mg–Al–Ca, Mg–Al–Sr) are eutectic. These phases strengthen grain boundaries but provide no Orowan strengthening inside the grains.

Precipitation hardening

Precipitation hardening, another mechanism of second-phase strengthening in alloys, occurs when the solid solubility decreases with decreasing temperature; this allows the solutionizing of as-cast second phases to be re-precipitated in a controlled fashion. When rapidly cooled from a homogeneous solid solution at high temperature, the alloy produces an unstable supersaturated solid solution which, upon aging for a sufficient time, may form fine and dispersed precipitates for further hardening. The precipitation may not go directly into the formation of the non-coherent equilibrium phase because of kinetic limitations and may instead go through a sequence

Table 5.3 Precipitation sequence in certain Mg alloys[30]

Alloys	Precipitation
Mg–Al	SSS[1)]-$Mg_{17}Al_{12}$
Mg–Zn (-Cu)	SSS-GP zones-$MgZn_2$ (rods, coherent)-$MgZn_2$ (disc, semicoherent)-Mg_2Zn_3 (triagonal, incoherent)
Mg–RE (Nd)	SSS-GP zones-Mg_3Nd (?[2)], hep)-Mg_3Nd (fcc, semicoherent)-$Mg_{14}Nd$ (incoherent)
Mg–Y–Nd	SSS-β' ($Mg_{12}NdY$)- β_1($Mg_{14}Nd_2Y$)
Mg–Ag–RE (Nd)	SSS-GP zones (coherent)-γ (?, coherent)
	SSS-GP zones (coherent)- β (?, semicoherent)-$Mg_{12}Nd_2Ag$ (incoherent)

SSS: supersaturated solid solution

of metastable intermediate precipitates. Chapter 4 provides a detailed discussion of precipitation hardening. Table 5.3 is provided here to show some of the possible precipitation sequences in certain Mg alloys.[30]

Age hardening requires some solid solubility, which occurs for Mg at around 12–15% atom size difference (Fig. 5.1). This region in Fig. 5.1 is also where there is some respectable solute strengthening. In the region where size factors are above 15% at around 20–40% (Cu, Ca, Si), where very limited solid solution is observed, there are also good opportunities for strength-enhancing elements, because here intermetallics and dispersoids of high melting point and high hardness are obtained for dispersion strengthening.

Dynamic precipitation

Dynamic precipitation is a term that can be used to describe precipitation which is thermally or stress-induced. Dynamic precipitation is common in cast alloys which are in non-equilibrium in the as-cast condition; upon thermal exposure, the alloy moves towards equilibrium, which may result in second-phase precipitation from the supersaturated matrix or from the supersaturated second phases. A well-known example is the precipitation of $Mg_{17}Al_{12}$ from the supersaturated Mg matrix in die-cast Mg–Al based alloys.[27] Dynamic precipitation has been observed in the AJ62 (Mg–6Al–2Sr) alloy that has been developed by Noranda in 2000–2, where Al–Mn particles precipitated out of the α-Mg matrix during elevated temperature heating.[6] A more recent example of dynamic precipitation has been reported in studies on Mg–Mn–Ce alloys with α-Mn nano-particles precipitating out of the supersaturated α-Mg and supersaturated eutectic $Mg_{12}Ce$ phase.[28,51,82]

5.4.2 Effects of second phases on creep

Mg alloy research focused on developing creep-resistant Mg casting alloys for automotive powertrain applications in the time period between 1991 and 2006. Hence, alloy design for creep resistance has become important. Second phases play a key role in determining the creep resistance of Mg alloys.

a. *Thermally stable intermetallic compounds at the grain boundaries* of the cast material may strengthen the alloy against grain boundary migration by providing obstacles to grain boundary motion) and against grain boundary diffusion (by increasing the atomic packing). Examples are Al_4Sr in Mg–Al–Sr alloys (AJ62). The thermal stability of the Al_4Sr phase is one of the key attributes of AJ62 in attaining its creep resistance.
b. On the other hand, grain boundary intermetallics that show thermal/*metallurgical instability* via coarsening or phase transformation exhibit local strain which causes grain boundary migration (GBM) or grain boundary diffusion (GBD) at elevated temperatures. There are numerous examples of this in Mg alloys, which results in loss of creep resistance: (i) the decomposition of $Al_{11}RE_3$ into Al_2RE and Al to form $Mg_{17}Al_{12}$ in AE42 above 150°C,[52] (ii) the decomposition of Mg_9Al_3Sr in AJ63 alloy,[6] (iii) the transformation of $\beta_1(Mg_{12}NdY)$ to $\beta(Mg_{14}Nd_2Y)$ in WE alloys at 300°C,[83] (iv) the transformation of age-hardening precipitates in Mg–Y–RE alloys.[83]
c. *Intradendritic precipitates*, when fine, closely and uniformly spaced, and in a large quantity, can be effective against the most common creep deformation process in Mg alloys under the service conditions of powertrain components, that is, dislocation climb. Such precipitates can be produced in the as-cast alloys by peritectic reactions or by precipitation hardening. In order to be effective, these precipitates need to successfully pin dislocations and maintain thermal stability at the creep temperatures. Most age-hardening precipitates lose their effectiveness against dislocation climb at temperatures where they are prone to coarsening and phase transformations. An important contribution to creep resistance comes from the fine dispersion of Mn_2Sc disc precipitates in Mg–Gd–Sc–Mn alloys – precipitates in overaged Mg–Gd–Nd–Zr alloys, and α Mn precipitates formed via precipitation from the α-Mg matrix upon temperature exposure in Mg–Mn alloys.[82–86]
d. Creep-induced dynamic precipitation of second phases can either impart or be detrimental to creep resistance. It can be said that the resistance of Mg alloys to dislocation climb can largely be attributed to the dynamic precipitation of fine particles during creep. In AJ62 alloy, the precipitation

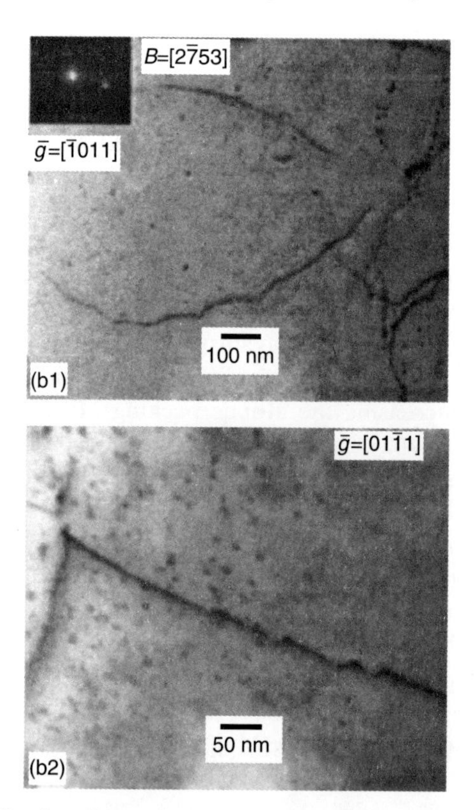

5.17 Effective dislocation pinning by Al–Mn precipitates in AJ alloys.[6]

of Al–Mn particles (Fig. 5.17) from the matrix leads to effective dislocation pinning.[6] In Mg–Ce–Mn alloys, the intradendritic co-precipitation of α-Mn with $Mg_{12}Ce$ slows down dislocation glide within the dendrites and delays the pile-up of dislocations at the intermetallic interface, preventing intermetallic cracking and the early onset of tertiary creep.[28,51,82] Alloys that are heat treatable can be used in solution heat-treated (T5) condition to allow for fine precipitation during creep. Gd and Y alloys with Mn and Sc in the T5 condition exhibit dynamic precipitation from the solid solution of stable Mg_5 Gd type phase (~2 μm in size) inside the grains during creep, which interact effectively with dislocations.[83–86]

e. *Interdendritic intermetallics* can also lead to the early onset of tertiary creep by cracking if the intradendritic and/or interdendritic region of the alloy is not able to prevent easy glide and strain accumulation at the matrix/intermetallic interface. Examples are the cracking of $Mg_{12}Ce$ interdendritic in Mg–Ce–Mn alloys when the dislocation pinning effect of α-Mn fine precipitates is lost leading to the formation of HDD zones at the matrix/intermetallic interface.[51]

5.4.3 Effects of second phases on the recrystallization of wrought Mg alloys

The second phases exert their influence on Mg alloy recrystallization in two ways: (i) they influence grain growth and (ii) they can nucleate new grains through a mechanism called particle-stimulated nucleation (PSN).

Second-phase interaction with grain growth

Fang *et al.*[87] studied the microstructures and mechanical properties of rolled AZ61 alloys containing different levels of Y (0–1.4%Y). Y refined the average grain size; Al_2Y second phase which had been finely broken during rolling was claimed to suppress grain growth. Mg–Li–Al–Zn–RE alloy (LAZ532–2RE), when extruded, recrystallized to a fine grain size[88] which was attributed to grain boundary pinning and to the retardation of grain growth by the fine Al_2Y precipitates. He *et al.*[89] observed grain refinement in the extruded/heat-treated (T6) structure of 1.3%Gd alloyed ZK60. The grain refining effect of Gd was attributed to grain boundary pinning by Mg-Zn-Gd phase particles.

Particle-stimulated nucleation

PSN is a mechanism whereby the recrystallization of the deformed metal is activated at the interface of second-phase particles. The important attributes of the second-phase particles in nucleating recrystallization are known to be optimal size and non-coherency, to ensure sufficient accumulation of dislocations to nucleate recrystallization.

The thermal stability of the precipitates is also of prime importance in maintaining the effectiveness of PSN at high temperatures. Competing mechanisms that occur at low temperatures (twinning, grain boundary bulging) and at high temperatures (non-basal slip) may overshadow PSN even when there are appropriate second phases.[90] The size of the distributed particles is also important. Clusters of particles, even when they are clusters of fine (<1 μm) particles, are known to be more effective than single particles.[91] Different alloys exhibiting PSN during recrystallization are presented in Table 5.4.

5.5 Alloying with surface-active elements

Surface-active elements are specific to a metal system and are those solutes that segregate to the surface because they decrease its surface energy. The surface could be liquid metal surface, solid surface, grain boundaries or the interface between the matrix and the second phase. In the case of liquid surfaces,

Table 5.4 PSN in different Mg alloys

Alloy	Precipitate	Deformation process	Effect	References
Mg + Mn	Mn particles	Hot compression	Matrix rotation at Mn particles	92
AZ41 + Mn,Ca	$Mg_{17}Al_{12}$, Al_8Mn_5, $A1_2Ca$	Hot compression	Large particles promote recrystallization. The geometry, size and distribution of the particles influence DRX	91, 93
AZ31	$Mg_{17}Al_{12}$ AlMn	Hot compression	Some evidence of PSN	94
AZ31	$Mg_{17}Al_{12}$, AlMn	Hot compression	PSN took place during hot deformation and was facilitated by the fragmentation of the $Mg_{17}Al_{12}$	95
AZ31	$Mg_{17}Al_{12}$ AlMn	Twin roll casting (TRC)	PSN occurs due to high amount of second phase in the as-TRC metastable structure and contributes to texture weakening	96, 97
AZ31 + Sr	Al_4Sr	Extrusion	Al_4Sr stringers lead to PSN and texture weakening at 350°C but not at 250°C	98, 99
Mg–1Mn + Sr	$Mg_{17}Sr_2$, Sr_5Si_3	Rolled sheet	Stringers cause PSN and texture weakening	100, 101
Mg–1Zn + Sr	Sr–Si compounds (SrSi, Sr_5Si_3)	Rolled sheet	Stringers cause PSN and texture weakening	100
WE43	Mg–Nd. Mg–Y Mg–Nd–Y	Extrusion	Recrystallization at second-phase particles were observed	102
WE54	Mg–Nd, Mg–Y Mg–Nd–Y	Extrusion	PSN took place and has weakened the basal texture	103

Table 5.5 Wigner-Seitz radii, *rSW*, of selected elements (nm)

Al	Mg	Sr	Ca	Ce	Ni	Nd	Si	Sb	Sn	Fe	Se	Te	Mn	Li
0.15	0.18	0.19	0.21	0.20	0.14	0.20	0.16	0.11	0.19	0.14	0.19	0.21	0.14	0.17

solute segregation can influence the surface tension and alter the chemical activity (e.g., oxidation) or the evaporation[104] of the alloy melt. Yamauchi[105] has performed the calculations using electron density profiles for a number of binary substitutional solid solutions and has formulated the rule that the element having the larger Wigner-Seitz radius, r_{WS} (*lower* average electron density), segregates to the surface. It implies that the driving force for surface segregation is reduction in the surface energy of the segregated alloy. This was reconfirmed by Kiejna *et al.*[106,107] for binary alkali-metal alloys (e.g., Na–K, Na–Cs). In other systems, Sb, Sn and Se have a strong tendency to surface-segregate on Fe and Fe–Si alloys (Table 5.1). Sb and Sn can alter the texture and improve the electrical properties of silicon steels, and Se and Te are surface-active in molten Fe and decrease wettability.[108] Micro-levels of surface-active Nd, Pr, Ce, La in nickel alloys, and of Ca in tool steels, poisons the grain boundaries and disperses carbides within the grains.[109] Ce is also surface-active in Al.[110] According to Table 5.5, Sr, Ca Ce, Nd, Sn, Se and Te would be surface-active in Mg based on their larger r_{WS} than Mg. These elements would reduce the surface energy and the surface tension of Mg.

5.5.1 Surface tension

It has been observed[111] that a number of elements alter the surface tension of liquid Mg (Fig. 5.18). As noted, Sr has a significant effect wherein the surface tension continues to decrease with increasing Sr. The surface activity gives a good guideline for surface segregation, but exceptions have been observed: for example, Zn surface-segregates in Mg but has a lower r_{WS} than Mg.[17] Sb also reduces the surface tension despite the smaller r_{WS}. It can be predicted from Fig. 5.18 that Sr would increase interdendritic feeding, thereby minimizing or altering shrinkage microporosity during casting. In the gravity cast AZ91 Mg alloy, Sr removes from the bulk the microporosity that results from the solidification shrinkage during dendritic freezing (which is usually present in long-freezing-range alloys such as AZ91) and concentrates it into the hot spot;[113] in the high-pressure die-cast AZ91, a porosity reduction is also observed[112] resulting in an improvement in pressure tightness (Fig. 5.19).

The reduction in the surface tension, in this case (whereby solutes segregate to the surface) is accompanied by a decrease in the number of Mg atoms occupying the surface. This decreases the vapor pressure and increases the

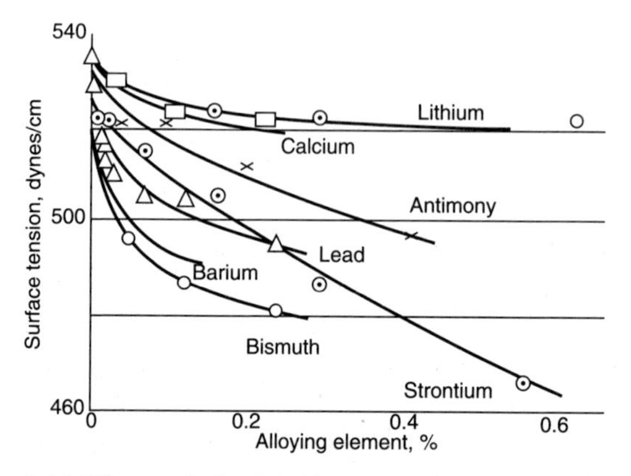

5.18 Effects of alloying elements on the surface tension of Mg.[111]

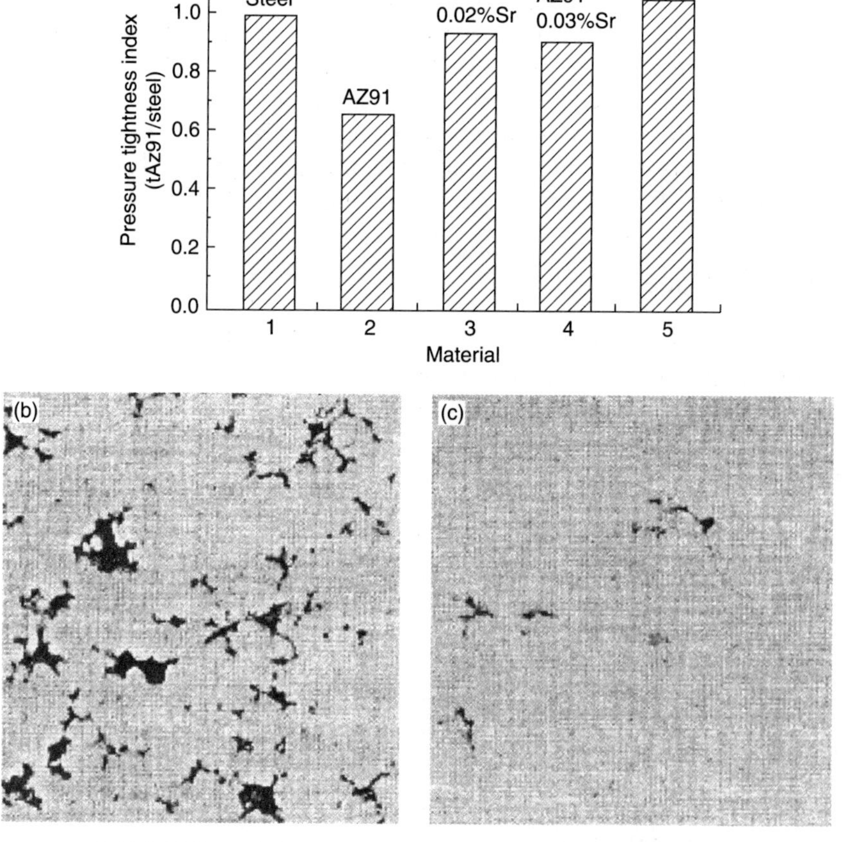

5.19 (a) Effect of Sr on the pressure tightness of die-cast AZ91D. Die-cast AZ91D (b) with no Sr; and (c) with 0.03 wt%Sr.[112]

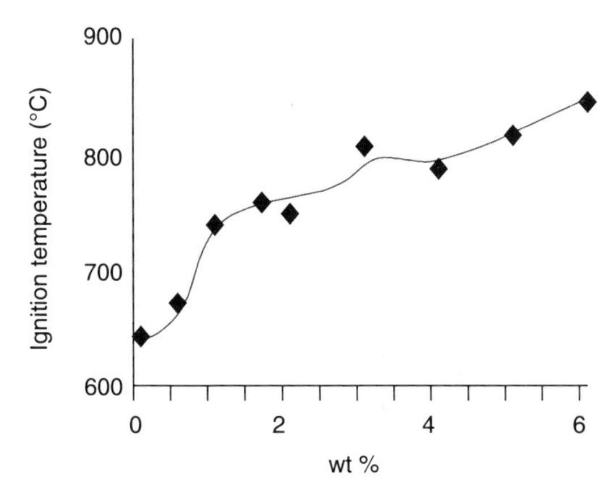

5.20 Ignition temperature vs. Sr.[114]

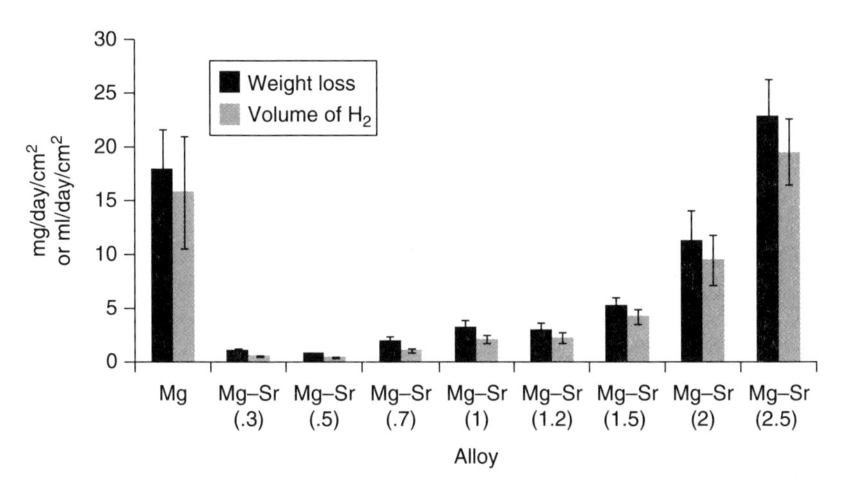

5.21 Biodegradation (weight loss; H_2 evolution) of Mg–Sr alloys in SBF (alloy compositions in wt%).[115]

ignition temperature of Mg. These effects have been observed in Mg–Sr, wherein the ignition temperature of the Mg–Sr alloy increases (Fig. 5.20) with increasing Sr content from 680°C at 0%Sr to 850°C at 6 wt%Sr.[114]

5.5.2 Bio-corrosion

Ca and Sr also improve (slow down) the bio-corrosion rate of Mg, paving the way for effective bio-absorbable cardiac and bone implants. Pekguleryuz and coworkers[115] have evaluated the biodegradation of Mg–Ca and Mg–Sr alloys; it was observed that compositions below 0.5% Sr and Mg-0.6%Ca

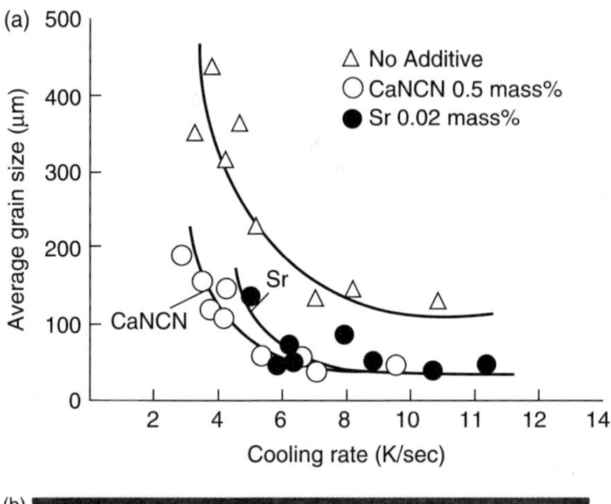

(b)

5.22 (a) The effect of Ca and Sr on grain refining; and (b) the Mazda wheel.[116]

show low rates of weight loss and hydrogen evolution (Fig. 5.21) in simulated body fluid (SBF). The improvement is related to the surface-active nature of Sr and Ca, which slows down the corrosion rate and improves the hydroxyapatite that forms on the surface.

5.5.3 Microstructural refinement

The criteria and rules governing the occurrence of fine grain boundary precipitates are not always clear. However, the selection of elements with a

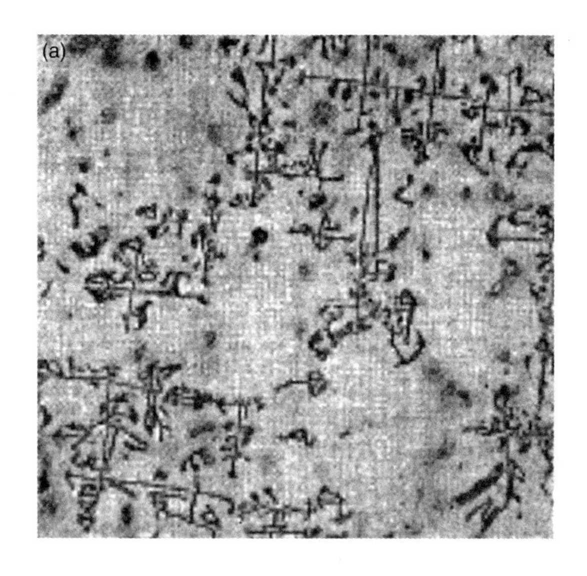

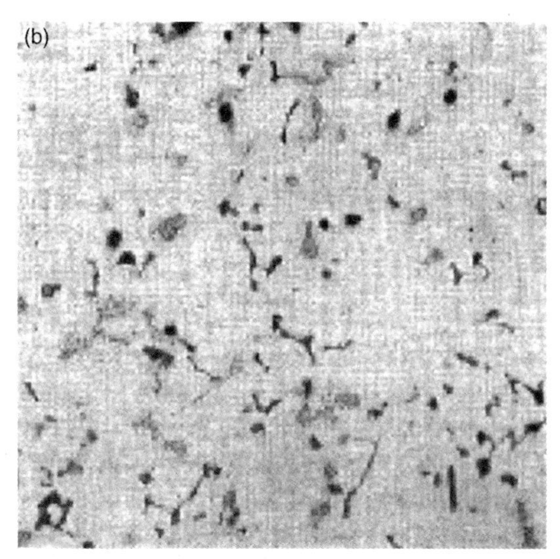

5.23 Microstructure of gravity cast (a) AS41 alloy with Chinese script Mg_2Si phase; (b) same alloy with 0.1 wt% Ca with refined Mg_2Si phase.[41]

probability of forming fine precipitates or refining existing phases can be facilitated by tapping into the knowledge base of other metal systems where these phenomena have occurred. Some of the elements, such as Ce, Ca or Sr, Ba and Sb, can be used in the microstructural design of many Mg alloys.

In Mg, the refining effect of Sr and Ca on the grain size and the $Mg_{17}Al_{12}$ phase in Mg–Al alloys had been first identified by Koubichek[111] in 1959.

Table 5.6 Mechanical properties of sandcast Ca-modified AS41 (parentheses give the sample size)[41]

Alloy	UTS (MPa)	YS (MPa)	Elongation (%)	Elastic modulus (GPa)	Mg_2Si morphology
AS41	152 ± 7 (14)	84 ± 3 (14)	1.9 ± 0.3 (9)	45 ± 3 (14)	Chinese script
AS41 + 0.06%Ca	155 ± 2 (16)	85 ± 2 (16)	4.5 ± 0.7 (15)	46 ± 2 (16)	Semi-dispersed
AS41 + 0.11%Ca	163 ± 7 (18)	85 ± 2 (18)	4.2 ± 0.7 (18)	49 ± 2 (18)	Dispersed
AZ91C	165	93	3.5	–	N/A

Comparing the Wiegner–Seitz ratio (Table 5.5), it can be seen that Sr and Ca would segregate to the surface or interfaces in Mg. This effect had not been explored for many years until Aliravci *et al.*[113] observed the grain refining effect of Sr in AZ91 (Mg–9Al–1Zn) alloy. This led to the use of Sr to grain-refine Mazda AZ80 forging wheels (Fig. 5.22).[116] In recent years more work has been carried out on the grain refining of Mg alloys via Sr.[117,118] Pekguleryuz *et al.*[41] used Ca in the 1990s to refine the Mg_2Si Chinese script morphology (Section 5.3.2) in the AS41 (Mg–4Al–1Si) alloy (Fig. 5.23) and a significant improvement in ductility of the gravity castings was obtained at 0.1%Ca additions (Table 5.6). In 2006, the Mg_2Si phase was refined by 0.1%Sr and Sb additions.[119,120]

5.6 Alloying elements and their effects

5.6.1 Al, Zn and Mn[8,26]

The most widely used alloys in this group are the AZ (Mg–Al–Zn) and AM (Mg–Al–Mn) series. Figure 5.24a illustrates the effects of Al and Zn on the strength and ductility of binary Mg alloys. The 3% Zn alloy is shown to be the most ductile, and the 9% Al alloy gives the maximum strength. Where a compromise needs to be made between strength and ductility, a binary 6% Al alloy (AM60, AZ61) offers acceptable strength, while a 5% Zn offers acceptable elongation. The combined effect of Al and Zn on Mg is not only on mechanical properties but also on castability (Fig. 5.24b). Al and Zn (up to 3%) improve the fluidity of Mg, making it castable. Al also influences corrosion resistance: Al decreases the corrosion resistance until ~8%, after which a reversal in behavior is seen.[121] For the AZ91 alloy where Al content ranges from 8.3% to 9.7%, great variations in corrosion resistance are inevitable. Tighter control of the Al limits may be a preferable measure in the future to avoid these variations.

Various problems related to molten metal handling and corrosion of these alloys due to heavy metal contamination have been solved over the last two decades. The most serious handicap with these alloys is poor high temperature strength, especially in cast parts. Their microstructure is characterized by $Mg_{17}Al_{12}$ (γ in the Al–Mg phase diagram, but also called β in the Mg alloy literature) intermetallic precipitates in a matrix of primary Mg. The β-precipitate has a cubic crystal structure incoherent with the HCP lattice of the matrix (Section 5.3.1). It exhibits covalent as well as metallic bonding and is prone to thermal instability.

5.6.2 Rare earth (RE) elements

REs impart both room and elevated temperature (200°C) strength to Mg. The strengthening is due to solid solution of the borderline sized RE atoms and second-phase hardening due to the $Mg_{12}X$ intermetallic (where X denotes RE element). The intermetallics (Table 5.2) have relatively low diffusivity (line compounds) at moderate temperatures and good matrix coherency.[122] Additional elements, such as Ag, may further improve the age-hardening response (such as in QE22 alloys). Die-casting alloy, AE44 has been utilized in the Corvette engine cradle. It takes advantage of the fluidity imparted by Al and the creep resistance imparted by REs. Al decreases somewhat the effect of REs due to the preferential formation of Al_2X which removes RE from solution. The alloy is costly but relatively easy to cast and has good creep properties. $Al_{11}RE$ is prone to phase transformation at moderate temperatures, making the alloys such as AE42 thermally unstable.

The major beneficial effect of REs on Mg is on creep strength up to 200–250°C. In Mg–Ce alloys it has been discovered that there is additional nucleation of a fine $Mg_{12}Ce$ precipitate at the grain boundaries during creep at 200–300°C, which is the decisive factor in improving the creep performance of Mg at 300°C by strengthening the grain boundaries. Furthermore, the additions that inhibit the coarsening of such fine precipitates contribute further to creep resistance. Mn further increases the creep strength of Mg–Ce alloys by restricting or altering the kinetics of the growth and altering the habit planes of $Mg_{12}Ce$ to achieve and maintain a fine dynamic precipitation.[51] The creep mechanism of REs can shed light on the development of creep-resistant Mg alloys. REs are surface-active in Mg and they can be used to modify and refine microstructural features, and for improving the oxidation and corrosion resistance. The beneficial effects of REs on the strength of Mg had been discovered as early as the 1930s and many commercial casting alloys suitable for gravity casting and some wrought alloys have been developed. Because of the higher cost associated with the RE alloys they have

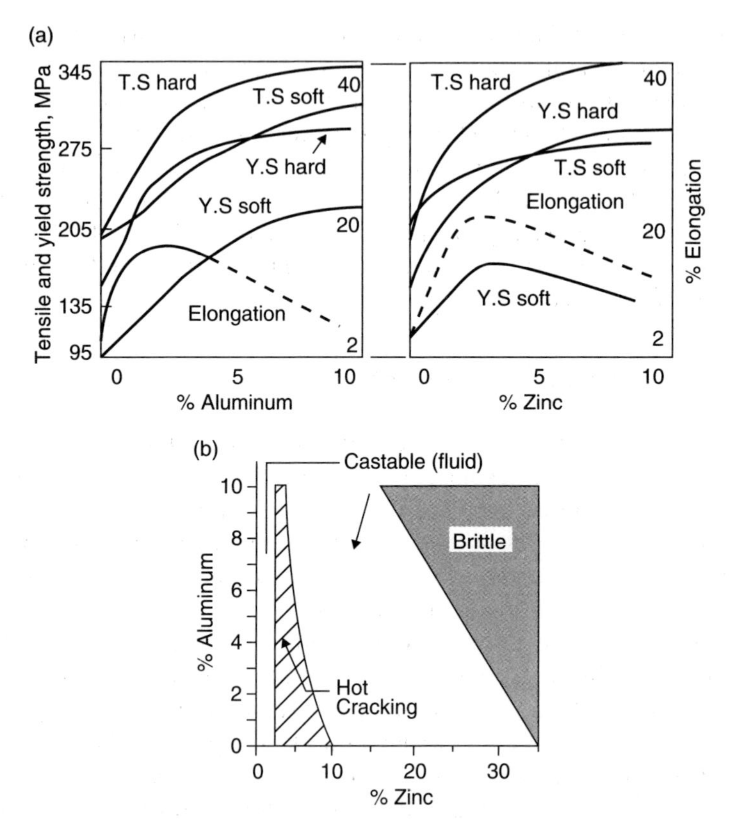

5.24 The effects of Al and Zn on (a) the mechanical properties of Mg and (b) castability.[8,26]

been mainly used for aerospace applications. A number of ternary and quaternary alloys have been commercialized, such as the ternary Mg–RE–Zr (EK30) and the quaternary alloys Mg–Zn–RE–Zr (ZE63, EZ33).

Selection of alternate lower cost solutes that produce similar intermetallic compounds in combination with solutes that stabilize them could be the key consideration for future work. However, most RE intermetallic compounds in Mg alloys have moderate melting points and most are subject to phase transformations, making the alloys metallurgically and thermally unstable for certain elevated temperature service. Other shortcomings are the high cost and the lack of worldwide supply of REs, posing a concern for high volume cost sensitive applications.

5.6.3 Alkaline earth elements

Alkaline earth elements Ba, Ca, Sr, like Mg, belong to Group II of the periodic table and exert effects similar to the REs when alloyed into Mg. Ba is not recommended for use in commercial alloys since the storage of Ba as a raw material poses some safety problems.

Alkaline earths are surface-active and, like REs, improve the corrosion and oxidation resistance of Mg (including bio-corrosion resistance) and also refine the grain size and the size of the second phases.[111-120] They form intermetallic compounds with Mg and with Al in Mg–Al alloys,[55–72] which strengthen the grain boundaries and thereby increase creep resistance. These intermetallics are thermally more stable than the RE compounds: due to the negligible solubility of the alkaline earths in Mg at the elevated temperatures, they do not dissolve in the matrix or transform to other compounds by reacting with aluminum in Mg–Al alloys like some of the Al–RE intermetallics. The Sr compounds, due to their thermal stability, play a more effective role in PSN of recrystallization during rolling and extrusion than do the Mg–Re intermetallics at elevated temperatures.[10,70–78]

5.6.4 Group IV and group V elements[7]

Another area for further investigation is the effect of elements such as Sn, Sb and Bi on strength.[7] These elements form line compounds of high melting point. Because of the low ductility of the binary systems, they should be investigated in combination with elements that impart ductility such as Zn or Al. An important element addition to replace the REs partially, or to take the place of Ag, may be Sb. Sb has been used to produce creep-resistant solder alloys such as the 95/5 Sn/Sb.

The precipitates of the Ag containing alloys have not been completely determined yet but it is known that in the case of the Mg–Ag system, 'electron compounds' of simple lattice structures (BCC AgMg at an electron: atom ratio of 1.5 and HCP $AgMg_3$ at a ratio of 1.8) form.[8] These simple phases may more readily maintain coherency with the HCP Mg matrix.

Sb is also known to refine the grain size and the second phases in Mg alloys,[119] even though it is not predicted to be surface-active based on Wigner–Seitz radii (Table 5.5). The refining effect may be related to nucleant effects rather than to growth inhibiting effects related to surface activity.

5.7 Summary: magnesium alloy design to enhance properties

5.7.1 Strength and ductility

When the mechanical properties of a series of binary Mg alloys are examined under comparable conditions, it is seen that alloying elements fall into three main categories[7,8]

i. Those that increase both strength and ductility. These elements in order of effectiveness are:
   Al, Zn, Ca, Ag, Ce, Ga, Ni, Cu, Th (strength criteria)
   Th, Ga, Zn, Ag, Ce, Ca, Al, Ni, Cu (ductility criteria).
ii. Those which provide little strengthening, but increase ductility: Cd, Tl and Li.
iii. Those which may confer considerable strengthening but at the cost of ductility. These are: Sn, Pb, Bi and Sb.

Based on the above information, it is not surprising that most common alloy systems have been developed in the Mg–Al, Mg–Zn and Mg–Al–Zn systems.

5.7.2 Creep resistance in cast alloys[122]

Alloying elements exert their influence on creep through solute effect or second-phase formation.

i. REs and Y reduce GBD, pipe diffusion, and decrease primary creep rate. Ce produces Mg_{12} Ce or $Al_{11}Ce_3$, which are unstable at elevated temperatures, but Nd and Y produce interdendritic and intradendritic intermetallics, some of which are thermally stable.
ii. Sr produces the thermally stable Al4Sr against GBD and GBM in Mg–Al alloys. Sr also reduces Al solubility in Mg, leading to the dynamic precipitation of Al–Mn in Mg–Al–Sr (Mn) alloys.
iii. Mn produces fine intradendritic α-Mn phase which co-precipitates dynamically with fine $Mg_{12}Ce$ particles, refining its size and leading to effective dislocation pinning.[51,82]

5.7.3 Formability[123]

Since the main cause of low formability of Mg at room temperature (lack of at least five slip systems that activate) is its HCP crystal structure with a *c*/*a* of 1.624. Li additions, which influence the *c*/*a* (for Li < 2 wt%) or change the crystal structure to BCC(Li > 14 wt%), are perhaps the most potent alloying elements for improving the yield symmetry and the uniform deformation in Mg alloys. Li may, however, adversely affect strength; furthermore, alloying, melting and casting processes are very challenging.

A further difficulty in forming Mg wrought alloys is the preferred orientation (basal texture) that develops when Mg alloys are rolled or extruded. Alloying elements that have contributed to weakening of the strong basal texture in wrought Mg are RE elements, Y, and Sr.

Table 5.7 E_{corr} (V_{SCE}) for intermetallic phases and impurities in 0.1 M NaCl.[125]

	Mg_2Ca	Mg	$Mg_{24}Y_5$	$Mg_{12}La$	$Mg_{12}Nd$	Mg_2Si	$Mg_{12}Ce$	$Mg_{17}Al_{12}$	Mn	$MgZn_2$	Fe	Ni	Cu
E_{corr}	−1.75	−1.65	1.60	−1.60	−1.55	−1.54	−1.50	−1.35	−1.28	−1.03	−0.60	−0.22	−0.15

5.7.4 Corrosion resistance[124]

Magnesium has low corrosion resistance due to its highly negative corrosion potential of −2.37 V. It also ignites at ~680°C due to its high vapor pressure following surface oxidation which increases the local surface temperature very rapidly. Bulk magnesium, because of its high thermal conductivity, can dissipate the heat from the surface quickly; consequently it is difficult to ignite bulk magnesium, but chips will ignite rapidly.

Magnesium alloys are also susceptible to micro-galvanic corrosion which results from the difference between the corrosion potentials of the Mg matrix and the intermetallic second phases. Table 5.7 gives the corrosion potential (E_{corr}) of some of the second phases found in Mg alloys.

Alloying elements influence the corrosion behavior of Mg in various ways. Al improves the corrosion resistance up to a limit of ~8 wt% after which the corrosion resistance falls likely due to enhanced micro-galvanic corrosion.[121] Mn added to Mg–Al alloys removes the Fe impurity during molten state and settles it to the bottom of the crucible as sludge; when the solubility is reduced with the addition of Al, the separation of the Fe–Mn sludge compounds in the molten state becomes possible. Mn is not as effective in removing Fe from other Mg alloys, where it has some solid solubility and does not settle out in the molten state.

Surface-active elements, Ca, Sr and REs change the surface composition of the Mg alloy and modify the driving force for corrosion, and oxidation. Second phases that have corrosion potential quite different from Mg (Table 5.7), and impurities such as Fe, Cu and Ni, are detrimental to the corrosion resistance of Mg because they lead to micro-galvanic corrosion.

5.8 References

1. M.R. Stoudt, 'Magnesium: applications and advanced processing in the automotive industry', *JOM*, **60**(11), 2008, 56–56.
2. M. Easton, 'Magnesium alloy applications in automotive structures', *JOM* **60**(11), 2008, 57–62.
3. K. Johnson, 'Magnesium automotive applications,' *Adv. Mater. Proc.*, **160**(6), 2002, 62–65.
4. M. Pekguleryuz, E. Baril, P. Labelle, D. Argo, 'Creep resistant Mg-Al-Sr alloys', *J. Adv. Mater. Proc.*, **35**(3), 2003, 32–38.

5. Magnesium Fosters Rebirth of an Automotive Engine: http://www.intlmag.org/files/mg001.pdf.
6. M. Kunst, A. Fischersworring-Bunk, G.L' Esperance, P. Plamondon, U. Glatzel, 'Microstructure and dislocation analysis after creep deformation of die-cast Mg–Al–Sr (AJ) alloy', *Mat. Sci. Eng. A*, **510–511**, 2009, 387–392.
7. M.O. Pekguleryuz, M.M. Avedesian, 'Magnesium alloying – some metallurgical aspects', *Magnesium Alloys and their Applications*, B.L. Mordike, F. Hehman (Eds). DGM, Germany, 1992, 213–220.
8. G.V. Raynor, *The Physical Metallurgy of Mg and Its Alloys*, Pergamon Press, London, 1959, 103.
9. J.L. Walter, M.R. Jackson, C.T. Sims (Eds), *Alloying*, ASM, Metals Park, Ohio, 1988, 424.
10. M. Masoumi, M. Hoseini, M. Pekguleryuz, 'The influence of Ce on the microstructure and rolling texture of Mg–1%Mn alloy', *Mat. Sci. Eng A*, **528**, 2011, 3122–3129.
11. S.R. Agnew, M.H. Yoo, C.N. Tome, 'Application of texture simulation to understanding mechanical behavior of Mg and solid solution alloys containing Li or Y', *Acta Mater*, **49**, 2001, 4277–4242.
12. F.E. Hauser, P.R. Landon, J.E. Dorn, 'Deformation and fracture of alpha solid solutions of lithium in magnesium', *Trans. ASM*, **50,** 1958, 856–883.
13. A. Becerra, M. Pekguleryuz, 'Effects of lithium, indium and zinc on the lattice parameters of magnesium', *J. Mater. Res.*, **23**(12), 2008, 3379–3386.
14. R.S. Busk, 'Lattice parameters of magnesium alloys' *J. Met.*, **188**, 1950, 1460–1464.
15. S.R. Agnew, M.H. Yoo, C.N. Tome, 'TEM investigation of dislocation structures in Mg and Mg-Li α-solid solution alloys', *Metall. Mater. Trans.*, **33A**, 2002, 851–858.
16. J.F. Stohr, J.P. Poirier, 'Etude en microscopie electronique du glissement pyramidal (1122) 1123) dans le magnesium', *Phil. Mag.*, **25**, 1972, 1313–1329.
17. A. Becerra, M. Pekguleryuz, 'Effects of lithium, indium and zinc on the grain size of magnesium', *J. Mater. Res.*, **24**(5), 2009, 1722–1729.
18. M. Pekguleryuz, M. Celikin, M. Hoseini, A. Becerra, L. Mackenzie, 'Study on edge cracking and texture evolution during 150°C rolling of magnesium alloys: The effects of axial ratio and grain size', *J. Alloy Compd.*, **510**(1), 2011, 15–25.
19. T. Uesugi, M. Kohyama, M. Kohzu, K. Higasi, 'Generalized stacking fault energy and dislocation properties for various slip systems in magnesium: a first principles study', *Mater. Sci. Forum, Trans Tech.,* **419–422**, 2003, 225–230.
20. M.H. Yoo, 'Slip, twinning, and fracture in hexagonal close-packed metals', *Metall. Trans. A*, **12A**, 1981, 409–418.
21. M.H. Yoo, J.K. Lee, 'Deformation twinning in h.c.p. metals and alloys', *Phil. Mag. A*, **63**(5), 1991, 987–1000.
22. J. Wang, J.P. Hirth, C.N. Tome, 'Twinning nucleation mechanisms in hexagonal-close-packed crystals', *Acta Mater.*, **57**, 2009, 5521–5530.
23. M.F. Ashby, D.R.H. Jones, *Engineering Materials* 1, Pergamon Press, 1989, 177.
24. P. Moreno, T.K. Nandy, J.W. Jones, J.E. Allison, T.M Pollock, 'Microstructural stability and creep of rare-earth containing magnesium alloys', *Scr. Mater.*, **48**(8), 2003, 1029–1034.

25. S.M. Zhu, M.A. Gibson, J.F. Nie, M.A. Easton, G.L. Dunlop, 'Primary creep of die-cast magnesium-rare earth based alloys', *Metall. Mater. Trans. A*, **40A**(9), 2009, 2036–2041.
26. E.F. Emley, *Principles of Magnesium Technology,* Pergamon Press, London, 1966, 1032.
27. M.S. Dargusch, G.L. Dunlop, *Magnesium Alloys and Their Applications*, Wolfsburg, Werkstoff -Informationsgesellschaft mbH., 1998, 277–282.
28. M. Celikin, A.A. Kaya, M. Pekguleryuz, 'Microstructural investigation and the creep behavior of Mg-Sr-Mn alloys', *Mat. sci. Eng. A*, **550**, 2012, 39–50. doi:10.1016/j.msea.2012.03.117.
29. K.Y. Zheng, X.Q. Zeng, J. Dong, W.J. Ding, *Mater. Sci. Eng. A*, **A492**(1–2), 2008, 185–190.
30. N. Hort, Y.D. Huang, K.U. Kainer, 'Intermetallics in magnesium alloys', *Adv. Eng. Mat.*, **8**(4), 2006, 235–240.
31. D. Singh, C. Suryanarayana, L. Mertus, R-H. Chen, 'Extended homogeneity range of intermetallic phases in mechanically alloyed Mg-Al alloys', *Intermetallics*, **11**, 2003, 373–376.
32. H.W. Liu, J.W. Liu, L.Z. Ouyang, C.P. Luo, 'Multiple orientation relationship of $Mg/M_{g17}A_{l12}$', *J. Appl. Cryst.*, **45**, 2012, 224–233.
33. K.N. Braszczyńska-Malik, 'Precipitates of γ-$Mg_{17}Al_{12}$ phase in AZ91 alloy', *Magnesium Alloys – Design, Processing and Properties*, F. Czerwinski (Ed.), InTech, Croatia, 2011, 95–112.
34. C.M. Wayman, *Introduction to the Crystallography of Martensitic Transformations,* Macmillan, New York, 1964.
35. R.M. Wang, D. Eliezer, A. Gutman, 'Microstructures and dislocations in the stressed AZ91D magnesium alloys', *Mat. Sci. Eng.*, **A344**, 2002, 279–287.
36. D.-W. Zhou, P. Peng, H.-L. Zhuang, Y.-J. Hu, J.-Sh. Liu, 'First-principle study on structural stability of Ca alloying $Mg_{17}Al_{12}$ phase', *Chinese J. Nonferrous Met.*, **15**(4), 2005, 546–551.
37. Z.W. Huang, Y.H. Zhao, H. Hou, P.D. Han, 'Electronic structural, elastic properties and thermodynamics of $Mg_{17}Al_{12}$, Mg_2Si and Al_2Y phases from first-principles calculations', *Physica B*, **407**, 2012, 1075–1081.
38. G.Y. Yuan, Z.L. Liu, Q.D. Wang, W.J. Ding, 'Microstructure refinement of Mg–Al–Zn–Si alloys', *Mater. Lett.*, **56**, 2002, 53–58.
39. M. Cabibbo, E. Evangelista, S. Spigarelli, Microstructure studies on a Mg-Al-Si alloy (AS21x)', *Metall. Sci. Tech.*, **22**, 2004, 9–13.
40. B. Bronfin, M. Katsir, E. Aghion, 'Preparation and solidification features of a series magnesium alloys', *Magnesium Technology*, H. Kaplan , J. Hryn , B. Clow (Eds), TMS, 2000, 253–259.
41. M.O. Pekguleryuz, A. Luo, C. Aliravci, 'Modification of sandcast AS41 alloy', *Light Metals Proc. Appl.*, CIM, Quebec City, Aug–Sep, 1993, 409–416.
42. G.Y. Yuan, Z.L. Liu, Q.D. Wang, 'Microstructure refinement of Mg-Al-Zn-Si alloys', *Mater. Lett.*, **56**, 2002, 53–58.
43. A. Srinivasan, U. T.S. Pillai, J. Swaminathan, S.K. Das, B.C. Pai, 'Observations of microstructural refinement in Mg–Al–Si alloys containing strontium', *J. Mater. Sci.*, **41**, 2006, 6087–6089.
44. D.H. Wood, E.M. Cramer, J. *Less-Common Met.*, **9**, 1965, 321–337.

45. Q.C. Johnson, G.S. Smith, D.H. Wood, E.M. Cramer, 'A new structure in the magnesium-rich region of the cerium-magnesium system', *Nature*, **201**, 1964, 600.
46. X. Zhang, D. Kevorkov, M. Pekguleryuz, 'Stoichiometry study on the binary compounds in the Mg-Ce system-part I', *J. Alloy Compd.*, **475**(1–2), 2009, 361–367.
47. X. Zhang, D. Kevorkov, M. Pekguleryuz, 'Study on the binary intermetallic compounds in the Mg-Ce system', *Intermetallics*, **17**(7), 2009, 496–503.
48. X. Zhang, D. Kevorkov, M. O. Pekguleryuz, 'Study on the intermetallic phases in the Mg-Ce system, Part II: Diffusion couple investigation', *J. Alloy Compd.*, **501**, 2010, 366–370.
49. X. Zhang, D. Kevorkov, I-H. Jung, M. Pekguleryuz, 'Phase equilibria on the ternary Mg-Mn-Ce system at the Mg-rich corner', *J. Alloy Compd.*, **482**, 2009, 420–428 (doi:10.1016/j.jallcom.2009.04.042, 2009).
50. P. Moreno, T.K. Nandy, J.W. Jones, I.E. Allison, T.M. Pollock, 'Microstructure and creep behavior of a die cast magnesium-rare earth alloy', TMS, 2002.
51. M. Celikin, A.A. Kaya, R. Gauvin M. Pekguleryuz, 'Effects of manganese on the microstructure and dynamic precipitation of creep resistant cast Mg-Ce-Mn alloys', *Scripta Mater*, **66**, 2012, 737–740.
52. B.R. Powell, V. Rezhets, M.P. Balogh, R.A. Waldo, 'Microstructure and creep behavior in AE42 magnesium die-casting alloy', *JOM*, **54**, 2002, 34–38.
53. T. Rzychoń, A. Kiełbus, G. Dercz, 'Structural and quantitative analysis of die cast AE44 magnesium alloy', *J. Achiev. Mater Manufact. Eng.*, **22**(2), 2007, 43–46.
54. G. Pettersen, H. Westengen, R. Hoier, O. Lohne, 'Microstructure of a pressure die cast magnesium-4wt.% aluminum alloy modified with rare earth additions', *Mat. Sci. Eng. A*, **A207**, 1996, 115–120.
55. T.B. Massalski, *Binary Alloy Phase Diagrams*, **1**, 2nd Ed., Materials Park, OH, ASM, 1990.
56. S. Amerioun, S.I. Simak, U. Haussermann, 'Laves-phase structural changes in the system $CaAl_{2-x}Mg_x$', *Inorg. Chem.*, **42**, 2003, 1467–1474.
57. Y. Zhong, J. Liu, R.A. Witt, Y-h. Sohnb, Z.-K. Liua, 'Al2(Mg,Ca) phases in Mg–Al–Ca ternary system: first-principles prediction and experimental identification', *Scripta Mater.*, **55**, 2006, 573–576.
58. R. Ninomiya, T. Ojiro, K. Kubota, 'Improved heat resistance of Mg-A1 alloys by the Ca addition', *Acta Metall. Mater.*, **43**(2), 1995, 669–674.
59. A.A. Luo, M.P. Balogh, B.R. Powell, 'Creep and microstructure of magnesium-aluminum-calcium based alloys', *Metall. Mater. Trans.*, **33A**, 2002, 567–574.
60. K. Ozturk, Y. Zhong, A A. Luo, Z-K. Liu, 'Creep resistant Mg-Al-Ca alloys: computational thermodynamics and experimental investigation', *JOM-US*, **55**, 2003, 40–44.
61. A. Suzuki, N.D. Saddock, J.W. Jones, T.M. Pollock, 'Solidification paths and eutectic intermetallic phases in Mg–Al–Ca ternary alloys', *Acta Mater.*, **53**, 2005, 2823–2834.
62. J. Grobner, D. Kevorkov, I. Chumak, R. Schmid-Fetzer, 'Experimental investigation and thermodynamic calculation of ternary Al-Ca-Mg phase equilibria', *Z. Materialkunde*, **94**, 2003, 976–982.

63. D.W. Zhou, J.S. Liu, P. Peng, L. Chen, Y.J. Hu, 'A first-principles study on the structural stability of Al_2Ca Al_4Ca and Mg2Ca phases', *Mater. Lett.*, **62**, 2008, 206–210.
64. M. Pekguleryuz, 'Creep resistant magnesium diecasting alloys based on Mg–Al–(alkaline earth element) systems', *Mater. Trans., JIM*, **42**(7), 2001, 1258–1267.
65. M. Pekguleryuz, 'Development of creep resistant magnesium diecasting alloys', *Mater. Sci. Forum*, **350–351**, 2000, 130–131.
66. E. Baril, P. Labelle, M.O. Pekguleryuz, 'Elevated temperature Mg–Al–Sr: creep resistance, mechanical properties, and microstructure', *JOM*, **55**(11), 2003, 34–39.
67. M. Pekguleryuz, E. Baril, P. Labelle, D. Argo, 'Creep resistant Mg–Al–Sr alloys', *J. Adv. Mater. Process.*, **35**(3), 2003, 32–38.
68. M. Pekguleryuz, P. Labelle, E. Baril, D. Argo, 'Magnesium diecasting alloy AJ62x with superior creep resistance,ductility and castability', *SAE Trans.*, **112**(5), 2003, 24–29.
69. M. Celikin, *PhD Thesis*. McGill, Montreal, 2012.
70. M. Masoumi, M. Pekguleryuz, 'The influence of Sr on the microstructure and texture evolution of rolled Mg-1%Zn alloy', *Mat. Sci. Eng. A*, **529**, 2011, 207–214.
71. M. Masoumi, M. Pekguleryuz, 'Effect of Sr on the texture of rolled Mg–Mn-based alloys', *Mater. Lett.*, **71**, 2012, 104–107.
72. H. Borkar, M. Hoseini, M. Pekguleryuz, 'Effect of strontium on flow behavior and texture evolution during the hot deformation of Mg-1wt%Mn alloy', *Mat. Sci. Eng A*, **537**, 2012, 49–57.
73. A. Sadeghi, M. Hoseini, M. Pekguleryuz, 'Effect of Sr addition on texture evolution of Mg–3Al–1Zn (AZ31) alloy during extrusion', *Mat. Sci. Eng. A*, **528**, 2011, 3096–3104.
74. A. Sadeghi, M. Pekguleryuz, 'Microstructural investigation and thermodynamic calculations on the precipitation of Mg–Al–Zn–Sr alloys', *J. Mater. Res.*, **26**(7), 2011, 1–8.
75. A. Sadeghi, M. Pekguleryuz, 'Microstructure, mechanical properties and texture evolution of AZ31 alloy containing trace levels of strontium', *Mater. Charact.*, **62**, 2011, 742–750.
76. A. Sadeghi, M. Pekguleryuz, 'Recrystallization and texture evolution of Mg–3%Al–1%Zn–(0.4–0.8)%Sr alloys during extrusion', *Mat. Sci. Eng. A*, **528**, 2011, 1678–1685.
77. A. Sadeghi, S. Shook, M. Pekguleryuz, 'Yield asymmetry and fracture behavior of Mg-3%Al-1%Zn-(0–1)%Sr alloys extruded at elevated temperatures', *Mat. Sci. Eng. A*, **528**(25–26), 2011, 7529–7536.
78. A. Sadeghi, M. Pekguleryuz, 'Tube extrusion of AZ31 alloy with Sr additions', *Mat. Sci. Eng. A*, **544**, 2012, 70–79.
79. D. Zhou, J.-S. Liu, P. Peng, 'A first principles study on electronic structure and elastic properties of Al_4Sr, Mg_2Sr and Mg_23Sr_6 phases', *Trans. Nonferrous Met. 2678, Soc. China*, **21**, 2011, 2677–2683.
80. L.L. Rokhlin, T.V. Dobatkina, I.E. Tarytina, V.N. Timofeev, E.E. Balakhchi, 'Peculiarities of the phase relations in Mg-rich alloys of the Mg–Nd–Y system', *J. Alloys Compds*, **367**, 2004, 17–19.
81. T. Rzychoñ, A. Kiebus, *J. Achiev. Mater. Manufact. Eng.*, **21**(1), 2007, 32–34.

82. M. Celikin, A.A. Kaya, M. Pekguleryuz, 'Effect of manganese on the creep behavior of magnesium and the role of α-Mn precipitation during creep', *Mat. Sci. Eng A*, **534**, 2012, 129–141.
83. F. Hnilica, V. Janik, B. Smola, I. Stulikova, V. Ocenasek, 'Creep behaviour of the creep resistant $MgY_3Nd_2Zn_1Mn_1$ alloy', *Mater. Sci. Eng. A*, 2008, **A489**(1–2), 93–98.
84. I. Stulikova, B. Smola, B.L. Mordike, 'New high temperature creep resistant Mg–YNd–Sc–Mn alloy', *Phys. Stat. Sol. A*, **190A**(2), 2002, R5–R7.
85. B.L. Mordike, I. Stulikova, B. Smola, 'Mechanisms of creep deformation in Mg-Sc-based alloy', *Metall. Mater. Trans. A*, **36A**, 2005, 1729–1736.
86. L. Jianping, G.W. Lorimer, J. Robson, B. Davis, 'The recrystallization behavior of AZ31 and WE43', *Mater. Sci. Forum*, **488–489**, 2005, 329–332.
87. X.– Y. Fang, D.-Q. Yi, B. Wang, W.-H. Luo, W. Gu, 'Effect of yttrium on microstructures and mechanical properties of hot rolled AZ61 wrought magnesium alloy', *Trans. Nonferrous Met. Soc. China*, **16**(5), 2006, 1053–1058.
88. R.Z. Wu, Y.S. Deng, M.L. Zhang, 'Microstructure and mechanical properties of Mg–5Li–3Al–2Zn–xRE alloys', *J. Mater Sci.*, **44**, 2009, 4132–4139.
89. S.M. He, L.M. Peng, X.Q. Zeng, W.J. Ding, Y.P. Zhu, 'Comparison of the microstructure and mechanical properties of a ZK60 alloy with and without 1.3 wt.% Gd addition', *Mat. Sci. Eng. A*, **433**, 2006, 175–181.
90. F.J. Humphreys, M. Hatherly, *Recrystallization and Related Annealing Phenomena*, Elsevier, Oxford, UK, 2004.
91. L. Wang, F. Pyczak, J. Zhang, L.H. Lou, R.F. Singer, 'Effect of eutectics on plastic deformation and subsequent recrystallization in the single crystal nickel base superalloy CMSX-4', *Mater. Sci. Eng A*, **532**, 2012, 487–492.
92. J.D. Robson, D.T. Henry, B. Davis, 'Particle effects on recrystallization in magnesium–manganese alloys: particle pinning', *Mat. Sci.Eng. A*, **528**, 2011, 4239–4247.
93. L. Wang, F. Pyczak, J. Zhang, R.F. Singer, 'On the role of eutectics during recrystallization in a single crystal nickel-base superalloy CMSX-4', *Int. J. Mater. Res.*, **100**, 2009, 1046–1051.
94. A.A. Mwembela, *Hot Workability of Magnesium Alloys*, PhD Thesis, Concordia University, 1997.
95. J. Jiang, A. Godfrey, Q. Liu, 'Influence of grain orientation on twinning during warm compression of wrought Mg-3Al-1Zn', *Mater. Sci. Technol.*, **21**, 2005, 1417–1422.
96. M. Masoumi, M. Pekguleryuz, 'Effect of cooling rate on the microstructure of AZ31 alloy', *AFS Trans.*, **117**, 2009, 617–626.
97. M. Masoumi, F. Zarandi, M. Pekguleryuz, 'Alleviation of basal texture in twin roll cast AZ31 magnesium alloy', *Scripta Mater.*, **62**, 2010, 823–826.
98. A. Sadeghi, M. Pekguleryuz, 'Precipitation during solidification of Mg-3wt%Al-1wt%Zn-(0.001–1%)Sr alloys', *AFS Trans*, **18,** 2010, 363–368.
99. A. Sadeghi, M. Pekguleryuz, 'Effect of pre-deformation anneal on the microstructure and texture evolution of Mg-3Al-1Zn-0.7Sr alloy during hot extrusion', *J. Mat. Science*, **47**(14), 5374–5384, 2012, doi: 10.1007/s10853–012–6416–0.
100. M. Masoumi, *Microstructure and Texture Studies on Magnesium Sheet alloys*, PhD Thesis, McGill, 2011.
101. H. Borkar, M. Pekguleryuz, 'Effect of strontium on the properties and extrusion behavior of Mg-Mn alloys', *Light Metals* 2010, *Advances in Materials and Processes*, D. Gallienne, M. Bilodeau (Eds), 231–238.

102. L. Mackenzie, G.W. Lorimer, J.F. Humphreys, T. Wilks, 'Recrystallization behavior of two magnesium alloys', *Mater. Sci. Forum*, **467–470**, 2004, 477–482.
103. E.A. Ball, P.B. Prangnell, 'Tensile-compressive yield asymmetries in high-strength wrought magnesium alloys', *Scripta Metall. Mater.*, **31**, 1994, 111.
104. P. Sahoo, T. Debroy, M.J. Mcnallan, 'Surface-tension of binary metal – surface-active solute systems under conditions relevant to welding metallurgy', *Metall. Trans. B*, **198**, 1988, 483.
105. H. Yamauchi, 'Surface segregation in jellium binary solid solutions', *Phys. Rev. B*, **31**, 1985, 7668.
106. AKiejna, K.F. Wojciechowski, 'Surface properties of alkali-metal alloys', *J. Phys. C, Solid Slate Phys.*, **16**, 1983, 6883.
107. A. Kiejna, 'Comment on the surface segregation in alkali- metal alloys', *J. Phys. Condens. Matter.*, **2**, 1990, 6331.
108. K. Ogino, K. Nogi, O. Yamase, 'Effects of selenium and tellurium on the surface tension of molten iron and the wettability of alumina by molten iron', *Trans. ISIJ*, **23**, 1983, 235.
109. Olshanet, A. D. Koval, 'Influence of microadditions of rare-earth metals on energy of nickel grain-boundaries', *Russian Metallurgy*, **1**, 1972, 90–96.
110. V.G. Shevchenko, V.I. Kononenko, I.A. Chupova, I.N. Latosh, N.V. Lukin, 'Oxidation of powdered alloys of aluminum and cerium during heating in air', *Combustion, Explosion, Shock Waves*, **37**(4), 2001, 413.
111. L. Kubichek, M.B. Mal'tsev, 'The influence of certain elements on the surface tension of ML5 alloy', *Izv. Akade. Nauk SSSR, Otdel Tekh. Nauk. Met. I Toplivo*, **3**, 1959, 144.
112. M. Pekguleryuz, *Inst. Magnesium Tech. (ITM) Research Report*, Quebec City, Canada, 1993.
113. C.A. Aliravci, J.E. Gruzleski, F.C. Dimayuga, 'Effect of strontium on the shrinkage microporosity in magnesium sand castings', *AFS Trans.*, **92**, 1992, 353–362.
114. A. Temur, M. Pekguleryuz, Light Metals-Aluminum, Magnesium, Titanium, M. Bilodeau, D. Gallienne, 'The Effect of Neodymium and Strontium on the High Temperature Oxidation and Ignition Behavior of Magnesium' in *Light Metals-Advances in Materials and Processes*, Eds. D. Gallienne and M. Bilodeau CIM, Westmount, QC, Canada, 2010, 191–198
115. M. Bornapour, M. Pekguleryuz, *The Effect of Sr on Corrosion Behavior of Magnesium as Biodegradable Implant*, TMS, Orlando, 2012.
116. M. Fujita, N. Sakate, S. Hirahara, Y. Yamamoto, 'Development of Magnesium Forged Wheel', *SAE Trans.* (# 950422), **1**, 1995.
117. Y. Pan, X. Liu, H. Yang, J. Wuhan, 'Sr microalloying for refining grain size of AZ91D magnesium alloy', *Univ. Technol.-Mater. Sci. Ed.*, **22**(1), 2007, 74.
118. J. Dua, J. Yang, M. Kuwabara, W. Li, J. Penga, 'Effect of Strontium on the Grain Refining Efficiency of Mg-3Al Alloy Refined by Carbon Inoculation', *J. Alloys Compd.*, **470**, 2009, 228.
119. K.Y. Nam, D.H. Song, C.W. Lee, S.W. Lee, Y.H. Park, K.M. Cho, I.M. Park, 'Modification of Mg_2Si morphology in As-cast Mg-Al-Si alloys with strontium and antimony', *Mater. Sci. Forum*, **510–511**, 2006, 238–241.
120. A. Srinivasan, U. T.S. Pillai, J. Swaminathan, S.K. Das, B.C. Pai., *J. Mater. Sci.*, **41**, 608, 2006.
121. O. Lunder, J.E. Lein, T. Kr.Aune, K. Nisancioglu, 'The role of $Mg_{17}Al_{12}$ phase in the corrosion of Mg alloy AZ', *Corrosion*, **45**, 1989, 741.

122. L.Y. Wei, *Development of Microstructure in Cast Magnesium Alloys*, PhD Thesis, Chalmers University of Tech., Sweden (1990).
123. M. Pekguleryuz, M. Celikin, 'Creep mechanisms in magnesium alloys', *Int. Mater. Rev.*, **55**(4), 2010, 197–217.
124. Chapter 1, 'Alloy development II-current and future developments', *Advances in Processing, Properties and Applications of Wrought Magnesium Alloys, Part I Types and Properties of Magnesium Alloys*, Eds C. Bettles, M. Barnett, Woodhead Publishing, (in press, 2012).
125. S.D. Cramer, S. Bernard (Eds), *Corrosion: Fundamentals, Testing, and Protection ASM Handbook, 13A*, ASM International, 2003.
126. A.D.Südholz, N.T. Kirkland, R.G. Buchheit, N. Birbili, 'Electrochemical properties of intermetallic phases and common impurity elements in magnesium alloys', *Electrochem. Solid-State Lett.*, **14**(2), 2011, C5–C7.
127. A. Kaya, P. Uzan, D. Eliezer, E. Aghion, 'Electron microscopic investigation of as cast AZ91D alloy', *Mater. Sci. Technol.*, **16**(9), 2000, 1001–1006.
128. J.F. Nie, X.L. Xiao, C.P. Luo, B.C. Muddle, 'Characterisation of precipitate phases in magnesium alloys using electron Microdiffraction', *Micron*, **32**, 2001, 857–863.

6
Forming of magnesium and its alloys

M.R. BARNETT, Deakin University, Australia

DOI: 10.1533/9780857097293.197

Abstract: The present chapter examines the formability of magnesium base alloys through the prism of the tensile test. Despite its inadequacies, this test is remarkably useful and it is used here to help highlight key differences between magnesium and its sister light metal, aluminium. Magnesium and its alloys display yield strengths comparable to aluminium base alloys. However, there is asymmetry and far greater anisotropy to yielding. Despite Lankford *r*-values that can be as high as 4, magnesium and its alloys typically display deep drawing ratios considerably less than aluminium and other metals. In room temperature tensile tests, wrought magnesium alloys typically display strain hardening exponents and forming limit curves marginally lower than comparable aluminium base alloys. However, the strains to failure seen in both tension and compression fall considerably below those observed in aluminium alloys. Formability is markedly improved by an increase in the forming temperature. Hot forming of magnesium alloys can be carried out using much the same equipment and conditions employed for aluminium base alloys.

Key words: magnesium, forming, tensile test, ductility, *r*-value, anisotropy.

6.1 Introduction

As discussed elsewhere in this book, magnesium and its alloys show particular promise in applications where light weight is an advantage. The optimal material for a light stiff beam has a high value of $E^{1/2}/\rho$, where E is Young's modulus and ρ is the density. For a light stiff panel the relevant material index is $E^{1/3}/\rho$. The values of these indices exceed those calculated for aluminium by one fifth and one third, respectively. While providing one illustration of the desirability of magnesium, this example also makes another point, which is that magnesium is frequently viewed in terms of how it compares to the metal it aims to replace – aluminium. Where practicable, the present chapter will follow this practice.

To obtain benefit from the favourable combination of lightness and stiffness in a panel undergoing flexure one must first produce the panel. This brings us to the topic of the current chapter – forming. Panels and beams, that is, long, broad and frequently thin components, are produced by metal shaping processes such as extrusion, rolling and drawing. Such products are often subsequently shaped further by bending, stretching and drawing. Forging is also sometimes employed to make more compact but integral parts. All these processes involve metal plasticity. In this regard, magnesium has a poor reputation. Authors of scientific papers on this topic have repeatedly pointed out that magnesium struggles, in one way or another, to provide the five independent shears required to satisfy von Mises'[1] requirement for arbitrary shape change. However, struggle though she does, and magnesium and her alloys all display some degree of plasticity under ambient conditions.

An obvious first point of call, in a search to unearth the origin of magnesium's poor reputation in relation to plasticity, is to consider its tensile ductility. In Roberts' book[2] published in 1960, he addresses the 'frequent engineering misunderstanding' in relation to magnesium's supposed tendency to brittleness. To do this, he quotes from Toaz and Ripling's earlier study,[3] where it is concluded that magnesium is simply at its poorest under typical tensile testing conditions. To put this to trial, the total tensile elongation reported in a popular on-line materials database (MatWeb) is plotted against tensile yield strength in Fig. 6.1 for the entire range of wrought alloys contained in the database. In the spirit indicated above, data for the full extent of available wrought aluminium alloys are also shown. A number of points can be made from this figure, but for now we note only one: while

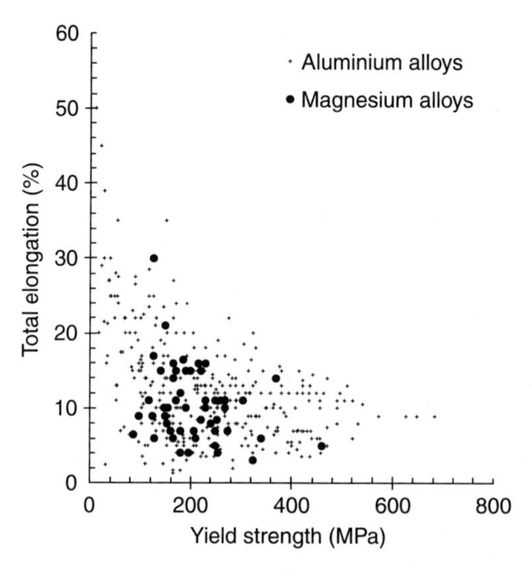

6.1 Plot of tensile total elongation against tensile yield strength obtained for the entirety of the wrought aluminium and magnesium wrought alloys contained in the popular MatWeb on-line material property database.

aluminium alloys certainly extend both to higher strengths and to higher ductilities, the tensile elongation of magnesium base alloys appears *not* to be, in the main, significantly lower than that seen in many aluminium alloys of comparable strength.

Clearly an important issue to be considered is the relevance of tensile ductility as a measure of formability. This is taken up in the first section of the present chapter. The subsequent sections then address key issues in the formability of magnesium with the aim of establishing a firm base for the metal's 'reputation'. The sections are organized as follows. The basic deformation mechanisms are very briefly reviewed and then separate subsections are dedicated to yielding, work hardening, failure strain, superplasticity and hot cracking. Where it is possible to do so, recent understanding with respect to the roles of temperature, grain size and alloying addition are considered. The chapter will present what the author believes is the most widely understood, or at least the most compelling, of current understanding. The aim is to present some rules-of-thumb for understanding, rather than to explore the details of all the debates that still continue across many aspects of the plasticity of magnesium base alloys.

6.2 Testing for formability

Despite its reputation in some circles, the tensile test is an excellent method for characterizing the forming properties of a metal. However, the parameters extracted from the procedure correlate differently with the different forming processes. In some cases no measure is appropriate and other tests are required. For example, Hosford[4] notes, 'rarely does the formability in sheet forming processes correlate well with… reduction in area or elongation at fracture'. A summary of the main correlations between testing and forming processes is presented in Table 6.1. Naturally, forming loads correlate well with the tensile stress–strain curve and this point is not included in the table. The correlations shown are of a general nature, and are not specific to magnesium base alloys. They have not all been verified to hold for the present alloys, though in the main they ought to because their basis is of a mechanical rather than material nature. In this regard it will be shown shortly below that the r-value measured in a tensile test is actually not as good a predictor of room temperature deep drawability for magnesium base alloys as it is for many other metal alloys. Nevertheless, there are two particularly important links for magnesium that are indicated in the table: (i) the connection between both the uniform elongation in a tensile test (via the work hardening exponent n) and the rate sensitivity of the flow stress (m) and the forming limit diagram and, consequently, stamping, and (ii) the correlation between tensile reduction in area with both bendability and rollability.

Table 6.1 Correlations between tests and forming processes/defects (courtesy of Professors W.B Hutchinson and J. Duncan)

Process	Simulation tests	Relevant tensile test parameters
Stamping	Erickson punch height FLD	Total elongation (%) Uniform elongation (%), n-value Strain rate sensitivity, m-value (+ve) Plastic strain ratio, r
Cylindrical deep drawing	Swift cup drawing	Plastic strain ratio, r Average normal anisotropy r_m (for LDR) Planar anisotropy Δr (for earing height)
Stretch flanging	Hole expansion test	Post-uniform elongation (%) Reduction of area (%) Plastic strain ratio, r
Bending	Bendability (r/t) Handkerchief fold test	Fracture strain, reduction of area (%)
Roll forming		Post-uniform elongation (%) Reduction of area (%) (Low) Tensile strength/yield stress ratio
High pressure hydro-forming	Erickson punch height FLD	Total elongation (%) Uniform elongation (%)
Warm forming	Hot stretch test FLD	Total elongation (%) Uniform elongation (%) High strain rate sensitivity, m-value
Blanking, punching		High strength (Low) Tensile strength/yield stress ratio Strain rate sensitivity, m-value (−ve)
Springback	Cup drawing Strip drawing	Tensile strength (%) Young's modulus
Cold rolling		Failure strain, reduction in area (%)
Wrinkling		Plastic strain ratio, r

6.3 Deformation mechanisms and formability

The plastic deformation mechanisms commonly encountered in magnesium base alloys include dislocation glide, deformation twinning and grain boundary sliding. Dynamic recrystallization also occurs in this material (e.g., Reference 5). This is in marked contrast to aluminium, which does not readily dynamically recrystallize, an important contrast that is quite possibly due to a difference of an order of magnitude in the rate of grain boundary diffusion[6] in addition to differences in tendency for recovery. However,

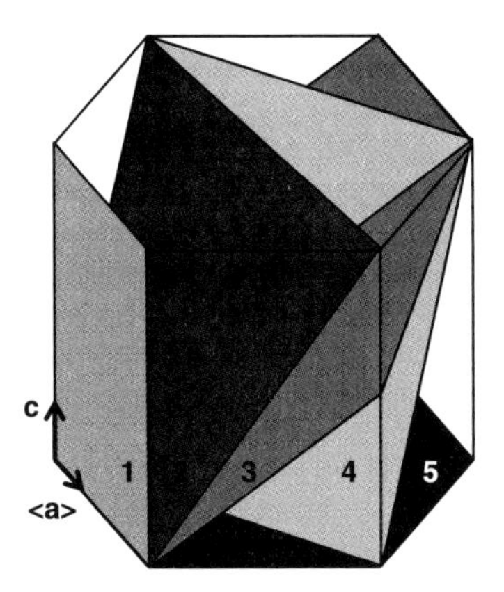

1- Prismatic: $\{10\bar{1}0\} < 1\bar{2}10 >$ slip

2- Pyramidal (I) plane: $\{10\bar{1}1\} < \bar{1}012 >$ twinning

3- $\{10\bar{1}2\} < \bar{1}011 >$ twinning

4- Pyramidal (II) plane: $\{11\bar{2}2\} < \bar{1}\bar{1}23 >$ slip

5- Basal plane: $(0001) < 1\bar{2}10 >$

6.2 Important planes in the hcp unit cell along with the deformation modes that occur on them.

while dynamic recrystallization does impact on the way in which temperature affects deformation, it does not contribute directly to the plastic strain. Here, we are more interested in processes that do.

Reviews of the main deformation modes in magnesium and its alloys can be found in References. 2, 7, 8 Dislocation glide of <*a*> dislocations occurs readily on the close packed basal plane. These dislocations also glide on the prismatic plane, after cross slipping out of the basal plane.[9] It is also possible that these dislocations cross-slip onto a pyramidal plane. There is some suggestion that this may be important,[10] but there is only little support for the contention and so it will not be considered directly in what follows. Observation of <*c* + *a*> dislocations have also been made (e.g., References 11–13). Glide of these dislocations provides deformation along the *c*-axis. This has been most frequently considered to take place on the $\{11\bar{2}2\}<\bar{1}\bar{1}23>$ slip mode (e.g., Reference 11) though there is some debate on the matter.[14] These three slip modes are shown schematically in Fig. 6.2. Also shown in this figure are the two main twinning habit planes: $\{10\bar{1}2\}$ and $\{10\bar{1}1\}$. Twins form most readily on the former, as with all *hcp* metals, and when this occurs, the *c*-axis is extended. This twinning mode is sometimes called tension, tensile or extension twinning and the characteristic shear is 0.13. The $\{10\bar{1}1\}$ twin has a similar characteristic shear, 0.14, but acts so as to contract the *c*-axis. It is sometimes called compression or contraction twinning.

More twinning modes than those shown in Fig. 6.2 have been reported (e.g., Reference 15) but those shown are observed most frequently in electron back scattering diffraction (EBSD) studies of relatively fine-grained polycrystalline alloys. However, there is one additional complication that is important to note. Twins can form in the interiors of other twins. In single crystals[16] and in polycrystals,[2,17] this occurrence may be important for understanding forming behaviour, and further discussion of this will be taken up below in the sections on work hardening and failure strain.

The importance of grain boundary sliding in the deformation of magnesium is a more controversial topic. Suffice it here to note that it seems to be important for creep and for superplastic deformation (e.g., Reference 18). It is also possible that deformation confined to the near boundary regions occurs under more conventional forming conditions and that this appears as grain boundary 'sliding' on sample surfaces.[19,20] If the deformation is mediated by grain boundaries then the boundaries must enter into considerations with regards to formability, and anisotropy, in a different way from what they would otherwise (e.g., Reference 19).

The determination of which deformation modes operate in any given forming event depends on the extent to which each mode is stressed and on the magnitude of the critical stresses that must be exceeded for activation to occur. In regards to the former, the texture is important, and this will be considered shortly. First, though, we consider the critical resolved shear stresses (CRSS) for the different deformation modes. Much has been written on this point, due to the considerable discrepancies seen between different alloys and between single and polycrystals.[7,21–23] These differences can be understood in part in terms of the following: complex solute softening and hardening effects for the different deformation modes, potential differences in the sensitivities to grain size of each deformation mode, and to the additive effects of the different hardening mechanisms. For the case of CRSS values determined for polycrystals, the experimental approach and the model employed impacts on the magnitude of the values obtained.

One particularly compelling method employed to establish values for the CRSS for different deformation modes is to use neutron diffraction to determine the internal stresses that correspond to the observable onset of relaxation in different orientation classes.[24] The experiment can be contrived so that each separate orientation class – characterized by a common diffraction vector – is dominated by a single deformation mode. Elastoplastic self-consistent crystal plasticity modelling can then be used to firm up the interpretation.[10,24] This modelling approach permits sequential activation of the different modes, which is particularly useful for the present material where some modes are considerably 'harder' than others.

Before examining data obtained in experiment, the common wrought alloys will be mentioned. These are given in Table 6.2. The most common

Table 6.2 Common wrought alloys and their compositions in weight percent

Alloy	Al	Mn (min)	Zn	Zr
AZ10	1.2	0.2	0.4	
AZ31	3.0	0.20	1.0	
AZ61	6.5	0.15	1.0	
AZ80	8.5	0.12	0.5	
M1A		1.2		
ZK60			5.5	0.45 (min)
ZM21		1	2	

Source: From the *ASM Handbook on Magnesium Alloys.*

Table 6.3 Relative values of the critical resolved shear stress for alloy AZ31 determined using neutron diffraction combined with elastoplastic modelling and subjected first to an additive normalization to bring the CRSS values for basal slip to a common value

	AZ31- sample A	AZ31- sample B	AZ31- sample C
Basal slip	1	1	1
Prismatic slip	3.25	4.5	3.75
Pyramidal II slip	3.5	4.75	4.25
Tensile twinning	2	1.5	1.25

Source: From References 10 and 24.

wrought alloy – both in rolled and extruded products – is alloy AZ31, which contains nominally 3 wt% Al, 1 wt% Zn and typically ~0.4 wt% Mn along with normal impurities. This is the alloy that has been subjected to the most study. It is also worth mentioning that rare-earth containing alloys are increasingly being scrutinized for their forming properties but there is no dominating widely accepted class or type for these alloys as yet.

In Table 6.3 the relative 'initial' CRSS values obtained by combining elastoplastic modelling with neutron diffraction are shown for three batches of alloy AZ31 – two from Reference 24 and one from Reference 10. These were determined only for simulations in which the deformation modes above were considered, bar compressive twinning. Two normalizations were performed to construct the table. First, an additive normalization was performed by adding (or subtracting) a common amount to (or from) each of the CRSS values for each material so that the values for basal slip were all brought into coincidence. The justification for this can be found in Reference 21 and is based on the observation that hardening is often found to be additive. This may not be completely true for the different deformation modes but we adopt it here as a reasonable first approximation. Second, all the CRSS values are then divided by the common value for basal slip. It can be seen that the CRSS ratios are reasonably consistent and that they fall in increasing order according to the following: basal slip, tensile twinning, prismatic slip and pyramidal II slip.

Some important distinctions need to be made as to the sensitivity of the CRSS values to temperature. First, we need to recall that the CRSS values referred to above are 'effective' values, in that they represent the resolved stresses required for appreciable slip to be detected in a deformation test. They therefore represent the stresses required to overcome the Peierls barrier and to activate a significant number of sources. They also represent stresses required to overcome any other obstacles that present or develop over the increment of strain that occurs prior to the attainment of a detectable level of slip. It turns out that the effective CRSS for the non-basal slip modes falls sharply with temperature above ambient (see single crystal data reviewed in Reference 25). In contrast, the CRSS for basal slip and tension twinning changes only moderately with increasing temperature, if at all.[25,26] One effect of this is to bring the CRSS values all closer together with increasing temperature. Another effect is to make non-basal slip more favourable than tension twinning, so that twinning is not seen once the temperature rises above a critical value (e.g., Reference 27).

Temperature has a greater effect on the slip of <*a*> dislocations on non-basal planes, compared to the basal plane, because cross-slip is required for the former. In the range of ambient to 350–400°C a cross-slip mechanism seems to dominate whereby constrictions in <*a*> dislocations dissociated in the basal plane bow out into non-basal planes.[9,28,29] The segments in the non-basal plane do not dissociate in that plane but they do dissociate back into the basal plane. Thus a configuration is created which is similar to a jog-pair or kink-pair seen in bcc metals where the Peierls stress is dominant. This mechanism of cross- slip in magnesium has been termed the Friedel–Escaig mechanism,[29] the pseudo-Peierls mechanism or the jog-pair or kink-pair mechanism.[9] The reader is alerted to the fact that in the later studies,[9,28] the term 'Friedel–Escaig mechanism' has been restricted in definition to cases where the cross-slip segment dissociates in the cross-slip plane, something that is *not* believed to occur in the present case.

Couret and Caillard[29] give a temperature of 430°C as the athermal temperature for the kink-pair mechanism and note that high activation energies point to an important role of diffusion and possibly to the climb of *c* dislocations at this and higher temperatures. Here we repeat that in polycrystals dynamic recrystallization occurs and this is seen at temperatures greater than ~ 200°C, depending on strain, strain rate and alloy. Clearly, diffusion is important in such a case. To a first approximation the effect of recrystallization on the effective CRSS values for the different slip modes is likely to be similar; dynamic recrystallization 'sweeps' out the forest dislocations that provide hardening. The role of temperature on yielding stresses will be considered briefly again further below.

The extent to which the applied stress resolves onto the systems that comprise each deformation mode is controlled by the texture. Wrought magnesium products have particularly strong fibre textures. This is an important point of demarcation with aluminium wrought products, which are seldom so strongly textured. In general, the basal pole in wrought magnesium is found to be aligned near to parallel to the direction of predominant material contraction. Thus, in rolling, the basal plane is found close to the plane of the sheet. The fibre texture is not perfect, however, and there is frequently a tilt of basal poles towards the rolling direction (e.g., References 30, 31). This can be ascribed to the action of $<c + a>$ slip[30] but it may also relate to 'compression' banding/shear banding, which also tends to rotate basal poles towards the rolling direction.[32,33] In extrusion of common alloys, the basal poles are typically found to lie perpendicular to the extrusion direction.[34] Having made these generalizations, it must be noted that in both rolling and extrusion of alloys containing rare-earth elements, weaker and altered textures are frequently observed (e.g., References 31, 34). This is not fully understood but the effect seems to relate to a combined influence of rare-earth elements on deformation and recrystallization.[31,34–37]

Given the relative magnitudes of the CRSS values in Table 6.3 and the nature of the textures commonly found in conventional (i.e., AZ31) rolled and extruded material, one can draw some conclusions with regards to which deformation modes are likely to dominate under common testing and forming conditions. Thus, in Table 6.4 it is shown that tensile twinning is favoured when extruded rod or plate is compressed along the direction in which the material experienced extension during processing (or in the transverse direction of rolled sheet). It is also seen that prismatic slip is expected to dominate the yield stresses seen in tension along the rolling transverse and extrusion directions.

6.4 Yield characteristics and drawability

A distinct yield point is not commonly observed in magnesium alloys though yield elongations are sometimes seen in finer-grained samples (e.g., Reference 38). One reason for this is that despite the range of deformation modes available, and the sharp textures, micro-yielding often occurs in those grains that are more favourably aligned for basal slip. Thus, 'yielding' can reflect a rather prolonged elastic–plastic transition.[24] In the wrought state, the magnitude of the stress at which yielding occurs in tension is shown in Fig. 6.1 to overlap the lower end of the range seen for wrought aluminium alloys.

Table 6.4 Dominant deformation modes for different mechanical strain paths, as expected for common wrought textures, such as those found in alloy AZ31

	Direction of contraction	Direction of extension	Dominant mode(s)
Rolled plate/ sheet	Free	RD	Prismatic slip, basal slip
	"	TD	Prismatic slip
	"	ND	Tensile twinning
	RD	Free	Tensile twinning
	"	TD	Prismatic slip
	"	ND	Tensile twinning
	TD	Free	Tensile twinning
	"	RD	Prismatic slip
	"	ND	Tensile twinning
	ND	Free	<*c* + *a*> slip, basal slip, compression twinning
	"	RD	<*c* + *a*> slip, basal slip, compression twinning
	"	TD	<*c* + *a*> slip, compression twinning
Extruded bar	Free	ED	Prismatic slip
	ED	Free	Tensile twinning

Note: Rolling, transverse and normal directions are given as RD, TD and ND, and the extrusion direction is abbreviated to ED.

Unlike aluminium, the yield locus in magnesium base alloys is strongly asymmetric. This arises in consequence of the strong fibre textures frequently seen in magnesium sheet alloys, and the variation in CRSS among the available deformation modes. This asymmetry has important consequences for modelling, forming, deep drawing, bending and design with wrought magnesium products. The yield locus on the σ_1–σ_2 plane has been determined by a number of workers (e.g., References 39, 40) using the method employed by Backofen.[41] An example is given in Fig. 6.3, in which the asymmetry of yielding in an extruded flat bar is illustrated. In the plane of the sheet, the yield stresses in compression (along with either RD or TD) are not infrequently seen to fall to around half of those measured in tension. The same holds true for deformation in the ED. With reference to Table 6.4 above, these deformations tend to pit tensile twinning (in compression) against prismatic slip (in tension), the latter of which has a much higher CRSS.

More detailed experimental determination of yield surfaces has been carried out using biaxial tension[42] and compression tests.[43] In these instances, the portions of the curves have been quite successfully described mathematically using the Logan–Hosford[44] yield criterion for both the biaxial quadrants. However, different coefficients are required in each case and only a

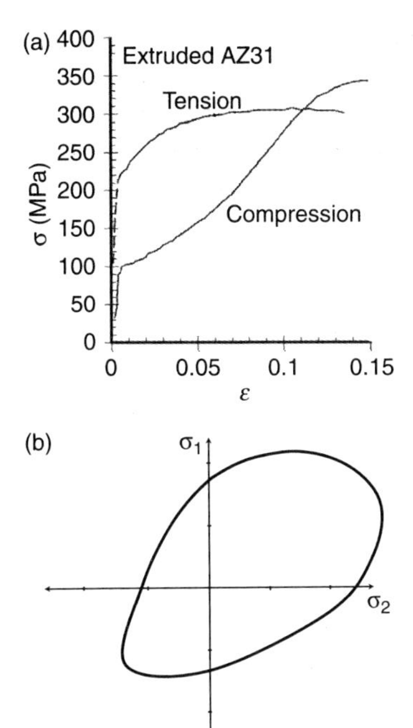

6.3 (a) True stress-strain curves obtained in tension and compression tests performed along the extrusion direction of a bar of magnesium alloy AZ31. (b) Tracing of a typically asymmetric experimental yield surface obtained for alloy ZM61 extruded into sheet shown. The axes refer to normal stresses in the plane of the sheet.[40] The bottom left corner is dominated by twinning.

small portion of the yield surface is fitted. Finite element modelling of forming requires more complete yet simply calculable expressions and this is a considerable obstacle. Lee *et al.*[45] have categorized the approaches tried so far into two main types. In one family of approaches, stresses are added to existing models, such as the work by Hosford.[46] Cazacu and co-workers[47–49] have added terms based on the third invariant of stress. In the other family, artificial 'back stress' terms are introduced in kinematic hardening models to create an asymmetric surface by 'pretending' the existence of some prior strain. Another promising approach is to employ a rapid crystal plasticity calculation based on an aggregate of a small number of representative grains.[50]

A curious outcome of the asymmetry of magnesium is that during bending of sheet, the sheet contracts in the direction perpendicular to the bend.[51] This arises because the yield stress in compression is lower than in tension, so the neutral plane in a bend tends to be closer to the tension side of the bend. The action of bending thus imparts some net contraction.

The Lankford r-value is a characteristic of the yield locus and is related to the slope of the yield surface on the σ_1–σ_2 plane where it intersects the positive axes. The r-value is written as the ratio of width to thickness strain in a tensile test, but is usually obtained by measuring the elongation and width contraction and assuming constancy of volume. The r-value is frequently found to correlate with the limiting cup draw height. This is sometimes termed the limiting draw ratio (LDR), which gives the maximum ratio of the blank diameter to the cup diameter achievable during the deep drawing of a cup. That is, high r-values typically give superior deep drawing depths (e.g., Reference 4). In magnesium alloy sheet it is common for the r-value to attain rather high levels in the transverse direction (i.e., up to r = 4, where r = 1–2 is more common in Al and Fe alloys).[19,52] However, in the rolling direction lower r-values are more typical[52] and, overall, the material is usually found to display disappointingly low levels of deep drawability at room temperature (e.g., References 53–55 – see Fig. 6.4). Furthermore, in magnesium alloys, higher r-values do not necessarily give superior deep drawability.[53] We will come back to these points below.

We now turn to the influence of temperature on yielding. In general, the decrease in yield stress that accompanies an increase in temperature is similar in magnesium and aluminium alloys (c.f., References 56–58). This is not too surprising, as both metals share similar melting temperatures. Forming can be accomplished under lower stresses at elevated temperature. As mentioned

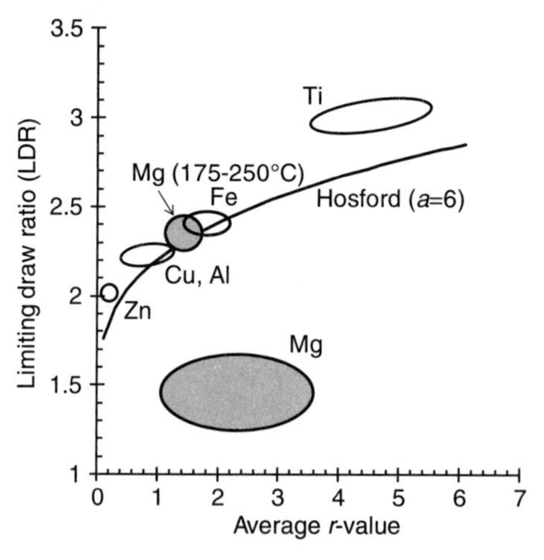

6.4 Influence of average r-value on the room temperature LDR for deep drawing of sheet. The data were obtained from Dieter's book for non-Mg alloys, and from Mori and Tanaka,[54] Iwanaga *et al.*[55] and Simon Jacob[84] for Mg (alloy AZ31). The curve is from Hosford and Cadell.[4] Warm deep drawing data for magnesium alloy AZ31 are also shown.[55,124]

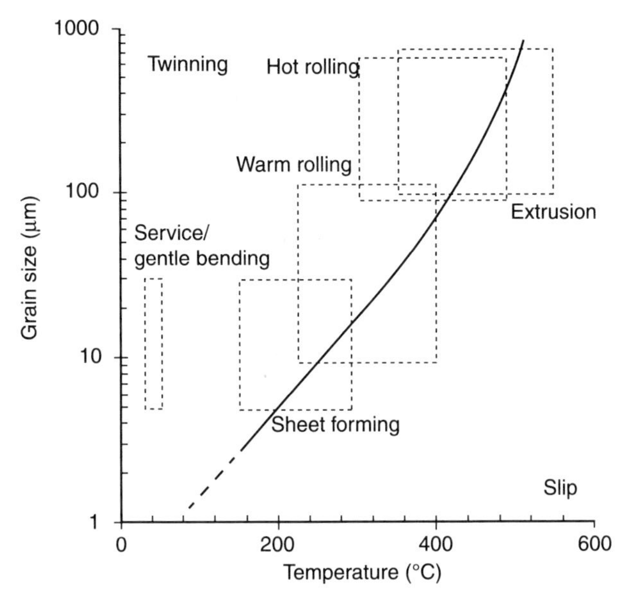

6.5 Plot of the transition between slip to twinning dominated deformation as a function of temperature and grain size for magnesium alloy AZ31. The relation is based on reference [27] and derives from compression tests performed on extrusions. Different deformations and texture may result in slip dominated deformation in the twinning region but not vice versa. Approximate 'windows' are shown for the main wrought processes.

above, the sensitivity of the different deformation modes to temperature is different. Thus, at ambient temperature the yielding of extrusion and plate in compression along the direction of prior extension occurs primarily by tensile twinning (Table 6.4). At higher temperatures, yielding under these strain modes is dominated by non-basal slip instead, simply because these systems are more easily activated. The conditions for this transition depend on grain size[27] and alloy (e.g., Reference 59). The slip-twinning transition for alloy AZ31 is shown in Fig. 6.5, along with the approximate conditions corresponding to the main forming processes. It is clear that there is scope to decide whether or not to form under conditions where yielding will occur by twinning.

In consequence of these effects, the influence of grain size on the yield stress in magnesium is a complex one (see Fig. 6.6). There are two key observations to note. One is that in deformation dominated by twinning, such as compression of an extrudate along the ED, the Hall–Petch slope tends to be higher than when slip is dominant. The other is that, with an increase in temperature, a marked drop is seen in such samples, once the critical temperature for the slip-twinning transition is met.[27] Samples not oriented for

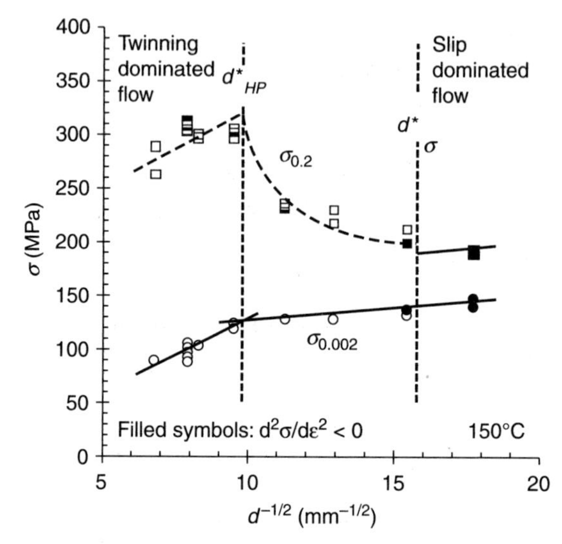

6.6 Influence of grain size on the yield and flow stress in magnesium AZ31 extrusions tested in compression at 150°C.[27]

twinning also experience a drop in Hall–Petch slope with increasing temperature, but in this case the effect is more gradual.[60]

Alloying addition alters the yielding behaviour of magnesium alloys through similar mechanisms to those seen in other metals. Thus, solute and precipitate hardening are employed to raise the strength. A detailed review is beyond the scope of the present work and on this note the reader is pointed to Reference 7. However, there is an interesting phenomenon whereby the addition of certain alloying elements, such as Zn and Al, can reduce the stresses required to activate prismatic slip.[61,62] Also, there are two important observations that can be made with respect to the addition of rare-earth elements (apart from the fact that these elements are commonly used to impart creep resistance – that is, they increase the high temperature yield strength[63]). One is that rare-earth containing alloys can display a dynamic strain ageing type phenomenon with the associated anomalies in stress–strain response and strain rate sensitivity.[36,64] The other was mentioned above and relates to the tendency of rare-earth alloys to display weaker textures. Sheets with weaker textures tend to yield at lower stresses in in-plane tension, all else being constant. But they also display greater deep drawability.[53]

There are a number of reasons why magnesium alloys tend to be less 'deep drawable' than other metals at equivalent *r*-values. One reason is that magnesium displays complex kinematic hardening, due largely to the fact that twinning only acts in one sense of strain and does so only for a limited duration of deformation. Thus, the bending and unbending that occur as the blank is drawn over the die corner and into the cup wall changes the flow behaviour compared to that of the unprocessed sheet. This seems to have received limited attention. However, with reference to Table 6.3, we can make some preliminary observations. As the material travels over the die profile radius, the upper surface will undergo tension. This tension is in the plane of the sheet and will be accompanied largely by slip processes. As the material is straightened as it enters the cup wall, the upper surface will undergo compression. Compression in the plane of the sheet is expected to induce twinning (Table 6.3). Any thickening of the blank prior to entry into the die will also be accompanied by twinning. Thus the material in the cup wall will be partially twinned and will thus tend to undergo yielding with the application of further tension, under lower stresses than it otherwise would. The full significance of this for deep drawing has yet to be determined.

Probably the most significant cause of the poor deep drawability of magnesium is that it is unable to withstand the strains required to:

i. form the bend radii at the corners of the punch and die, and/or
ii. reduce the outer circumference of the blank into that of the die.

With respect to the latter, the maximum compressive strain in the plane of the sheet corresponding to the different drawing ratios is simply ln(LDR). Magnesium alloys display LDR values at room temperature between 1.2 and 1.7 (Fig. 6.4) so this corresponds to compressive strains at the outer radius of the blank in the range of 0.18–0.53. The difficulty for magnesium to achieve such strains at room temperature without failure will be discussed further below. But it is worthy of note that an increase in the LDR can be achieved by modifying the hold-down plate used in the deep drawing process to increase the compressive hydrostatic stress in the flange just before it enters the cup wall.[54] This extends the strain that the rim of the blank can withstand prior to cracking.

Magnesium sheets with weaker textures tend to display greater deep drawability.[53] Increasing the temperature also improves the deep drawability, despite the drop in *r*-value.[53,55,58,65,66] Iwanaga *et al.*[55] report an LDR value of ~2.5 for deep drawing at 175 °C and Doege and Droder[67] give values between 2.2 and 2.5 for M1, AZ61 and AZ31 tested at 200–250 °C. These values are superimposed on Fig. 6.5, where it can be seen that they sit consistently with the values seen for other metals. The effect of temperature can be understood in terms of the increase in failure strain that occurs with temperature. This will be explained further below, but for now we note that an increase in failure strain will eliminate problems associated with failure in deep drawing due to bending around die radii and contraction of the blank rim.

Although the drop in *r*-value with temperature with alloy AZ31 seems not to be detrimental to deep drawability, it raises a number of questions as to why it occurs at all. Agnew and Duguly[65] and Stanford[66] make a strong argument that the effect arises from the increasing ease of <*c* + *a*> slip with increasing temperature. Others[19] have suggested that grain boundary sliding, or at least a deformation mechanism mediated in the near grain boundary regions, is important. However, the effect has not been seen in all alloys.[66]

6.5 Work hardening and stretching

As noted above, the propensity of a metal to display work hardening in a tensile test is a good predictor of formability in a number of forming processes. Work hardening stabilizes the material against localization of deformation. This is most simply captured in the Considère criterion, which predicts that the strain for diffuse necking in a tensile test in a power law hardening material (where $\sigma = k\varepsilon^n$) should equal n, the work hardening exponent. This strain is often termed the uniform strain and it corresponds to the peak engineering stress, the ultimate tensile, σ_{TS}. These are general comments but it turns out that they hold for magnesium alloy sheet and extrusions that display typical basal textures. That is, tension in the plane of the sheet and in the ED of conventional alloy AZ31 produces a stress–strain curve that can be approximated reasonably well (though not always perfectly[68]) with a power law fit and that the peak in engineering stress, that is, the uniform elongation, occurs at a strain near n.[69–72]

With a little algebraic manipulation, it can be shown that for a power law work hardening material, the ratio of the ultimate to the yield stress, σ_{TS}/σ_y, can be given by $n^n / \varepsilon_y^n (n+1)$, where ε_y is the yield strain, that is, the arbitrary strain at which the yield stress is determined. This expression can be used to estimate values of n from data of tensile and yield strengths for different magnesium alloys. Such a procedure was performed using the same dataset of tensile properties employed to construct Fig. 6.1 (for $\varepsilon_y = 0.002$) and the results are given in Fig. 6.7. As with Fig. 6.1, the magnesium data overlap those seen for aluminium but generally fall short of the best performing aluminium alloys, that is, those that display the highest n for a given strength. The range is $n = 0.05$–0.2. This agrees well with direct measurements.[68–71,73,74]

Although a detailed discussion of the mechanisms of work hardening is beyond the scope of the current work, it is important to note that the occurrence of deformation twinning, and therefore texture and stress state, impacts notably on the stress–strain curve. In Fig. 6.3a the apparent work hardening in the compression test, which is dominated by twinning, is considerable. (From here on 'work hardening' will be employed to refer simply to the increase in stress seen with strain during plastic deformation, not to denote any particular mechanism.) The phenomenon in Fig. 6.3a relates largely to the change in orientation that accompanies twinning. While $\{10\bar{1}2\}$

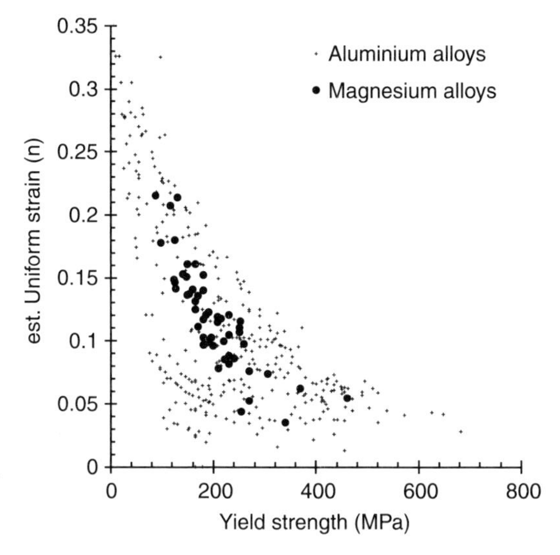

6.7 Plot of estimated tensile uniform strain against tensile yield strength obtained for the entirety of the wrought aluminium and magnesium wrought alloys contained in the popular MatWeb on-line material property database. The uniform strain was estimated using published values for the yield and ultimate tensile strengths assuming the validity of power law hardening and the Considère criterion (see text for details).

twinning is activated under relatively low stresses, considerably higher stresses are required for deformation to continue by slip in the twinned regions. There is ongoing debate as to the extent to which twin boundaries and transformed (twinned) dislocations play a role (e.g., Reference 75). Nevertheless, the consequence is that texture and the deformation sense (i.e., compression or tension) play important roles in apparent work hardening. Thus, in materials with weak and or non-typical textures it is quite common for higher work hardening exponents to be reported.

In these cases, the stress–strain curves do not always follow power law hardening and there may be other factors in addition to twinning at play (increased basal slip, for example). In Mg–Gd–Zn alloys, for example, which show weak textures (as is common in many RE containing alloys), Wu *et al.*,[76] report values of uniform elongation in the range 0.25–0.35. Tensile tests taken across the face of extruded AZ31 bar reveal uniform elongations of ~0.22 compared with ~0.07 when tested along the bar.[77] In material textured by equal channel angular extrusion, higher uniform strains are also seen (~0.3, compared to ~0.12 in a comparable 'conventional' sample of the same alloy).[78,79] Also, inspection of published tensile curves for cast material (e.g., in Reference 2) reveals high *n*-values (~0.5), similar to those seen in 'high' work hardening materials such as brass and stainless steel. However, more random textures do not always raise the uniform elongation,[31] due to the role of other competing factors, such as grain size and solute additions. Nevertheless, in general, it is apparent that the combination of typical basal textures with tensile testing in either the rolling plane or ED tends to produce low work hardening exponents. Weaker textures and different test directions often give higher values of uniform elongation.

A more subtle effect of deformation twinning will now be raised. Twinning on the $\{10\bar{1}1\}$ plane can occur with increasing strain in tensile tests carried out in the plane of rolled sheet or in the ED. This is often followed by $\{10\bar{1}2\}$ twinning inside the primary twin.[80] After these two twinning reorientations, the basal planes become favourably aligned for slip, and flow can localize in the doubly twinned region.[16,80,81] Although seen initially in single crystals, it has been proposed that this can lead to flow softening in polycrystals and that this can be detected in the macroscopic stress–strain curve, under appropriate conditions.[17,73] In these studies it is suggested that the n-value in room temperature tensile tests performed on extruded samples is lowered by twinning. Although the n-value typically drops with increasing temperature, in certain magnesium alloy extrusions the opposite occurs, at least initially.[17,72,73] The argument is that because twinning tends to decrease in prevalence with higher temperatures, its detrimental impact on the work hardening exponent vanishes as the temperature rises. This can account for the increase in n-value with temperature.[17,73]

Before considering the ramifications of work hardening and uniform elongation for sheet formability, the effect of grain size on work hardening will be discussed. It is common for many metals to display *decreasing* values of n as the grain size is refined.[82] The same result is obtained also from the observation that the stress–strain curve is often seen to move uniformly upward when the grain size is refined. That is, the grain size and work hardening terms are additive. In contrast to these observations, magnesium and its alloys can display values of n that *increase* with grain refinement. The uniform elongation measured in a series of extruded samples is plotted against grain size in Fig. 6.8. It can be seen that the uniform elongations, though quite low, generally increase as the grain size is refined. This can be readily rationalized if one accepts the idea that compressive type twinning can serve to lower the apparent work hardening rate at higher strains. With grain refinement, twinning is suppressed along with its detrimental effect on n and the uniform strain. However, more work is needed to verify this hypothesis.

Conversely, for weaker textures and textures in which off basal components are present, tension twinning impacts on the apparent work hardening rate. In these instances, grain refinement is expected to lead to a rather rapid drop in work hardening as twinning is suppressed.

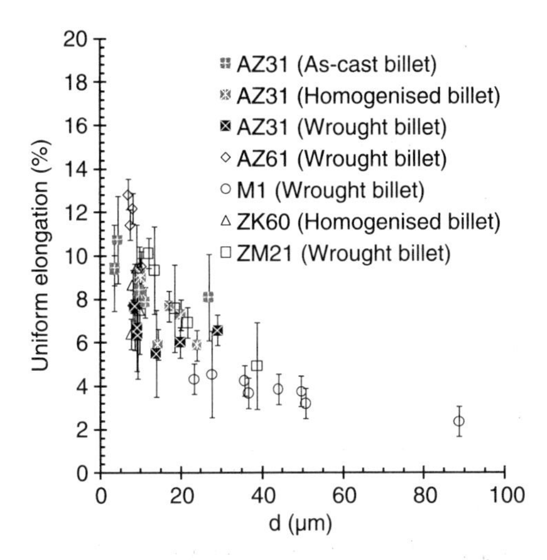

6.8 Impact of grain size on the uniform elongation seen in tensile tests performed on a range of magnesium alloys extruded uniaxially in a laboratory press from 30 mm billet to 5 mm rod. (Dale Atwell is acknowledged for his assistance with this figure, which contains unpublished work.)

The formability of metal sheet in stretching applications is often characterized by using a forming limit diagram, an example of which is provided in Fig. 6.9.[68,83–86] The axes refer to orthogonal principal strains in the plane of the sheet and the curves shown represent the strains at which local necking gives rise to failure. A simple local necking analysis assuming power law hardening predicts that the intersection of the limit curve with the vertical axis (which gives the value FLD_0) should occur at n.[4] The more sophisticated Marciniak–Kuczynski approach tends to give lower values (e.g., References 4, 87). Additionally, in practice, values of FLD_0 higher than n are often seen, due to thickness effects and to the fact that higher strains are required for necking to register at the resolution of the experiment.[4] Furthermore, in magnesium AZ31, there is indication that the work hardening rate drops off with strain at a greater rate than seen in power law hardening.[17,68] This will tend to give a value of FLD_0 less than an 'averaged' value for n. The upshot of these, and possibly other, effects is that for the limited data available for magnesium base alloys, the room temperature FLD_0 value falls below the value of n extracted from uniaxial tensile tests (Fig. 6.9).[68,83–86] However, it should be noted that experimental forming limit curves can vary considerably.[4] In any case, the sheet formability as expressed by FLD_0 of the common sheet alloy AZ31 is seen to be typically below that for sheet forming aluminium alloys. For example FLD_0 values reported in the literature for 5XXX series Al grades include ~0.19,[87] ~0.28[4] whereas the values in Fig. 6.9 span from 0.05 to 0.16.

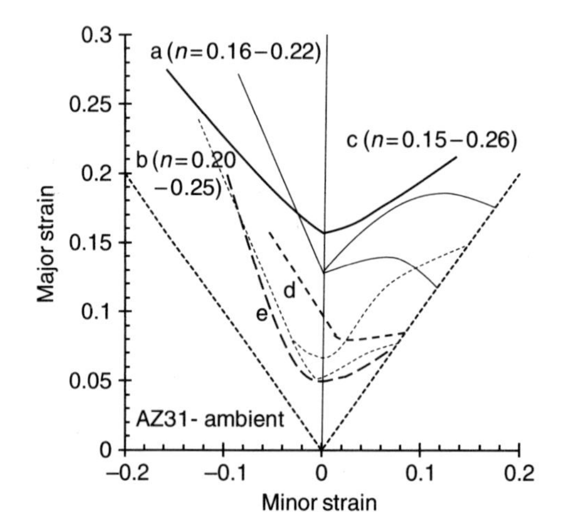

6.9 Forming limit diagrams for magnesium alloy AZ31 sheet tested at room temperature. The curves were obtained by different workers (a-[83], b-[68], c-[84], d-[85], e-[86]). The work hardening exponents measured in tensile tests of the same materials are also shown, where they were available.

Before turning to some of the approaches employed to remedy this problem, it is worth mentioning that some of the forming limit curves in Fig. 6.9 display a 'hump' on the right hand side. John Neil and Agnew[88] have predicted such a feature theoretically based on anisotropy considerations. Naka *et al.*[89] find a hump in their theoretical predictions, but in their case it arises from the intersection of the local necking locus with the fracture condition. In low ductile materials, the onset of rupture becomes important in the biaxial stress regime[90] and it is feasible that such an event is occurring here. Forming limit curves truncated by rupture show a 'hump' in the biaxial quadrant.

It was noted above that the *n*-value is raised by weakening the texture and diluting the typically strong sheet basal fibre. It follows that a concomitant improvement should be seen in the forming limit. This has been demonstrated theoretically in forming limit predictions[88] and, in practice, in a weakly textured rare-earth containing alloy.[85] For the latter, an increase in FLD_0 from ~0.1 to ~0.17 was seen at room temperature, as compared with a similarly grain-sized AZ31 sample. Erichsen cup heights, another measure of sheet stretching (see Reference 4), have also been found to increase from ~2.5 mm to ~7 mm with decreasing texture sharpness.[55] In this last case the alloy and microstructure are constant; the sheet was cut at different angles from an extrusion. If the lubrication is sufficient, such a test provides the fracture strain.

A significant increase in the level of the forming limit is seen with an increase in forming temperature. For this reason, the sheet forming of magnesium base alloys is typically carried out at temperatures above ambient. Naka *et al.*[89] report FLD_0 values in the range 0.25–0.4 at 200 °C for AZ31 and Stutz *et al.*[85] report ~0.22 and ~0.27 for the same alloy at 200 °C and 250 °C respectively. Forcellese *et al.*[91] report higher values of 0.35–0.5 for the same alloy and temperature.

Significant increases in the level of the forming limit diagram with temperature are also seen for aluminium base alloys (e.g., Reference 92). The underlying cause in both materials is probably the same; that is, with an increase in temperature there is an increase in the strain rate sensitivity of the flow stress, m.[71,74,93] Higher strain rate sensitivity provides resistance to local necking. This effect has been considered in the modelling work of John Neil and Agnew.[88] These workers also note that, as the strain rate sensitivity of the flow stress increases with temperature, n drops (e.g., References 71, 74) although, as mentioned above in extruded material an increase in n is sometimes observed with temperature. So, after an initial increase in formability with temperature, a plateau in FLD_0 arises.[89]

The effect of m on the forming limit curve of alloy AZ31 is also clearly illustrated in the modelling and experiments of Naka *et al.*[89] It has been shown that at 'warm' forming temperatures, the strain rate has a large effect on FLD_0.[89,91] The forming limit is higher when the forming rate is lower and the effect is a strong one. This may be due to simply to the lower levels of stress, a hypothesis that requires flow to depart from the usual Backofen type law ($\sigma = K\varepsilon^n\dot{\varepsilon}^m$), which is yet to be verified. Alternatively, it may be that with lowering strain rate the flow enters the superplastic regime, which assumes m rises as the rate drops. This is something we will return to briefly further below. Finally, it is worth noting that the strain rate sensitivity does not increase steadily with temperature in all alloys. In certain RE containing alloys, dynamic strain ageing occurs and this can impact upon m (e.g., Reference 36).

6.6 Failure strain behaviour, compression, rolling and bending

The strain required to rupture a tensile or compression sample can be usefully obtained from the sample dimensions in the region of the fracture. In a tensile test, this is well quantified by measuring the minimum in cross-section area after fracture. This area can then be employed to establish the 'reduction in area' at failure or the 'failure strain'. In the magnesium literature these values are not commonly reported. Instead, the total elongation is typically given. However, if failure occurs prior to diffuse necking, the two values are equivalent. Though not explicitly stated, such is clearly the case for the low tensile elongations seen in coarse-grained pure magnesium extrusions studied by Hauser *et al.* in their early work.[94] It is also true in material tested in the as-cast state, due to the abundance of defects, but we focus here on wrought materials. Similarly low ductilities have also been seen recently in 1.5 mm thick rolled AZ31 sheet with a mean grain size of 80 μm.[84,95] Here, tensile failure also preceded diffuse necking. This is in marked contrast to coarse-grained pure aluminium, which can neck down to effectively 100% reduction in area, strains far greater than the diffuse necking strain.

Chapman and Wilson[96] report reductions in area values for extruded pure magnesium over a range of grain sizes. At room temperature, their 8 μm and 60 μm grained samples failed at 30% and 15% reductions in area, respectively. Davidson *et al.*[97] report the low value of RA of ~8% for their pure magnesium extruded samples, but show that this can be increased to effectively 100% with the imposition of ~500 MPa of hydrostatic pressure. In both this and the early Hauser study, the fracture was attributed, at least in part, to an intergranular failure mode. However, in the AZ31 case mentioned above, the failure was attributed to twinning.[95] More work is required to fully elucidate the relative roles of these modes of failure, and in the following more focus will be placed on the roles of twinning and crystallographic phenomena.

It is clear from the forgoing that pure magnesium displays inherently lower tensile fracture strains than pure aluminium. We turn now more fully to its alloys. In this regard we notice that alloyed magnesium frequently displays higher reduction in areas than that seen in coarse-grained pure material. Values given by Beck[98] in his book for eleven wrought alloys attain values up to 47% (though the lowest value is 3%). One reason for higher tensile failure strains in alloys is that the failure strain in magnesium base alloys is strongly sensitive to grain size,[96] as alluded to above. And alloying frequently serves to refine the grain size. This observation is illustrated in Fig. 6.10a where the reduction in area values for the same samples used for Fig. 6.8 are presented. It can be seen that the range of values is the same as that reported by Beck and that the higher values are always attained at finer grain sizes.

The contrast between magnesium and aluminium alloys becomes striking when one considers the failure strain in compression (measured along the extrusion direction e.g., References 25, 27, 99), or in the plane of rolled plate (e.g., Reference 100)). Figure 6.10b presents the compressive strains at failure that correspond to the same samples employed for Fig. 6.10a (see also images in Fig. 6.11). The grain size is not so important in Fig. 6.10b and the absolute values do not exceed 25%. Thus, it can be seen that the reduction in area in tensile tests frequently exceeds that in compression, when measured along the ED. This is in distinction to most metals, where hydrostatic compressive stresses in compression tests permit considerably greater strains to failure to be achieved. As noted above, it is not that the failure strain in magnesium is

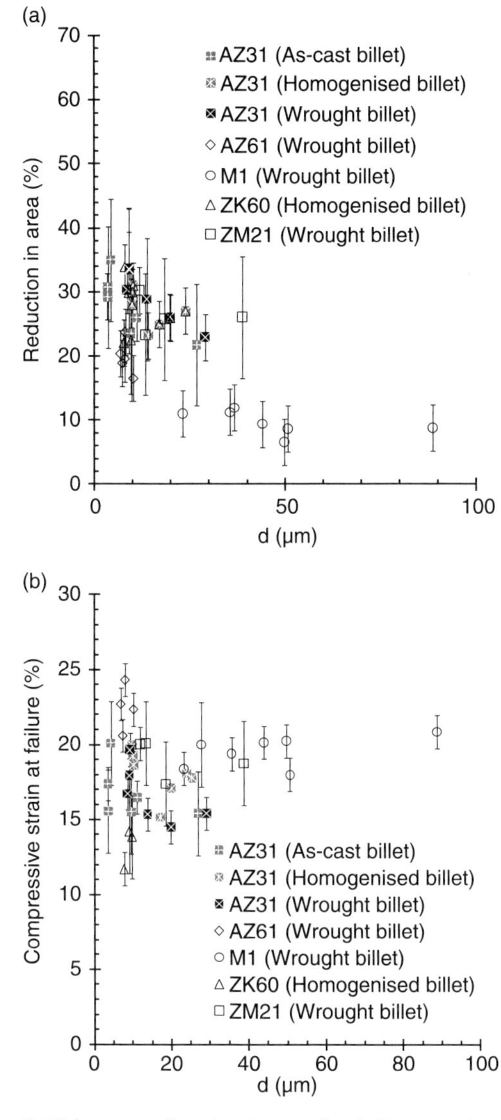

6.10 Impact of grain size on the failure strains in (a) tension (reduction in area) and (b) compression seen in tests performed on the same range of extruded magnesium alloys used for Fig. 6.7. (Dale Atwell is acknowledged for his assistance with this figure, which contains unpublished work.)

not sensitive to the hydrostatic pressure; it is. Rather, the effect of the different deformation modes, combined with a strong texture, dominates.

The 'strongest' combination of stress and orientation in magnesium arises when contraction occurs along the *c*-axis. Single crystal experiments

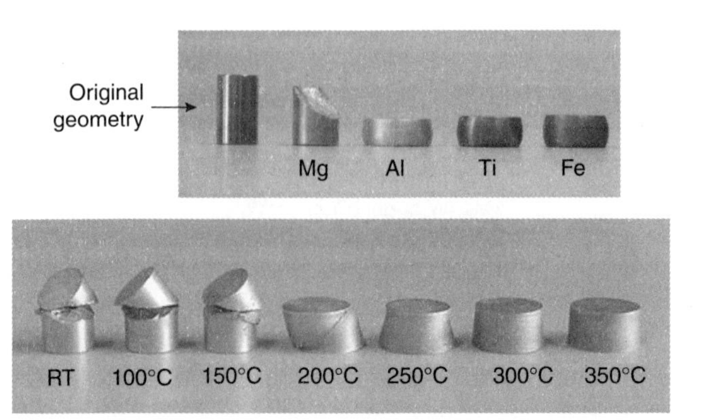

6.11 Examples of the shear failure seen in magnesium extrusions subjected to compression along the extrusion direction.

also reveal that this combination displays very poor ductility.[101] In tension of sheet and extrusions that display typical textures, the tension direction lies predominantly near to the basal plane. Contraction is then expected to be shared by the *c* directions and other directions in the basal plane. That is, contraction is not 'forced' to occur along the *c*-axis (although such deformation may occur once sufficiently high stresses are attained). On the other hand, in compression of extrusions, the initial compression lies near to the basal plane and $\{10\bar{1}2\}$ twinning is activated. Once this mode of twinning saturates, which occurs after a strain of ~10%,[102] the *c*-axes are largely aligned with the compression direction. Compression testing of other textures leads to a similar result; the basal poles tend to rotate into the compression direction. Thus, the application of compression invariably eventually results in *c*-axis compression, which is the least ductile of the orientation–stress combinations. For this reason, failure strains in compression can frequently fall below those in tension. This is despite the tendency for the opposite due to the hydrostatic component of the applied stress.

We consider now some of the reasons compression along the *c*-axis is rapidly followed by failure. Firstly, as mentioned above, this is the hardest orientation and higher stresses mean higher levels of stored elastic energy to 'drive' failure. Secondly, single crystals tested in this manner fail by parting within doubly twinned volumes.[16,80,81,103,104] As mentioned above, $\{10\bar{1}1\}$ contraction twinning followed by $\{10\bar{1}2\}$ tensile twinning creates a twinned volume that is crystallographically soft; after twinning induced reorientation, the basal plane becomes well aligned for slip. This mechanism plays an important role in polycrystalline failure too[17,95,105,106] (see Fig. 6.12). Contraction along the *c*-axis is inherently unstable, in the sense that any perturbation in orientation or flow will provide a 'softer' solution. Indeed,

6.12 Examples of shear failure and void formation in deformation twins seen in magnesium alloy (a) AZ31 and (b) ZK60 subject to tension along the extrusion direction[17] (used with permission).

any other mechanism able to generate orientation change has the potential to be favoured. Shear band formation is expected under such conditions. Indeed, shear localization is ubiquitous in tensile and compressive (including rolling) failure (see Figs 6.11 and 6.12).[2,32,97,107] It is particularly obvious in tension tests carried out under high hydrostatic pressures.[97]

The tensile and compressive failure strains increase dramatically with temperature (e.g., Fig. 6.11 and References 17, 77, 96, 108). This mirrors single crystal results for tension in the basal plane.[16] In that work, the failure strain shows an inverse correlation with the stress for non-basal slip. With lower stresses for non-basal slip, more strain can accumulate before the stress is attained to initiate double twinning and subsequent failure in the twinned volume. Also, more generally, lower stresses for non-basal slip will mean lower levels of elastic stored energy, lower driving forces for shear banding, and less stress inhomogeneity over a polycrystal. In the case of compression, increasing the temperature raises the failure strain also, because of the suppression of tension twinning, which slows the development of c-axis compression orientations. As the temperature is raised further, failure strains also rise, due to the lowering of flow stresses through dynamic recovery and recrystallization.

As noted above, alloying raises the tensile failure strain in many cases, due to its impact on the grain size. The grain size is important most probably because of its impact on twinning; as noted above, finer grain sizes tend to suppress twinning and favour slip. Finer grain sizes can also suppress shear banding (e.g., Reference 109). The grain size is not so important for the failure strains in compression, because here there is not the same scope for competition between twinning and slip. Initially, tension twinning occurs and at room temperature this is not suppressed with grain refinement unless very fine grains are obtained (e.g., Reference 27). Once tension twinning is complete, compressive twinning 'competes' only with $<c + a>$ slip; these are the only two modes that provide *c*-axis compression. In contrast, in tension, compressive twinning competes with both $<c + a>$ slip and prismatic slip, because all of these modes can provide extension along directions in the basal plane.

There are two important direct roles of solute addition that will now be mentioned. One is that softening accompanies the addition of certain alloying elements, see for example, Li, on prismatic slip. This leads to higher strains to failure.[61] The other is that hard brittle second phase particles have a detrimental effect on failure, something which is not unique to magnesium alloys. In the early work of McDonald, the favourable effect of alloying addition on tensile ductility ceases at composition levels that correspond reasonably well with the solid solution limit.[110,111]

We turn now to the impact of the present discussion on forming processes. The connections are fairly obvious so we will not labour them – the favourable roles of temperature, texture weakening and grain refinement are all expected to hold, although they have not all been fully investigated in isolation. It was already noted in the first section of this chapter that it appears that deep drawing is limited by the failure strain. In the present section we have shown how little the failure strains are for compression of extrusions along the ED or compression in the plane of the plate (i.e., ~20%). The latter is the deformation mode that occurs at the periphery of a blank as it is being deep drawn. This is quite clear in thickness increases that are measured in this region (e.g., Reference 54). Above, it was shown that the limiting drawing ratio at room temperature for alloy AZ31 falls between 1.2 and 1.7 and that this corresponds to compressive strains at the outer radius of the blank in the range of 0.18–0.53. These are of a similar order to the failure strains seen in compression testing.

The bending of sheet is also limited by tensile and compressive strains to failure. The tensile strain to failure has been found to correlate well with the minimum bend ratio.[4] However, in the case of magnesium, failure is sometimes seen to begin on the inner compressive side of the bend.[98] It can be seen that here too the inability of magnesium to withstand significant strains in compression is important. If we assume a limiting compressive strain of ~0.2, we can estimate the limiting bend radii to be ~ 2.5 t. But the drift of the neutral plane to the tension side of the bend needs to be taken into account, so values greater than this will be expected. Reported limiting bend radii for wrought alloys fall around 3–5 t.[98]

Finally, we return to the case of rolling. Here, the deformation zone geometry, temperature gradient and lubrication are all likely to be significant, but these factors appear not to have been examined yet in a comprehensive study. It is also not easy to compare studies, because one or other of these variables is often not quoted. However, there are some clear trends. Cold rollability of magnesium base alloys is limited. The most common wrought alloy, AZ31, cannot withstand much more than ~20% reduction before failure (e.g., Reference 33) and pure magnesium has been reported to have been cold rolled up to 50% in reduction[2] although other studies observed failure at lower reductions.[33,112] Because of this, rolling is typically performed at elevated temperatures, usually in the range 300–400°C. However, one notable phenomenon is that of apparently unlimited cold rollability in certain alloys (e.g., Mg–Mn, Mg–Ce, Mg–Mn–Misch Metal).[2,33,107,112,113] It appears that the way in which shear bands ('compression bands' in some studies[2]) develop in these grades is such that instead of leading to failure they provide a means for continued deformation. The effect is complicated by simultaneous changes in texture, grain size, stacking fault energy and non-basal slip.[33,112] A full study on the formability of such cold rolled alloys appears to be lacking.

6.7 Superplastic deformation and hot forming

The factors that impact on the total elongation achieved in a tensile test are many and varied. The parameter itself is also of limited use in predicting formability, despite its widespread use. However, large values of total elongation usually do correspond to superior levels of formability. Although it depends on the factors that affect both the localization of flow and the tensile strain, it is particularly sensitive to the strain rate sensitivity of the flow stress. High values of m give high post-uniform elongations.[114] As is the case for aluminium alloys, if a sufficiently stable fine-grained structure can be developed, total elongations in the superplastic regime (i.e., $\varepsilon_{total} > \sim 500\%$) can be achieved (e.g., Reference 115). Much has been written on this topic so it will not be considered in detail here; rather we will point out some methods for producing the requisite grain size and give an indication of the conditions under which superplasticity can be attained.

Superplastic deformation in magnesium base alloys is seen when the grain size is in the range of 1–3 μm and finer.[116] Such grain refinement is achieved in the laboratory typically by the application of severe plastic deformation. A very common method is to use equal angle channel extrusion (ECAP) (e.g., References 117, 118). However, it should be mentioned that in magnesium base alloys such grain refinement can be readily achieved with conventional approaches (e.g., References 96, 119, 120) and it is bemusing that more studies are not directed towards exploiting this. In this application, the ready occurrence in magnesium of dynamic recrystallization is of particular use. It can be used to generate fine grains. However, it must not be allowed to coarsen the structure during forming if superplasticity is to occur.

With the achievement of a suitably fine and stable grain structure, the conditions of deformation required for superplasticity are slow hot forming. For obvious reasons, research effort has been directed towards increasing the required strain rate and decreasing the required temperature. One recent study reports optimal conditions for superplasticity in a magnesium alloy of 240°C and 0.01 s^{-1}.[116] Despite falling into the category of high rate superplasticity, a part requiring strains of, say 500%, would still take ~3 min to form. Furthermore, the excessive thinning that accompanies superplastic forming frequently makes it unsuitable for tensile processes such as stamping. It is more attractive for precision forging of components such as housings for electronic devices.

6.8 Hot cracking and extrusion

Finally, we return to a formability measure that is not usually studied using tension testing, though it can be; and that is hot cracking. We will restrict our short discussion here to the contribution of hot shortness; that is, to the occurrence of incipient melting and the concomitant formation of cracks. This is important because it places a ceiling on the temperatures employed for rolling and extrusion. Probably more importantly, it places an important constraint on the speed of extrusion.

It has been pointed out that the speeds used for the hot extrusion of magnesium base alloys can be three times slower than for aluminium alloys.[121] The equipment required is the same, so this contrast is a serious impediment for an extruder considering taking up the production of magnesium alloy products. However, it is apparent that the effect does not relate to magnesium metal itself but rather to the hot strength and solidus values of the alloys that are commonly employed.[122] Magnesium alloys with lower hot strengths and high solidus values can be extruded at speeds approaching that of 6XXX series aluminium alloys.[122] The low hot strength permits higher speeds to be attained at lower extrusion temperatures where the press limit becomes important. The high solidus helps prevent hot shortness under the local heating which can occur in extrusion. These considerations are important in the development of new wrought alloys but are surprisingly frequently overlooked in the effort to generate improved performance in service. These matters are more fully considered in a recent review.[123]

6.9 Conclusions: key issues affecting the formability of magnesium

The present chapter has examined the formability of magnesium base alloys largely through the prism of the tensile test. Despite its inadequacies, this useful test has permitted us to consider the main elements of the formability of magnesium alloys. We have done this in a manner that highlights the differences between magnesium and its sister metal, aluminium. The most salient points are recapped below.

It has been shown that magnesium and its alloys display yield strengths comparable to aluminium base alloys but that there is a far greater asymmetry and anisotropy to yielding. This is due chiefly to the development of textures that strongly favour twinning when sheet or bar is loaded in compression in the direction of prior extension. Despite Lankford r-values that can be as high as 4, magnesium and its alloys typically display deep drawing ratios considerably less than other metals. This is due to their low fracture strains.

In room temperature tensile tests, wrought magnesium alloys typically display strain hardening exponents and forming limit curves only marginally lower than comparable aluminium base alloys. Under certain circumstances deformation twinning can raise or lower the extent of effective work hardening, depending upon which type of twinning is activated.

The strains to failure seen in both tension and compression fall considerably below those observed in aluminium alloys. The effect is most stark in compression, where magnesium can show failure strains less than it can attain in tension. The reason for this is that magnesium readily undergoes shear type failure due to either twinning related events or to more 'conventional' shear bands. The textures developed are unstable in that they favour the formation of these crystallographic instabilities. The low strain to failure limits deep drawing, bending and rolling.

In all of the above cases, formability is markedly improved by an increase in the forming temperature. Temperature does many things: it decreases anisotropy, it lowers the flow stress (particularly for non-basal slip), it suppresses twinning, it raises the rate sensitivity of the flow stress, and it suppresses flow localization, to name some.

Elevated temperature forming can be further enhanced by grain refinement, which permits superplastic strains to be achieved. Such grain refinement is favoured in magnesium base alloys by the occurrence of dynamic recrystallization. The extrusion of magnesium alloys can be carried out using the same equipment employed for aluminium base alloys and at similar temperatures. However, the speeds are typically considerably less for the common wrought alloy AZ31. Newer alloys will display extrusion speeds similar to that of aluminium extrusion alloys.

6.10 Future trends

Although rare-earth containing alloys have been around ever since magnesium alloys have been of interest, it is probably this class of alloys that will provide the future wrought alloys. Future forming alloys will at least display the behaviour and exploit the mechanisms that operate in the current and prototype rare-earth alloys. These alloys will contain the bare minimum of alloying elements in order to maintain a high solidus temperature and to ensure that the hot strength is low. (This of course assumes low temperature applications.) These alloys exhibit and will exhibit weak non-basal textures and fine grains. Both of these features favour nearly all of the formability measures considered in the present chapter.

The forming alloys of the future will probably be strip cast, in the case of sheet, and extruded in the case of bar. Narrow sheet (< ~700 mm) wide will probably be extruded using 'spreader' dies. These technologies provide the highest throughput for minimum cost assuming, that is, that magnesium tonnages stay shy of the exceptionally high tonnages seen for steel and aluminium sheet, which demand tandem rolling mills. Novel technologies for the application of temperature to forming processes with minimum of fuss will continue to be developed.

The author would like to acknowledge advice on the text and topic he received from Professors S. Agnew, W.B. Hutchinson and J. Duncan.

6.11 References

1. von Mises, R., Z. Agnew, *Math. Mech.*, 1928. **8**: 161.
2. Roberts, C. S., in *Magnesium and Its Alloys*. 1960, John Wiley & Sons, Inc.: New York. p. 81–107.
3. Toaz, M. W. and E.J., Ripling, *Transactions of AIME*, 1956. **206**: 936.
4. Hosford, J., W. F. and R.M. Cadell, *Mechanics and Metallurgy*, 3rd Edition, 1993, Prentice Hall: Upper Saddle River.
5. Ion, S. E., F.J. Humphreys, and S.H. White, *Acta Metall. Mater.*, 1982. **30**: 1909–1919.
6. Frost, H. and M. Ashby, *Deformation-Mechanism Maps*. 1982, Oxford: Pergamon.
7. Agnew, S. R., in *Advances in Wrought Magnesium Alloys*. 2011, Edited by C.J. Bettles and M.R. Barnett, Woodhead, Cambridge, chapter 2.
8. Barnett, M. R., in *Advances in Wrought Magnesium Alloys.* 2011, Edited by C.J. Bettles and M.R. Barnett, Woodhead, Cambridge, chapter 3.
9. Püschl, W., *Progress in Materials Science*, 2002. **47**(4): 415–461.
10. Muránsky, O., D.G. Carr, M.R. Barnett, E.C. Oliver, and P. Sittner, *Materials Science and Engineering: A*, 2008. **496**(1–2): 14–24.
11. Obara, T., H. Yoshinga, and S. Morozumi, *Acta Metallurgical* , 1973. **21**: 845–853.
12. Stohr, J. F. and J.P. Poirier, *Philosophical Magazine A*, 1972. **25**(6): 1313–1319.
13. Agnew, S. R., J.A. Horton, and M.H. Yoo, *Metallurgical and Materials Transactions A*, 2002. **33A**: 851–858.

14. Li, B. and E. Ma, *Acta Materialia*, 2009. **57**(6): 1734–1743.
15. Reed-Hill, R. E. and W.D. Robertson, *Acta Metallurgica*, 1957. **5**: 717–727.
16. Reed-Hill, R. E. and W.D. Robertson, *Acta Metallurgica*, 1957. **5**: 728–737.
17. Barnett, M. R., *Materials Science and Engineering*, 2007. **464**(1–2): 8–16.
18. Panicker, R., A.H. Chokshi, R.K. Mishra, R. Verma, and P.E. Krajewski, *Acta Materialia*, 2009. **57**(13): 3683–3693.
19. Barnett, M. R., A. Ghaderi, I. Sabirov, and B. Hutchinson, *Scripta Materialia*, 2009. **61**(3): 277–280.
20. Koike, J., R. Ohyama, T. Kobayashi, M. Suzuki, and K. Maruyama, *Materials Transactions*, 2003. **44**(4): 445–451.
21. Hutchinson, W. B. and M.R. Barnett, *Scripta Materialia*, 2010. **63**(7): 737–740.
22. Lou, X. Y., M. Li, R.K. Boger, S.R. Agnew, and R.H. Wagoner, *International Journal of Plasticity*, 2007. **23**: 44–86.
23. Raeisinia, B., S.R. Agnew, and A. Akhtar, *Metallurgical and Materials Transactions A: Physical Metallurgy and Materials Science*, 2011. **42**(5): 1418–1430.
24. Agnew, S. R., D.W. Brown, and C.N. Tomé, *Acta Materialia*, 2006. **54**(18): 4841–4852.
25. Barnett, M. R., *Metallurgical and Materials Transactions A*, 2003. **34A**: 1799–1806.
26. Chapuis, A. and J.H. Driver, *Acta Materialia*, 2010. In Press, Corrected Proof.
27. Barnett, M. R., Z. Keshavarz, A.G. Beer, and D. Atwell, *Acta Materialia*, 2004. **52**(17): 5093–5103.
28. Caillard, D. and J.L. Martin, Thermally Activated Mechanisms in Crystal Plasticity. *Pergamon Materials Series*, ed. R.W. Cahn. 2003, Oxford: Pergamon.
29. Couret, A. and D. Caillard, *Acta Metall. Mater.*, 1985. **33**(8): 1447–1454.
30. Agnew, S. R., M.H. Yoo, and C.N. Tome, *Acta Materialia*, 2001. **49**: 4277–4289.
31. Bohlen, J., M.R. Nurnberg, J.W. Senn, D. Letzig, and S.R. Agnew, *Acta Materialia*, 2007. **55**: 2101–2112.
32. Barnett, M. R. and N. Stanford, *Scripta Materialia*, 2007. **57**(12): 1125–1128.
33. Barnett, M. R., M.D. Nave, and C.J. Bettles, *Materials Science and Engineering A*, 2004. **386**(1–2): 205–211.
34. Stanford, N. and M.R. Barnett, *Materials Science and Engineering: A*, 2008. **496**(1–2): 399–408.
35. Mackenzie, L. W.F. and M.O. Pekguleryuz, *Scripta Materialia*, 2008. **59**(6): 665–668.
36. Stanford, N., I. Sabirov, G. Sha, A. La Fontaine, S.P. Ringer, and M.R. Barnett, *Metallurgical and Materials Transactions a-Physical Metallurgy and Materials Science*, 2010. **41A**(3): 734–743.
37. Barnett, M. R., A. Sullivan, N. Stanford, N. Ross, and A. Beer, *Scripta Materialia*, 2010. **63**(7): 721–724.
38. Muránsky, O., M.R. Barnett, V. Luzin, and S. Vogel, *Materials Science and Engineering: A*, 2010. **527**(6): 1383–1394.
39. Kelley, E. W. and J. Hosford, W. F., *Transactions of the Metallurgical Society of AIME*, 1968. **242**: 654–660.
40. Safi-Naqvi, S. H., W.B. Hutchinson, and M.R. Barnett, *Materials Science and Technology*, 2008. **24**(10): 1283–1292.
41. Lee, D. and W.A. Backofen, *Transactions of the Metallurgical Society of AIME*, 1966. **236**: 1077–1084.

42. Naka, T., T. Uemori, R. Hino, M. Kohzu, K. Higashi, and F. Yoshida, *Journal of Material Processing Technology*, 2008. **201**: 395–400.
43. Shimizu, I., N. Tada, and K. Nakayama, *International Journal of Modern Physics B: Condensed Matter Physics; Statistical Physics; Applied Physics*, 2008. **22**(31/32): 5844–5849.
44. Logan, R. W. and W.F. Hosford, *International Journal of Mechanical Sciences*, 1980. **22**(7): 419–430.
45. Lee, M.-G., R.H. Wagoner, J.K. Lee, K. Chung, and H.Y. Kim, *International Journal of Plasticity*, 2008. **24**(4): 545–582.
46. Hosford, W. F., *Transactions on ASME Journal of Applied Mechanics*, 1972. **39**: 607.
47. Cazacu, O. and F. Barlat, *International Journal of Plasticity*, 2004. **20**(11): 2027–2045.
48. Plunkett, B., O. Cazacu, and F. Barlat, *International Journal of Plasticity*, 2008. **24**(5): 847–866.
49. Cazacu, O., B. Plunkett, and F. Barlat, *International Journal of Plasticity*, 2006. **22**(7): 1171–1194.
50. Rousselier, G., F. Barlat, and J.W. Yoon, *International Journal of Plasticity*, 2009. **25**(12): 2383–2409.
51. Handbook, A., in Vol. 14: *Forming and Forging*, 9th edition. 1988, ASM International: Metals Park, Ohio.
52. Agnew, S. R. and O. Duygulu, *International Journal of Plasticity*, 2005. **21**(6): 1161–1193.
53. Yi, S., J. Bohlen, F. Heinemann, and D. Letzig, *Acta Materialia*, 2011. **58**(2): 592–605.
54. Mori, K. and H. Tsuji, *CIRP Annals – Manufacturing Technology*, 2007. **56**(1): 285–288.
55. Iwanaga, K., H. Tashiro, H. Okamoto, and K. Shimizu, *Journal of Materials Processing Technology*, 2004. **155–156**: 1313–1316.
56. Beer, A. G. and M.R. Barnett, *Materials Science and Engineering A*, 2006. **423**(Structural Materials: Properties, Microstructures and Processing; Elsevier SA, Switzerland): 292–299.
57. Barnett, M. R., *Journal of Light Metals*, 2001. **1**: 167.
58. Atwell, D., B. Hutchinson, and M.R. Barnett, submitted to *Materials Science and Engineering A*, 2011. **549**: 1–6.
59. Beladi, H. and M.R. Barnett, *Materials Science and Engineering: A*, 2007. **452–453**: 306–312.
60. Ono, N., R. Nowak, and S. Miura, *Materials Letters*, 2003. **58**(1–2): 39–43.
61. Hauser, F. E., P.R. Landon, and J.E. Dorn, *Transactions of the American Society of Metals*, 1958. **50**: 856–883.
62. Akhtar, A. and E. Teghtsoonian, *Acta Metallurgical*, 1969. **17**: 1351–1356.
63. Rokhlin, L. L., Magnesium alloys containing rare earth metals: structure and properties. *Advances in Metallic Alloys*, ed. J.N. Fridlyander and D.G. Eskin. 2003, London: Talyor and Francis.
64. Zhu, S. M. and J.F. Nie, *Scripta Materialia*, 2004. **50**(1): 51–55.
65. Agnew, S. R. and Ö. Duygulu, *International Journal of Plasticity*, 2005. **21**(6): 1161–1193.
66. Stanford, N., K. Sotoudeh, and P.S. Bate, *Acta Materialia*, 2011. **59**(12): 4866–4874.

67. Doege, E. and K. Dröder, *Journal of Materials Processing Technology*, 2001. **115**(1): 14–19.
68. Osada, N., K. Ohtoshi, M. Katsuta, S. Takahashi, and T. Yamada, *Keikinzoku/ Journal of Japan Institute of Light Metals*, 2000. **50**(2): 60–64.
69. Luo, A. A. and A.K. Sachdev, *Magnesium Technology*, 2004: 79–85.
70. Jiang, L., J. Jonas, A. Luo, A. Sachdev, and S. Godet, *Scripta Materialia*, 2006. **54**(5): 771–775.
71. Agnew, S. R., C.E. Dreyer, F. J. Polesak Iii, W.V. Chiu, C.J. Neil, and M. Rodriguez. in *Magnesium Technology*. 2008.
72. Luo, A. A. and A.K. Sachdev. in *Magnesium Technology*. 2004: TMS, Warrendale.
73. Jiang, L., J.J. Jonas, A.A. Luo, A.K. Sachdev, and S. Godet, *Scripta Materialia*, 2006. **54**(5): 771–775.
74. Kim, W. J., H.K. Kim, W.Y. Kim, and S.W. Han, *Materials Science and Engineering: A*, 2008. **488**(1–2): 468–474.
75. Oppedal, A. L., H. El Kadiri, C.N. Tome, J.C. Baird, S.C. Vogel, and M.F. Horstemeyer. in *Magnesium Technology*. 2011: TMS.
76. Wu, D., R.S. Chen, and E.H. Han, *Journal of Alloys and Compounds*, 2011. **509**(6): 2856–2863.
77. Barnett, M. R., *Materials Science and Engineering*, 2007. **464**(1–2): 1–8.
78. Mukai, T., M. Yamanoi, H. Watanabe, and K. Higashi, *Scripta Materialia*, 2001. **45**: 89–94.
79. Agnew, S. R., J.A. Horton, T.M. Lillo, and D.W. Brown, *Scripta Materialia*, 2004. **50**(3): 377–381.
80. Wonsiewicz, B. C. and W.A. Backofen, *Transactions of the Metallurgical Society of AIME*, 1967. **239**(9): 1422–1431.
81. Reed-Hill, R. E., *Transactions of the Metallurgical Society of AIME*, 1960. **218**: 554–558.
82. Morrison, W. B. and R.L. Miller. *Ultrafine-grain Metals: Proceedings of the 16th Sagamore Army Materials Research Conference*. Sagamore Conference Center, Raquette Lake, New York, August 19–22, 1969: Syracuse University Press.
83. Ohtoshi, K., T. Nagayama, and M. Katsuta, *Keikinzoku/Journal of Japan Institute of Light Metals*, 2003. **53**(6): 239–244.
84. Jacob, S., *Low Temperature Formability of Differently Processed AZ31B Magnesium Alloy Sheets*. 2007, Deakin University: Geelong.
85. Stutz, L., J. Bohlen, D. Letzig, and K.U. Kainer. *Magnesium Technology*. 2011, Ed. Sillekens, W. H. and Agnew, S. R., TMS, Warrendale, 373–378.
86. Huang, G.-s., H. Zhang, X.-y. Gao, B. Song, and L. Zhang, *Transactions of Nonferrous Metals Society of China*, 2011. **21**(4): 836–843.
87. Naka, T., G. Torikai, R. Hino, and F. Yoshida, *Journal of Materials Processing Technology*, 2001. **113**(1–3): 648–653.
88. John Neil, C. and S.R. Agnew, *International Journal of Plasticity*, 2009. **25**(3): 379–398.
89. Naka, T., Y. Nakayama, T. Uemori, R. Hino, F. Yoshida, M. Kohzu, and K. Higashi, *International Journal of Modern Physics B*, 2008. **22**(31–32): 6010–6015.
90. Embury, J. D. and J.L. Duncan, *Annual Review of Materials Science*, 1981. **11**: 501.
91. Forcellese, A., F. Gabrielli, and M. Simoncini, *Computational Materials Science*, 2011. **50**(11): 3184–3197.

92. Naka, T., Y. Nakayama, T. Uemori, R. Hino, and F. Yoshida, *Journal of Materials Processing Technology*, 2003. **140**(1–3): 494–499.
93. Montheillet, F. and J.J. Jonas, *Metallurgical and Materials Transactions A*, 1996. **27**(10): 3346–3348.
94. Hauser, F. E., P.R. Landon, and J.E. Dorn, *Journal of Metals, AIME Transactions*, 1956. **206**: 589–593.
95. Barnett, M. R., S. Jacob, B.F. Gerard, and J.G. Mullins, *Scripta Materialia*, 2008. **59**(10): 1035–1038.
96. Chapman, J. A. and D.V. Wilson, *Journal of the Institute of Metals*, 1962. **91**: 39–40.
97. Davidson, T. E., J.C. Uy, and A.P. Lee, *Acta Metallurgica*, 1966. **14**: 937–947.
98. Beck, A., *The Technology of Magnesium and its Alloys*. 1943, London: Hughes.
99. Klimanek, P. and A. Pötzsch, *Materials Science and Engineering*, 2002. **A324**(1–2 Special Issue SI): 145–150.
100. Jain, A. and S.R. Agnew, *Materials Science and Engineering*, 2007. **462**(1–2): 29–36.
101. Kelley, E. W. and J. Hosford, W. F., *Transactions of the Metallurgical Society of AIME*, 1968. **242**(1): 5–13.
102. Brown, D. W., S.R. Agnew, M.A.M. Bourke, T.M. Holden, M. Vogel, and C. Tome, *Materials Science and Engineering A*, 2005. **399**: 1–12.
103. Hartt, W. H. and R.E. Reed-Hill, *Transactions of the Metallurgical Society of AIME*, 1968. **242**: 1127–1133.
104. Reed-Hill, R. E., *Transactions of the ASM*, 1959. **51**: 105–107.
105. Koike, J., *Metallurgical and Materials Transactions A*, 2005. **36A**: 1689–1696.
106. Ando, D., J. Koike, and Y. Sutou, *Acta Materialia*, 2011. **58**(13): 4316–4324.
107. Couling, S. L., J.F. Pashak, and L. Sturkey, *Transactions of the ASM*, 1959. **51**: 94–107.
108. Wilson, D. V. and J.A. Chapman, *Philosophical Magazine*, 1963. **8**: 1543–1551.
109. Ridha, A. A. and W.B. Hutchinson, *Acta Metallurgica*, 1982. **30**(10): 1929–1939.
110. McDonald, J. C., *Transactions of the AIME*, 1940. **137**: 430–441.
111. McDonald, J. C., *Transactions of the AIME*, 1941. **138**: 179–182.
112. Sandlöbes, S., S. Zaefferer, I. Schestakow, S. Yi, and R. Gonzalez-Martinez, *Acta Materialia*, 2011. **59**(2): 429–439.
113. McDonald, J. C., *Transactions of the Metallurgical Society of AIME*, 1958: 45–46.
114. Langdon, T. G., *Journal of Materials Science*, 2009. **44**(22): 5998–6010.
115. Kawasaki, M., N. Balasubramanian, and T.G. Langdon, *Materials Science and Engineering A*, 2011. **528**(21): 6624–6629.
116. Figueiredo, R. B. and T.G. Langdon, *Scripta Materialia*, 2009. **61**(1): 84–87.
117. del Valle, J. A., F. Carreño, and O.A. Ruano, *Acta Materialia*, 2006. **54**(16): 4247–4259.
118. Miyahara, Y., Z. Horita, and T.G. Langdon, *Materials Science and Engineering: A*, 2006. **420**(1–2): 240–244.
119. Stanford, N. and M.R. Barnett, *Journal of Alloys and Compounds*, 2008. **466**:182–188.
120. Tan, J. C. and M.J. Tan, *Scripta Materialia*, 2002. **47**: 101–106.
121. Katrak, F. E., J.C. Agarwal, F.C. Brown, M. Loreth, and D.L. Chin., in *Magnesium Technology 2000*, Warrendale, TMS Annual Meeting, 2000, 351–354.

122. Atwell, D. and M.R. Barnett, *Metallurgical and Materials Transactions A*, 2007. **38**(12): 3032–3041.
123. Beer, A. G., in *Advances in Wrought Magnesium Alloys*. 2012, Edited by C.J. Bettles and M.R. Barnett, Woodhead, Cambridge, chapter 8.
124. Dröder, K. and S. Janssen. in International Body Engineering Conference and Exposition. 1999. Detriot, Michigan, SAE Technical Paper 1999–01–3172.

7
Corrosion and surface finishing of magnesium and its alloys

S. BENDER, iLF Institut für Lacke und Farben e.V, Germany, J. GÖLLNER, Otto von-Guericke University of Magdeburg, Germany, A. HEYN, Federal Institute for Materials Research and Testing (BAM), Germany, C. BLAWERT and P. BALA SRINIVASAN, Helmholtz-Zentrum Geesthacht, Germany

DOI: 10.1533/9780857097293.232

Abstract: While the possibilities of improving corrosion resistance (especially galvanic corrosion) by alloying are limited, surface finishing of magnesium alloys is the alternative for improving corrosion resistance. Due to the low corrosion potential of Mg and the danger of galvanic corrosion in the case of a coating defect, the choice of coatings is limited. This chapter will summarize the corrosion behavior of magnesium alloys and the metallurgical possibilities to improve the corrosion resistance of the alloys and reviewing critically the most commonly used surface treatments and coatings for magnesium.

Key words: corrosion, corrosion testing, negative difference effect (NDE), conversion coatings, plasma electrolytic oxidation, organic coatings, galvanic coatings.

7.1 Introduction

Magnesium is reactive and its alloys are not corrosion resistant under various aqueous or humid conditions. The relatively low corrosion resistance of magnesium alloys limits their applications. The use of magnesium alloys in the automotive industry, for example, is currently limited to components in the interior, or outside the field of view, in part due to poor corrosion behavior. Understanding and improving the corrosion behavior of magnesium alloys is important for increasing their potential application.

While the possibilities of improving the corrosion resistance (especially galvanic corrosion) by alloying are limited, surface finishing of magnesium alloys is an alternative method. Due to the low corrosion potential of Mg and the danger of galvanic corrosion in the case of coating defects, the choice of coatings is limited. This chapter will summarize the corrosion behavior of magnesium alloys and metallurgical options to improve their corrosion resistance. A critical review of the most commonly used surface treatments and coatings for magnesium is also provided.

7.2 Magnesium corrosion in aqueous media

Magnesium production is energy intensive; while, 6600 kJ/kg are required to produce iron from iron oxide, 23 520 kJ/kg are necessary for the extraction of magnesium from magnesium oxide. Since matter always tends to a low energy condition, the dwell time of the metallic state is naturally restricted. The corrosion behavior of magnesium alloys depends, in addition to alloy reactivity, on a variety of factors. The pH value of the surrounding medium and near the metal surface, the dependence of surface films on the alloy-composition, and the type and distribution of intermetallic phases all influence the corrosion rate of magnesium alloys. If the pH value is above 10.5, a stable and self-healing passive layer arises, which is responsible for the corrosion resistance. For pH in the range of 8–10, the layer varies between a stable and an unstable state. Reaction layers with relative thickness consisting of loosely adhering corrosion products are developed, but they only appear as passivity layers similar to high-alloyed steels. The corrosion rate is controlled by diffusion and depends mainly on the layer thickness. During dissolution and hydrogen evolution, the pH rises in the area near the surface. The result is a change in the formation of the reaction layer with higher stability.

The essential aspects of the corrosion mechanism of magnesium alloys are well described in literature (Song and Atrens, 1999, 2003). The corrosion process starts with the dissolution of magnesium. Mg^{2+} ions are dissolved and a corresponding number of free electrons are left behind on the metal surface (Equation [7.1]). To continue the dissolution process, an electron-consuming reaction is needed. This occurs by the reduction of hydronium (H_3O^+) to molecular hydrogen (acidic corrosion, Equation [7.2]). This reduction of hydronium ions is most important, and the formation of hydrogen bubbles is essential for the continuous dissolution of magnesium. Due to this cathodic reaction, the hydrogen development and active dissolution increase with decreasing pH values. The pH value in the surrounding media increases through the consumption of hydronium ions. As a secondary reaction, the magnesium ions react with hydroxide ions (OH^-) to magnesium hydroxide ($Mg(OH)_2$), which deposits on the metal surface (Equation [7.3]).

$$\text{Anodic reaction: } Mg \rightarrow Mg^{2+} + 2e \quad [7.1]$$

$$\text{Cathodic reaction: } 2H^+ + 2e \rightarrow H_2 \quad [7.2]$$

$$\text{Secondary reaction: } Mg^{2+} + 2OH^- \rightarrow Mg(OH)_2 \quad [7.3]$$

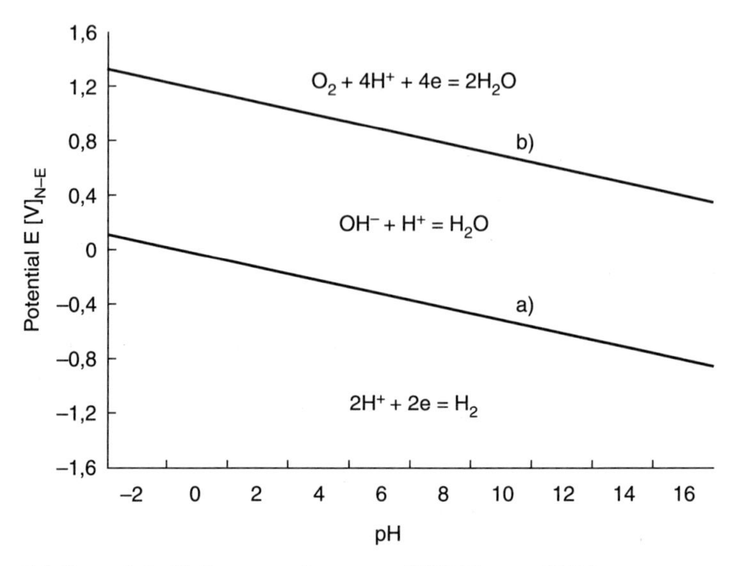

7.1 Potential-pH-diagram of water at 25°C (Jones, 1996).

As can be seen in Pourbaix diagram (Pourbaix, 1974), $Mg(OH)_2$ is stable at pH values above 10.5. Only under these specific conditions can a protective layer form on the surface. But the reaction layers of $Mg(OH)_2$, which form in neutral to slightly alkaline media, cannot be described as protective passive layers or even as bonding layers. In the active dissolution of magnesium, these layers are very porous.

Magnesium alloys are resistant to intergranular corrosion, since the grain boundaries are cathodic to the grain surfaces. Crevice corrosion does not occur in magnesium alloys because the differences in the oxygen concentration play no critical role in the mechanism of magnesium corrosion. The formation of hydrogen as cathodic reaction can be seen in Equation [7.2] and in the potential-pH diagram of water at 25°C (Fig. 7.1). In Fig. 7.1, the potential-pH relationships of water are shown by the cross lines (a) and (b), which were determined by the Nernst equation. Since the free corrosion potential of magnesium at –1.7 V in aqueous solutions is very negative, the area under the cross line (a) is of essential interest. Hydrogen is thermodynamically stable, therefore, hydrogen ions are spontaneously reduced to molecular hydrogen at the metal surface. This process corresponds to the cathodic process of corrosion in acids, where high hydrogen ion concentrations ($[H^+] \geq 1 \times 10^{-4}$ mol l^{-1}) are present. Additionally, under the cross line (a), water is unstable and is therefore reduced to hydrogen and hydroxide ions. The area under the cross line (a) covers the main corrosion process of magnesium, namely the acid corrosion, which underlines the importance of the electron-consuming process of hydrogen formation.

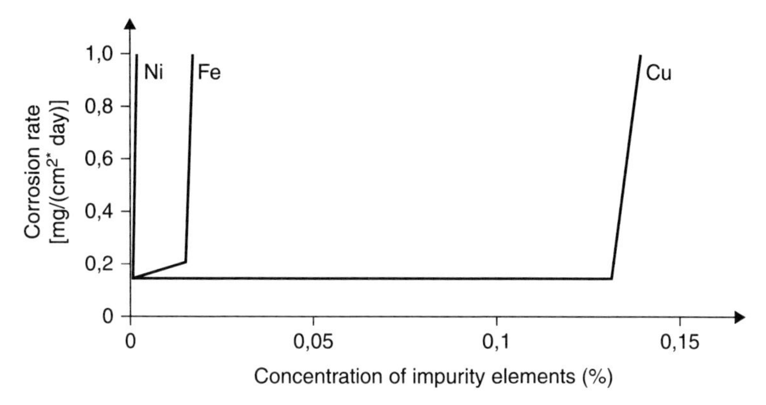

7.2 Schematic illustration of the dependence of the tolerance limit of magnesium for iron, nickel and copper (Hanawalt *et al.*, 1942).

7.2.1 Influence of alloying elements

The corrosion behavior of magnesium alloys can be influenced by alloying elements. The elements iron, nickel and copper can be dissolved only slightly in magnesium alloys during the manufacturing process. The precipitates that are formed in magnesium alloys are nobler than the magnesium matrix and have a lower hydrogen overvoltage. This favors the recombination of hydrogen (Ghali, 2000), so they are very effective local cathodes. The precipitation of some of the elements as impurities increases the corrosion rate by a factor of 10–100. The negative effects of these precipitates are reduced in the order nickel > iron > copper. Magnesium has, for each element, its own tolerance limit, up to which the corrosion rate is hardly influenced by the presence of the impurities and remains relatively low (Fig. 7.2).

If this tolerance limit is exceeded, the corrosion rate increases dramatically. For this reason, high-purity initial materials must be used for alloy development. Depending on the magnesium alloy, there are precise limits on the content of iron, nickel and copper impurities. Thus, the tolerance limit of pure magnesium is 5 ppm for nickel, 170 ppm for iron and 1300 ppm for copper (Ghali, 2000). The tolerance limit of the magnesium alloy AZ91 in salt water is 5 ppm for nickel, 50 ppm for iron und 300 ppm for copper (Song and Atrens, 1999). So far, there is no model that can explain the causes of the tolerance limits (Song and Atrens, 1999, 2003). Other metals such as aluminum, zinc, cadmium and tin in various media are also nobler than magnesium, but less effective cathodes. They avoid the recombination of hydrogen on their surface due to the high hydrogen overvoltage (Ghali, 2000). Basically, the addition of aluminum leads to an improvement of the corrosion resistance of magnesium alloys. The positive influence on the corrosion behavior is significant up to 4% aluminum. This observation correlates with the enrichment of Al_2O_3 in the reaction layer (Nordlien and Kemal Ono, 1996). It reaches its maximum of 35% aluminum in the reaction layer when the aluminum in the alloy reaches 4%. Thus, a completely closed network of passivating Al_2O_3 is formed in the reaction layer, which serves as a kind of skeletal structure in the amorphous mixture of aluminum and magnesium oxides as well as hydroxides. The addition of more aluminum does not increase the content of aluminum in the reaction layer.

The precipitation of $Mg_{17}Al_{12}$ begins in two-phase alloys at 2% aluminum (in an equilibrium state). The β-$Mg_{17}Al_{12}$ can act as a corrosion-protective network in the alloy, if the β-phase is distributed evenly in the magnesium alloy. The content of β-phase can be increased by the addition of more aluminum (Ghali, 2000). The addition of aluminum over 9% causes only a moderate improvement. It is shown that the distribution of the β-phase is finer in certain casting conditions; in the order of fineness (going from coarser to finer) the castings can be ranked as casting → die casting → thixomolding due to smaller grain size (Pieper, 2005). If the β-phase is distributed unevenly in the alloy due to a coarse microstructure or too low an aluminum content, the β-phase accelerates the corrosion as it works as an effective cathode and not as a protective network.

7.2.2 Investigation methods

To investigate the corrosion properties of magnesium alloys, the choice and conditions of test procedures are very important. It is therefore necessary to look critically at currently used methods (salt spray test), modify conventional methods on magnesium specifics (volume measurement of evolved hydrogen and polarization measurements in combination with the rotating-disc electrode) and particularly use newly developed methods (electrochemical noise measurements).

Methods such as the salt spray test (DIN-88, 1988) were developed originally for coated materials. They do not provide adequate characterization of the specific corrosion behavior of magnesium and magnesium alloys. Using this test, it is difficult to make a statement about the corrosion mechanism of magnesium alloys under different conditions. The conclusions from the salt spray test should be viewed critically. Practice shows that products can perform well in the salt spray test yet fail in use, or the reverse. Nevertheless, a limited amount of comparison between different magnesium alloys is possible by determination of weight loss in this test.

Another possibility for the determination of corrosion reactions is the volume measurement of evolved hydrogen during magnesium corrosion (Song *et al.*, 2001). The amount of hydrogen is directly proportional to the amount of corroded magnesium. For the formation of 1 mole of molecular hydrogen exactly 1 mole of magnesium must go into solution. Therefore, the weight loss of the magnesium specimen can be calculated from the resulting amount of hydrogen. The corrosion products formed during dissolution should not affect the evolution of hydrogen. A further condition for the determination of the corrosion rate is related to the fact that the contribution of oxygen reduction can be neglected. In comparison to the measurement of weight loss after the salt spray test, measuring the hydrogen evolution has the advantage that the change in corrosion rate by varying pH values can be measured over the course of corrosion. This allows a 'monitoring' of corrosion rate as a function of time. It is recommended to combine the measurement of hydrogen evolution with polarization measurements in order to obtain comprehensive information about the corrosion behavior of magnesium alloys and its evolution over time (Song *et al.*, 2005).

Electrochemical measurements are used to determine the free corrosion potential and the polarization behavior of the alloys. The corrosion behavior can be characterized based on the flowing current and the corrosion potential. But in neutral electrolytes, voluminous reaction products are formed and lead to a diffusion controlled dissolution of magnesium alloys. Thus, microstructural and alloying influences are strongly suppressed. In alkaline environments, passivation occurs as a result of the formation of a stable hydroxide layer on the magnesium surface. Therefore, differences in the corrosion behavior between alloys of different chemical compositions and structures are hardly detectable. A better suited approach for characterization is the utilization of a *rotating-disc electrode* (RDE). The rotation removes reaction products and adjusts defined flow conditions on the surface. Bender *et al.* (Bender *et al.*, 2007a) developed a material-specific method for corrosion testing of magnesium and magnesium alloys based on the commonly used polarization measurement method in combination with an RDE. The working electrode was an RDE in a classical three-electrode cell. After preliminary investigations, the following test parameters for polarization measurements were considered as most effective: 2000 rpm, 0.01 M NaCl and pH9. During the polarization measurements argon was bubbled into the electrolyte to avoid a decrease of pH value by adsorption of CO_2 from ambient air. Regarding the current density, polarization measurements with the RDE are suitable for characterizing and comparing different magnesium alloys.

Electrochemical noise measurements offer a simple, sensitive and virtually non-destructive approach for the assessment of the corrosion susceptibility of magnesium alloys and can be used for investigation of corrosion processes as well. A specific experimental set-up is necessary to record the potential, potential noise, current and current noise. These measurements are carried out in a three-electrode cell consisting of two working electrodes, which are macroscopically identical specimens (same size and preparation), and a reference electrode. Due to the high sensitivity of the measurements, a Faraday cage is required to suppress external influences. The experiments are carried out with a zero-resistance ammeter (ZRA) and a high-impedance potential measuring device. The direct and alternating parts of the current and potential must be filtered by a low-pass and a band-pass filter. One common method for the quantification of noise data is the calculation of the standard deviation (S) over fixed periods and the further calculation of a noise resistance (R_N = SU/Si). Another evaluation method is the calculation of the charge (Q) by integration of the rectified current noise over fixed time intervals. The electrochemical noise measurements without external polarization confirm the results of the polarization measurements with RDE. But the results are more convincing in relation to the high resolution of this method. The use of the electrochemical noise measurement in the corrosion examination makes it possible to evaluate the time dependence of corrosion events. Furthermore, it enables the differentiation of various magnesium alloys, their chemical composition, structures and different impurities within a short time. The advantages of the electrochemical noise technique are obvious: due to the short time, noise investigations can indicate tendencies and occurrences of corrosion before a macroscopic damage develops (Bender *et al.*, 2007b, 2008).

From the wide variety of corrosion testing possibilities and the different test conditions it is obvious that the investigation of the corrosion behavior of magnesium and its alloys is difficult. The reason can be seen in the very complex corrosion behavior of this light metal.

7.2.3 Phenomenology of NDE

NDE can be seen in the current-potential curves of Fig. 7.3. The anodic and cathodic reactions for a general metal are shown by the black solid curve marked I_a(Mg) and $I_c(H_2)$. This is the case for normal electrochemical polarization behavior and would be expected for magnesium. But experimentally the dashed line is found, which shows that with increasing potential, hydrogen evolution increases instead of decreasing along the expected curve $I_c(H_2)$.

Simultaneously, the anodic dissolution increases in a way that is stronger than expected. This effect, that the measured hydrogen evolution is greater than the expected hydrogen evolution, is designated as the NDE. Furthermore, for anodic dissolution of magnesium, the experimentally determined weight loss is greater than calculated by using Faraday's Law.

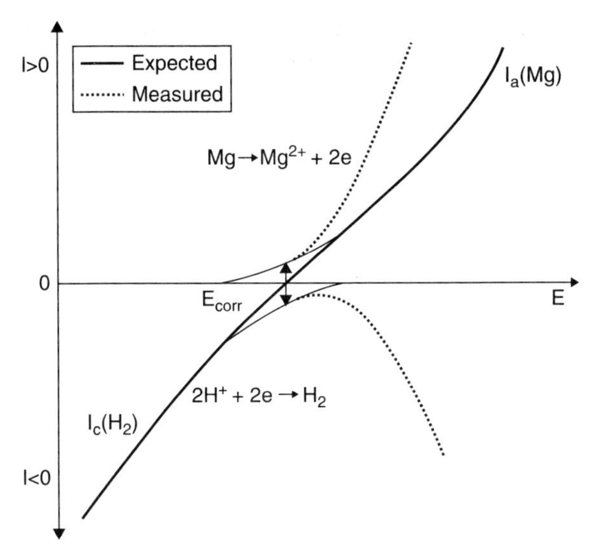

7.3 Schematic illustration of the anodic Ia(Mg) and the cathodic Ic(H_2) partial reaction curves for the expected (solid line) and for the experimentally found (dashed line) polarization behavior.

7.2.4 New model to explain the NDE

Several models have been presented in order to explain this phenomenon (Song, 2005). The main models published about the NDE of magnesium are based on the breakdown of a protective film (Robinson and King, 1961; King, 1966; Tunold *et al.*, 1977; Petrova and Krasnoyarskii, 1990; Kisza *et al.*, 1995) or the formation of magnesium hydride (Gulbrandsen, 1992). These models only explain the effect of the abnormal hydrogen evolution. No explanation is given for the difference between weight loss calculated with Faraday's Law and the gravimetrically determined weight loss. Another model tries to explain NDE by the presence of embedded secondary phase particles (Makar *et al.*, 1988; Makar and Kruger, 1990). This model describes the NDE exclusively for alloys of magnesium. But the 'strange' hydrogen evolution behavior arises with increasing positive polarization current density or potential only on pure magnesium. Therefore, this model gives no satisfactory explanation for the NDE. The dissolution of magnesium involving monovalent ions (Mg^+) is the most cited and discussed model (Song and Atrens, 1999; Bender *et al.*, 2008). Convincing evidence for Mg^+ has been forthcoming from the work of Atrens (Atrens and Dietzel, 2007) while a recent study by Williams (Williams and McMurray, 2008) and Swiatowska (Swiatowska *et al.*, 2010) showed no evidence for an Mg^+ intermediate. Resulting from further examinations and considerations (Bender, 2010) the following explanation of the

NDE arises: If the magnesium is in contact with water, it immediately sends ions (Mg^{++}) into solution, as shown in Equation [7.4]. The electrons ($2e^-$) remain in the metal.

$$\text{Anodic reaction: } Mg \rightarrow Mg^{++} + 2e^- \quad [7.4]$$

The magnesium ion could also leave monovalent its atomic place (Equation [7.5]). Then it would exist in an unstable intermediate state and possess an appropriately short life time. After giving up the second electron (Equation [7.6]), the magnesium ion would leave the metal surface. Till now, it has not been clear under which conditions the one or the other reaction has priority.

$$\text{Anodic reaction: } Mg \rightarrow Mg^{+} + e \quad [7.5]$$

$$\text{Anodic reaction: } Mg^{+} \rightarrow Mg^{++} + e \quad [7.6]$$

For this new theory, it is not important whether the magnesium ions generate first as monovalent metal ions and are oxidized electrochemically in a second step to the bivalent magnesium ion, or if the magnesium ions go as bivalent into the solution. In every case, a corresponding number of electrons stay in the metal. The reaction will stop since the electrostatic force (positive ions in the medium, negative charge carriers in the metal) prevents a further charge separation. An electron-consuming process is necessary for the continuing dissolution. The progress of electron consumption occurs by unloading of hydrogen ions, as can be seen in Equation [7.7].

$$\text{Cathodic reaction: } 2H^{+} + 2e^- \rightarrow H_2 \quad [7.7]$$

The source of the hydrogen ions (more exactly the hydronium ion H_3O^+) in neutral solutions (pH7) is the dissociated water, as shown in Equation [7.8].

$$\text{Dissociation: } H_2O \leftrightarrow OH^- + H^+ \quad [7.8]$$

In a secondary reaction, magnesium ions react with hydroxide ions to magnesium hydroxide, as seen in Equation [7.9].

$$\text{Product formation: } Mg^{++} + 2OH^- \rightarrow Mg(OH)_2 \quad [7.9]$$

Now, the equilibrium between water and its ions is disturbed and the reaction is stronger in one direction (principle of Le Chatelier and Braun (Grigull, 1963)), as seen in Equation [7.10].

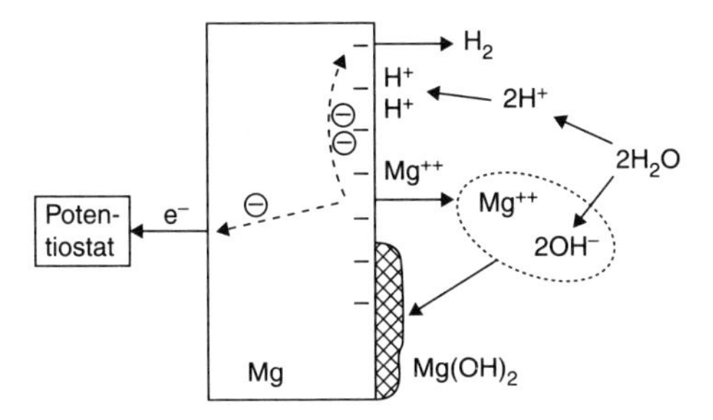

7.4 Schematic illustration of the new model for explanation of NDE.

$$H_2O \rightarrow OH^- + H^+ \quad [7.10]$$

An acid is produced which discharges the free electrons and the metal dissolution goes on further. That means for the process that the more magnesium ions that go into solution, the stronger is water dissociation and the higher the number of electrons consumed by hydrogen ions. The dissolution of magnesium increases rapidly. In addition, under potentiodynamic control, electrons are transported by potentiostat to the counter electrode and the magnesium dissolution increases. Additionally, this disturbs the non-equilibrium of the dissociation and an increasing level hydrogen evolution can be observed.

The new model is schematically illustrated in Fig. 7.4. The dashed arrow towards the left on the magnesium corresponds to the electron consumption by external polarization (potentiostat) and the other dashed arrow on the magnesium corresponds to the electron consumption by acidic corrosion. The aqueous electrolyte is dissociated into hydrogen and hydroxide ions. If magnesium is in contact with water, magnesium ions go into solution and react with the hydroxide ions of the dissociated water. The hydrogen ions are discharged on the metal surface.

This new model can explain the effects of NDE. The reason for strong hydrogen evolution by anodic polarization and the difference between calculated and measured weight loss, is the increased discharge of hydrogen ions. Due to anodic polarization of magnesium, more electrons were consumed and more magnesium ions go into solution. The electron consumption by potentiostat can be measured as a flowing current on the ammeter. Therefore, a great number of magnesium ions are available, which react with the hydroxide ions of the dissociated water to form magnesium hydroxide. Due to this secondary reaction, the equilibrium between water and its ions is disturbed. It should come to non-equilibrium in the dissociation of water, which would immediately be balanced due to new dissociation of the water into hydrogen and hydroxide ions. The chemical equilibrium is mainly disturbed on the phase boundary. So the hydrogen ions are reduced in the electrochemical double layer and no acidification in bulk electrolyte can be registered. It would always produce an acid on the phase boundary, which, additionally to the potentiostat, consumes the electrons. Therefore, more hydrogen develops on the magnesium surface. This second electron consumption process can be seen in the formation of gas bubbles and can only be measured by the simple hydrogen evolution technique. Therefore, under anodic polarization, the corrosion of magnesium alloys depend on two cathodic processes.

7.3 Surface finishing

With the development of high-purity magnesium alloys, an important step towards better corrosion performance was achieved. However, none of the metallurgical approaches towards better corrosion resistance is suitable to prevent galvanic corrosion of magnesium in contact with other metals. Furthermore, even the general corrosion resistance of magnesium and magnesium alloys still remain inadequate in more aggressive neutral and acidic media. In all these cases, only a coating may provide sufficient protection. The following section provides a short overview about coating technology suitable for magnesium. A more detailed overview is provided by Gray *et al.* (Gray and Luan, 2002).

7.3.1 Conversion coatings

As the name suggests, conversion coatings are intended to convert the surface of an alloy into a layer of thin protective film, comprising simple or complex compounds that are formed during the chemical and/or electrochemical treatments in specific electrolytes. The surfaces to be conversion-coated are subjected to rigorous cleaning to create a virgin and active surface for the conversion process. Even though conversion coatings are expected to provide an improved corrosion resistance, these coatings on their own may not provide great benefit, as they are usually very thin. In most circumstances, these coatings are used with other thick coatings, for example, lacquer, paint, etc., for which these films act as an effective barrier as well as promoting adhesion to the substrate.

As for aluminum and zinc alloys, for magnesium alloys, too, a variety of conversion coatings are used. The most common and well known is the chromate-conversion coating. Chromate films are said to provide good corrosion resistance to magnesium alloys, but with the major disadvantage of being toxic. The treatment that is performed in solutions containing chromate/dichromate ions is initiated by the metal dissolution, the consequent pH changes at the metal–solution interface, and the associated precipitation/formation of film with chromium compounds on the surface. The chromate-conversion film is reported to be constituted with a dense magnesium/chromium hydroxide inner layer and an exterior porous chromium hydroxide layer (Simaranov *et al.*, 1992). The corrosion protection comes from the barrier effect based on the thickness and the inhibiting effect of the film comprising chromate ions in corrosive environment.

The chromate coatings on magnesium alloys are produced from acidic electrolytes containing dichromate. The pH of the electrolyte can be in the range of 1.2 to 6.0, depending on the other ingredients in the electrolyte. The chromate-treatment duration is dependent on the chemistry, pH and temperature of the electrolyte. For example, the treatment duration in solutions containing sodium dichromate and nitric acid at ambient temperature is between 10 and 120 s, as against 2 h for that in solutions containing sodium dichromate and manganese–magnesium sulfates. The finishes and aesthetic appeal also vary with variations in the chemistries of electrolytes (Elektron, 2011). Chromate treatments can yield a range of colors from yellow to brown to black, depending on the alloy, type and processing condition of electrolyte.

Due to increasing concern over environmental issues and health awareness, chromate treatments are slowly becoming abolished, and non-chromate alternatives have been investigated since the 1990s as conversion coatings for magnesium alloys. A few of these are obtained chemically and electrochemically from solutions containing phosphate, permanganate, phosphate–permanganate, stannate and rare-earth salts. Numerous patents address the technology of phosphate coatings on magnesium and claim improvements in adhesion and corrosion behavior of additional coatings over the conversion coating (Heller, 1960; Makoto *et al.*, 1994; Naohiro *et al.*, 1999). Phosphate coatings on AZ91D alloy from acidic solutions containing phosphoric acid, zinc oxide and sodium fluoride as main ingredients, obtained at 40–45 °C, were reported to consist of zinc–magnesium and aluminum phosphates. The pH was reported to play a critical role in the film formation. More acidic conditions ($pH < 1.8$) resulted in inferior growth/poor coverage, owing to the excessive corrosion damage on the substrate during the treatment. It has been claimed that this coating had a better adhesion for the paint film compared to that offered by the chromate film on this magnesium alloy substrate, and the benefit was attributed to the anchoring microstructure. Phosphate coatings obtained from a slightly modified solution ($NaClO_3$ added as an accelerator to the solution mentioned above) on AZ91D alloy was found to have different morphology, but with a more or less similar film chemical composition (Li *et al.*, 2006). Nucleation of zinc–phosphate phase and metallic phase was reported in the early stages, the progress of which resulted in slab-like structure of only the zinc–phosphate phase (hopeite). The corrosion behavior of the phosphate surface was higher than that of the untreated alloy, as assessed by polarization tests in 5% NaCl solution.

The use of sodium dodecyl sulfate (SDS) as an additive/accelerator was reported to improve (a) coating weight/coverage, (b) compactness and (c) corrosion resistance of the conversion coatings formed on AZ31 magnesium alloy. The addition of SDS as a replacement for sodium nitrite was claimed to be beneficial in terms of the environmental perspective, in that the sludge handling in the treatment solutions containing nitrites can be avoided. The improved corrosion behavior is attributed to the reduction in the micro-cracks in the film, and also to the formation of higher amounts of uniform hopeite phase in the coating (Amini and Sarabi, 2011).

Permanganate-based coatings also containing phosphate have been developed recently (Zhao *et al.*, 2006) and are claimed to be on a par with, or better than, chromate films. Control of pH was found to be a key factor in achieving coatings of good quality without any powdery deposition. Phosphate–permanganate coatings on AZ91D magnesium alloy in acidic permanganate solutions in the pH range of 3–5 was reported to yield uniform, continuous coatings of 7–10 μm, with non-penetrating pores on the surface. Electrochemical measurements, *viz.*, potential and polarization, showed that the corrosion behavior of this coating with or without an additional organic film on the surface was better than that of the conventional Dow™ chromate-conversion coatings. Typical microstructures of the zinc–phosphate coatings with and without the chemical accelerator additive to the electrolyte are shown in Fig. 7.5.

Conversion coatings from stannate electrolytes have been attempted by researchers with claims of good success. Stannate electrolytes contain NaOH and K_2SnO_3 as major ingredients, meaning that the coatings from these alkali electrolytes are distinctly different from the conversion coatings that are produced from acidic electrolytes. The immersion chemical treatments are performed usually at higher temperatures, in the range of 45°C to 80°C. Lowering the pH, increasing the temperature, and increasing the stannate concentration in the electrolyte, were reported to lead to reduced incubation periods and increased growth rates. The coating grows as layers of hemispherical particles, and the recesses that are left at the contacting particles become defective sites in the coated surfaces. Providing congenial conditions for more nucleation and growth, especially by lowering the pH, is reported to help in achieving finer sized hemispherical particles, and thus minimizing the defects that arise from larger sized particles (Lin *et al.*, 2006).

Zucchi *et al.* (2007) investigated the formation and protective performance of stannate and phosphate–permanganate conversion coatings on AZ31 and AM60 alloys using *in situ* electrochemical impedance spectroscopy (EIS). The resistance of the phosphate–permanganate conversion film

7.5 Typical surface morphology of phosphate coated AZ91D alloy (a) without accelerator (b) with accelerator (Niu *et al.*, 2006; Li *et al.*, 2006).

as it grew on the magnesium alloy substrate during the coating evolution process was found to be lower than that of the stannate coatings, which was attributed to the cracks and connecting defects in the phosphate–permanganate coatings. Despite showing a higher initial resistance, the stannate coatings were found to degrade quickly in corrosive environments due to the easy permeability of electrolyte through the defects at the hemispherical particle interfaces. The characteristic surface morphology of phosphate-permanganate and stannate coatings on magnesium alloy substrate is shown in Fig. 7.6.

The role of chemical cleaning (pickling) in a dilute solution containing a mixture of hydrochloric and hydrofluoric acid was found to influence the coating coverage, morphology and corrosion resistance of stannate coating significantly (Elsentriecy *et al.*, 2007). Complete removal of the oxide film

7.6 Permanganate-phosphate and stannate coatings (Zucchi *et al.*, 2007).

on the magnesium alloy substrate by the pickling process was reported to facilitate the creation of virgin surface to promote uniform and dense coating on the surface. The better coverage with reduced defects in the resultant coatings was claimed to provide a better corrosion resistance than that observed on samples coated without pickling. Stannate coatings by potentiostatic condition (–1.1 V *vs.* SCE) was also reported to promote formation of good coatings, and in this case, too, a lower pH and higher temperature combination was found to provide a superior coverage and consequent higher corrosion resistance to the substrate (Elsentriecy *et al.*, 2008).

There has been a continued interest in developing conversion coatings on magnesium alloy substrates from molybdenum, vanadium, cerium, zirconium, lanthanum, and niobium containing salt solutions. Treatment of AZ63 magnesium alloy in solutions containing cerium chloride and hydrogen peroxide at ambient temperature was reported to result in a very thin film of the order of 0.2 μm, with mud cracks. SEM/EDS characterization revealed the presence of cerium oxide on the coated surface, and this conversion film was found to increase the pitting resistance of the alloy in chloride and sulfate environments (Dabala *et al.*, 2003). Pre-treatment in dilute HCl solutions was reported to improve the quality of cerium conversion coatings. The pre-treated/conversion-coated surfaces exhibited a nobler corrosion potential and a better corrosion resistance than that without the pre-treatment (Brunelli *et al.*, 2005).

The development of conversion coatings containing cerium, niobium and zirconium, both individually and in combination, was attempted by Ardelean *et al.* (2008). The electrolytes were based on nitrates, oxides and oxy-fluorides of the above elements. The authors reported that the long-term treatments in these solutions resulted in coatings that provided a nobler corrosion potential and a significant increase in corrosion resistance in dilute sodium sulfate solutions. These coatings were also claimed to have excellent adhesion to organic coatings. Similar benefits have been claimed for WE43 magnesium alloys coated with cerium, lanthanum and praseodymium based conversion coatings (Rudd *et al.*, 2000).

Cerium- and lanthanum-based conversion coatings on AZ31 magnesium alloy were investigated by Montemor *et al.* (2007). The films were reported to be constituted of cerium and lanthanum oxide/hydroxides, in addition to magnesium/aluminum hydroxides. Polarization studies in 0.005 M NaCl solution revealed the benefits of the conversion layers in terms of the corrosion resistance, as signified by the corrosion current density. However, contrary to the many other investigations on conversion films, more active corrosion potentials (than the uncoated substrate) for the coated surfaces have been reported (Fig. 7.7).

Conversion coatings on AZ61 magnesium alloy from slightly alkaline solutions containing various concentrations of sodium vanadate by immersion at different temperatures and durations were produced by Yang *et al.* (2007). It was reported that a concentration of 30 g/L of $NaVO_3$ was the optimum concentration for achieving an effective coating, and coatings produced at 80°C with gave the best corrosion resistance.

The available literature on the rare-earth and allied conversion coatings seems to be limited, and the claims of various research attempts on these coatings still require substantial validation with more systematic work. Nevertheless, the drive towards environmentally cleaner coating technologies is a welcome move and this could eventually end up in the elimination of toxic coatings from the industrial world.

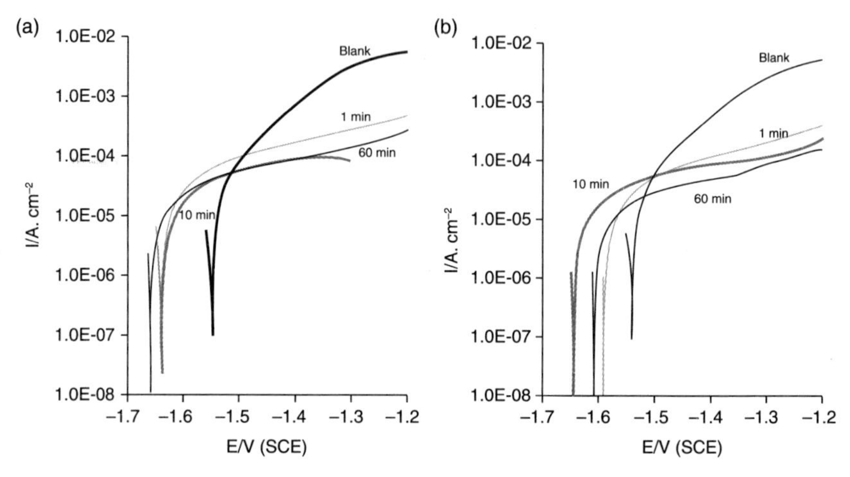

7.7 Polarization behavior of (a) cerium and (b) lanthanum conversion coated AZ31 magnesium alloy in 0.005 M NaCl solution (Montemor *et al.*, 2007).

Beside this overview of scientific research activities, there are presently a couple of successful chrome-free industrial conversion coatings available. However, their stand-alone performance is hard to judge as most tests performed are on combined conversion coating–paint systems. Pre-treatments appear to be more demanding, but they can reach at least the same performance compared to the chrome-containing conversion coatings. Magpass (AIMT, 2011), Gardobond (Chemetall, 2011), Alodine (Henkel, 2011) are some commercial conversion coating systems that work very well with magnesium alloys.

7.3.2 Plasma electrolytic oxidation

Anodic oxidation is an electrochemical process by which the surfaces of light alloys *viz.*, aluminum and magnesium, can be converted into a protective oxide film for improved corrosion resistance. The anodic oxidation of magnesium alloys can be grouped in two *viz.*, low-voltage processes and high-voltage processes. In the former case, the natural passive film growth and its characteristics on the magnesium alloy are governed by the processing voltage in appropriate electrolytes. The latter are called by different names *viz.*, plasma electrolytic oxidation (PEO), micro-arc oxidation (MAO), or plasma-chemical oxidation (PCO) processes, and involve very high voltages, well above the dielectric breakdown potentials of the passive films formed on magnesium alloys. The discharges that happen on the surface/electrolyte interface produce incipient melting and oxidation of magnesium alloy to form oxides in aqueous electrolytes (Yerokhin *et al.*, 1999; Blawert *et al.*, 2006). The species from the electrolyte also takes part in the reactions, forming complex compounds in addition to the magnesium oxide in the coatings. Thus, a ceramic coating is formed on the surface, with a high hardness compared to the soft magnesium alloy substrate, providing the twin benefits of wear and corrosion resistance. The following section gives a brief overview of the plasma electrolytic process and its variants.

The process of PEO of magnesium alloys, which is usually carried out in alkaline electrolytes, is reported to happen in four stages *viz.*, (a) the dissolution of magnesium alloy substrate with the associated passive film formation, (b) breakdown of the passive film, and the onset of fine discharges/sparks on the surface/electrolyte interface, (c) increase in the intensity of sparking with a steady increase in the process voltage and (d) the last stage, where the increase in voltage is marginal, with large sized relatively long-lived sparks of low density (Bala Srinivasan *et al.*, 2010). The breakdown potential of the passive film is governed more by the electrolyte chemistry, pH and conductivity. On the other hand, final processing voltage is a function of the chemical composition of the alloy, electrolyte composition, pH, power source, applied current density and treatment time. Due to the inherent processing features, especially the discharge intensity/density, the PEO coatings are porous, and pore size and morphology are dictated by the combined effect of processing conditions. Typical morphologies of a magnesium alloy surface PEO coated in silicate and phosphate-based electrolytes are shown Fig. 7.8 (Liang *et al.*, 2009).

Depending on the end application, the pores in the coating may be sealed with an appropriate sealing technique for improving the corrosion resistance (Tan *et al.*, 2005; Duan *et al.*, 2006; Malayoglu *et al.*, 2010). For tribological applications, the pores in the PEO coatings can be advantageously used as impregnating sites to hold lubricants, for improved lubricity.

Most of the alkaline electrolytes that are used for PEO processing contain silicate or phosphate as the major constituent (Cai *et al.*, 2006; Liang *et al.*, 2007; Arrabal *et al.*, 2008) and a variety of additives have been attempted by researchers for modifying the coating chemistry with a view to achieving improved tribological and corrosion performance (Liang *et al.*, 2005; Duan *et al.*, 2007; Chen *et al.*, 2010). The thickness of PEO coatings on magnesium alloys can vary from 5 to 200 μm, depending on various parameters including the chemical composition of the alloy. The pores that are developed in the PEO coatings in most cases spherical with mean diameter from <1 μm to about 50 μm. Many researchers claim the presence of a thin barrier/inner layer next to the substrate in the PEO coatings, and have reported a significant beneficial influence on the corrosion resistance of the coated magnesium alloys. It is not always necessary that the corrosion resistance would increase with increase in PEO coating thickness. It has been reported that

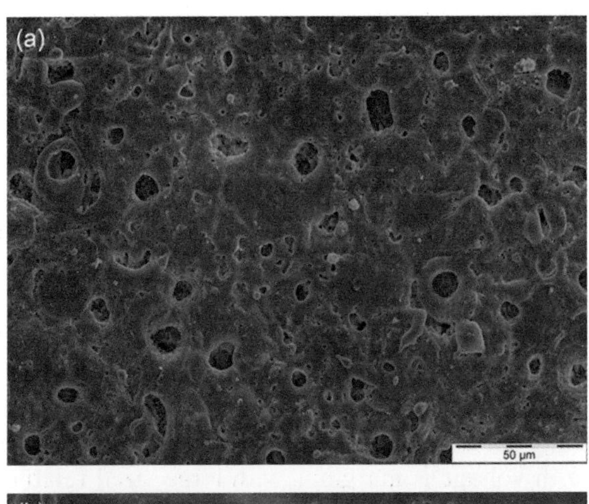

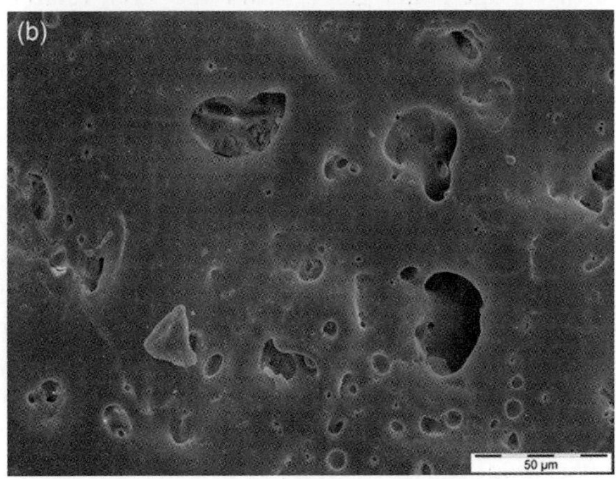

7.8 Scanning electron micrographs showing the surface morphology of PEO coatings on AM50 magnesium alloy obtained from (a) silicate and (b) phosphate based electrolytes (Liang *et al.*, 2009).

thick coatings produced at higher current densities have resulted in an inferior corrosion resistance compared to thin coatings, on account of the higher degree of defect levels, *viz*., larger sized pores and numerous interconnecting micro-cracks in the coating (Bala Srinivasan *et al.*, 2009). Figure 7.9 shows the typical cross-section micrographs revealing the thickness and features of PEO coatings from silicate- and phosphate-based electrolytes (Bala Srinivasan *et al.*, 2009). Arrows in the micrograph (b) indicates the cracks and pore channels in the coating produced at high current density.

The phase composition of the PEO coating is influenced by the chemical composition of the alloy and chemistry of electrolyte. Magnesium oxide, magnesium silicate and magnesium phosphate are the most common phases in PEO coatings, as most of the coatings are obtained from Si

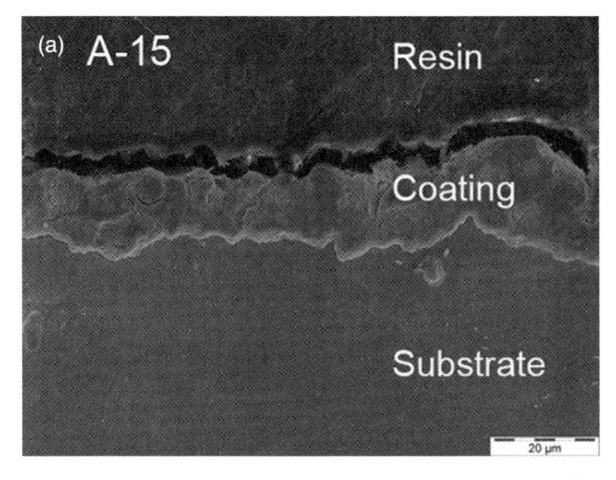

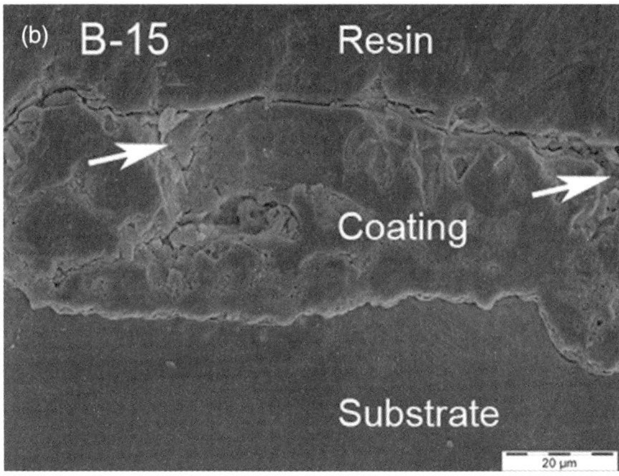

7.9 Micrographs showing the cross-section of silicate PEO coatings on AM50 magnesium alloy obtained at two different current densities in a 15 min treatment (Bala Srinivasan *et al.*, 2009).

and P electrolytes. However, coatings produced from zirconate, titania-sol containing electrolytes, were reported to be constituted with ZrO_2, TiO_2 in addition to the MgO and MgF_2 phases (Liang *et al.*, 2009a; Mu and Han, 2008; Luo *et al.*, 2009; Liang *et al.*, 2007). Coatings containing nano-ZrO_2 particles have also been produced successfully, demonstrating significant benefits for the corrosion resistance (Arrabal *et al.*, 2008a; Lee *et al.*, 2011). Addition of aluminate and fluoride to the alkaline PEO electrolytes facilitates faster growth of the coating and gives rise to the formation of magnesium aluminate and magnesium fluoride phases, which are said to be beneficial for improving the corrosion resistance of magnesium alloys.

The PEO-coated surfaces are hard compared to the substrate, and the hardness is dictated by the constituents (ceramic oxides/compounds) and the compactness of the coating. Low load hardness measurement techniques have been proved to be useful for understanding the properties of the coating, and that the hardness of PEO coatings can be as high as 650 HV. Coatings from silicate-based electrolytes have been reported to be harder than those from phosphate-based electrolytes. The roles of additives *viz*., tungstate, etc., have also been investigated, and beneficial effects on the hardness and the corrosion resistance have also been reported (Ding *et al*., 2007).

Harder magnesium alloy surfaces with the presence of ceramic compounds/oxides from the PEO process give the material improved resistance against sliding/ abrasive wear (Goretta *et al*., 2007; Bala Srinivasan *et al*., 2009a). Interestingly, the dry sliding-friction coefficient of the PEO-coated surfaces against AISI 52100 steel and silicon nitride balls have been reported to be very high compared to that for an uncoated magnesium alloy substrate. Nevertheless, the PEO coatings offer an excellent wear resistance (Liang *et al*., 2007; Bala Srinivasan *et al*., 2009a, 2009). In general, a hard, thick, compact coating provides superior wear resistance to thin coatings. In general, PEO coatings with more defects are likely to crumble under stress during sliding, resulting in high wear and may even have an additional adverse effect on the wear partner.

The corrosion resistance of the PEO coatings is better than that of the conversion coatings, essentially on account of the relatively thick coating and some of the stable phases in the coating. However, PEO coatings are suitable for situations where the environment is mild, and for short term protection. The corrosion resistance of the PEO-coated magnesium alloy substrate is dictated by (a) the chemical composition of the coating and (b) thickness and quality of coating. Electrochemical corrosion tests have been widely performed to understand the degradation of the coatings, and also to assess the relative performance of the PEO coatings. Liang *et al*. (2009; 2009a; 2010) have studied the silicate-, phosphate- and zirconate-based coatings for their short- and long-term corrosion behavior in neutral, acidic and alkaline chloride solutions. The stability of the PEO coatings and their degradation behavior have been documented (Liang *et al*., 2009, 2010). It is possible that some coatings may offer an excellent short term corrosion resistance, and that it may not give the same benefit in long-term exposure. The converse is true with some coatings, an example of this being the ZrO_2 coatings reported by Liang *et al*. (Liang *et al*., 2009a).

The corrosion resistance of some of the commercially available coatings, as claimed by the manufacturers, is very impressive. While the Magoxid PEO coatings were reported to withstand salt spray tests in accordance with DIN EN ISO 9227 for 80–100 h (AIMT, 2011), the Keronite and Anomag coatings have been claimed to withstand 1000 h in accordance with ASTM B117 tests (Ltd. Keronite, 2011).

7.3.3 Organic coatings

The main benefit of organic coatings in corrosion protection is their ability to prevent or at least to retard the access of electrolyte and oxygen to the metal surface. They can be used for permanent protection (e.g., paint, polymer, etc in combination with chemical conversion or anodized coatings, the latter usually offering excellent corrosion protection (Umehara *et al.*, 1999), or they are used, for example in the form of waxes and oil, as a temporary protection. The pre-treatment removes the natural outer corrosion layers and replaces it with a thin film firmly bound to the metal surface, improving the adhesion of the paint (Allsebrook, 1955). In any case, the selected organic compound has to be resistant to the highly alkaline reaction products forming during magnesium corrosion (Murray and Hills, 1990). However, the performance of organic paint systems depends on the corrosion resistance of the substrate alloy as well – the better the corrosion resistance of the alloy, the better the performance of the coating system (Wray, 1941).

State of the art for corrosion protection of magnesium alloys is usually a combination of conversion coating and paint. The paint can consist of several layers with different functions, for example, primer, filler and top coat. For applications which are not in the direct field of view normally a system of conversion coating, KTL and powder coat is used (Fig. 7.10). The corrosion performance is reported to be a >1000 h salt spray test DIN EN ISO 9227 or >10 cycles of VDA alternating corrosion test. Such a coating system can protect magnesium even if it is in contact with other metals. Well known examples are the tailgate from the 3-liter VW Lupo (Schreckenberger and Laudien, 2000) or from the recent Daimler E-class (Schreckenberger *et al.*, 2010). In the case of the Lupo, the multilayer protection system consists of yellow chromate pickling, KTL (20 microns) and

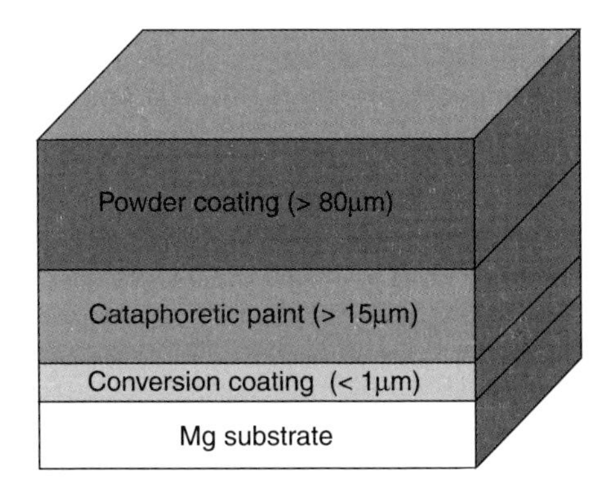

7.10 Typical coating system for magnesium alloys in automotive applications according to (Knoll, 2008).

powder coating (> 80 microns). This is similar to the protection system of the E-Class (Schreckenberger *et al.*, 2010). The application of such an automotive coating system generally combines multiple steps, such as degreasing/ acid pickling – AZ91/HNO_3 and AM50, AMZXX, AE44/ H_2SO_4 –/ Mn–Zr conversion treatment/ drying/ cooling/ KTL > 15 μm/ drying/powder coat >80 μm/ drying) (Knoll, 2008). Additionally, coating of the counterparts and/or constructive measures can be used to prevent galvanic corrosion between magnesium and the other more noble construction materials.

Apart from these industrial coating systems new polymer coatings (Yfantis *et al.*, 2002; Scharnagl *et al.*, 2009; Shao *et al.*, 2009; Conceicao *et al.*, 2010) with and without inhibitors are being developed and studied, mainly with the ultimate goal of creating coatings with self-healing ability.

7.3.4 Galvanic coatings

Magnesium can be coated electroless (without external current) with nickel-based, copper- or zinc-alloy coatings after the surface has been pre-treated accordingly. For magnesium alloys the standard coating bath compositions used for other metals were sometimes modified to the needs of the magnesium alloys; particularly acidic baths are changed to alkaline baths, reducing the attack of the magnesium substrate during coating formation, or acids/ions are used which do not attack or even protect magnesium to a great extent (e.g., HF/F^-). More information about the specific alloys and the pre-treatments can be found in the following selected references, which are far from complete – Ni–P (El Mahallawy *et al.*, 2008), Ni–W–P (Zhang *et al.*, 2007), Ni–Sn–P (Zhang *et al.*, 2008), Cu (DeLong, 1964; Yang *et al.*, 2006), Zn (DeLong, 1964). The formation of composite coatings reinforced with nano-particles is also possible (Song *et al.*, 2007). To improve corrosion resistance, sometimes multilayer Ni–P Coatings with different P content have been used (Gu *et al.*, 2005). The problem with harmful pre-treatments based on chromate or hydrofluoric acid is addressed by the use of organic interlayers on which the intended metal is electroless plated for example, Ni–P (Zhao *et al.*, 2007) or Ag (Zhao and Cui, 2007). Otherwise, with the use of Zn and/or Cu interlayers (Fig. 7.11) the galvanic deposition (electroplating) of nearly all metals is possible (DeLong, 1964; Haßenpflug, 2001).

The main application of galvanic metal layers is found in electronics and telecommunications. The problem with those layers is not only the high risk of galvanic-enhanced corrosion in the case of coating defects, but also the contamination and enrichment of secondary alloys especially with Cu and Ni during a recycling process if coated components are in the scrap fraction.

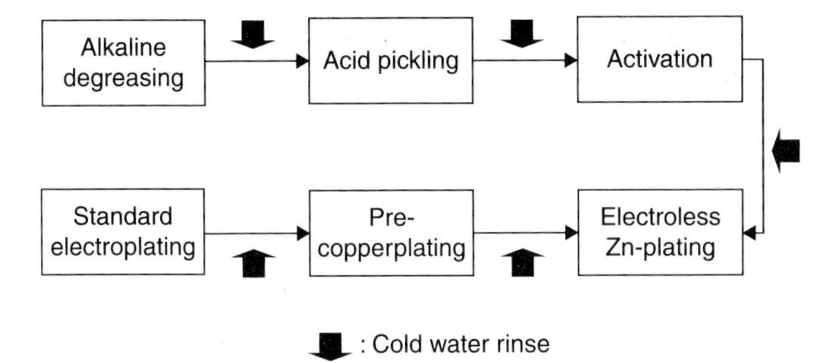

7.11 Typical treatment steps for the electroplating/galvanizing of magnesium alloys according to DeLong (1964).

7.3.5 Others

There are a couple of other coating technologies for magnesium alloys available. However, some of them are still under development, not industrially used or only interesting for niche products. Among them, sol–gel processes are the most promising technologies due to their ease of application and relative low investment cost. Corrosion and wear-resistant coatings are generated from organic–inorganic nano-composites via so-called cross-linking processes (Khramov *et al.*, 2006; Zaharescu *et al.*, 2009).

Physical methods, such as laser (Ignat *et al.*, 2004; Gao *et al.*, 2006; Mondal *et al.*, 2006), and electron beam (Hao *et al.*, 2005) technology, are mainly used for local surface engineering. By means of heat treatments, alloying, dispersing or coating the wear- and corrosion-resistance of magnesium surfaces can be modified. PVD (Lee *et al.*, 2003; Hoche *et al.*, 2005; Bohne *et al.*, 2007; Blawert *et al.*, 2008) and CVD techniques (including vacuum (Fracassi *et al.*, 2003; Yamauchi *et al.*, 2007) and atmospheric pressure treatments) can be used also to deposit a variety of metallic, organic and ceramic coatings on magnesium. Interesting is the ability of Mg-based PVD coatings to provide cathodic protection for magnesium substrates (Bohne *et al.*, 2007). Due to the low load-bearing capacity of the magnesium substrate, substrate heating effects during deposition, or process inherent coating defects, the use of PVD and CVD coatings is still limited.

Thermal spraying is another option for cladding Mg alloys with more corrosion- and/or wear-resistant materials for example, Al (Zhong *et al.*, 2004). With the development of new techniques with reduced heat input, thermal spraying might also be interesting for coating magnesium alloys. Use of high kinetic energies, high-speed flame spraying, and especially the newly developed cold spraying, offers the opportunity to avoid the oxidation of sprayed material and substrate substantially and thus produce very dense and corrosion-resistant protective coatings. However, the future will show whether one of those processes will find widespread industrial application.

7.4 Implications for improving corrosion resistance and future trends

7.4.1 Metallurgical

To a certain level, the corrosion resistance of magnesium can be improved by alloying and controlling the microstructure. Most important is that bulk and/or surface contaminations with heavy metal impurities are prevented. However, due to the fact that magnesium is the most electrochemically active construction material, it has to be protected against corrosion in harsh environments, or in direct contact with other materials if exposure to electrolytes is expected.

7.4.2 Surface finishing

Nowadays, corrosion protection of magnesium alloys is not really a major concern, as there are a number of coating techniques available (Gray and Luan, 2002; Zhang and Chaoyun, 2010). Most of the time the corrosion protection system is built layer by layer, combining several different coating layers. Which of the possible protection systems or combinations is used depends mainly on the application, the aggressiveness of the surrounding, the material combination and costs (Schreckenberger *et al.*, 2010; Elektron, 2006; Murray and Hillis, 1990). Normally, only the absolutely necessary coating system is used as long as the requirements of the application are fulfilled and sufficient protection throughout the lifetime of the component is guaranteed.

Nevertheless, besides protection with coatings, constructive measures are also important to minimize the environmental impact on the component. This includes the selection of more electrochemically compatible material counterparts (coatings), optimized draining of water, electrical isolation (or at least a resistance as high as possible) and optimized anode to cathode area ratios.

The compatibility between magnesium and a second metal is determined by the potential difference between anode (Mg) and cathode, which should be as low as possible, and by the polarization resistance of the cathode, which should be maximized. The materials most compatible to Mg are aluminum alloys of the 5XXX and 6XXX series, due to the relatively low potential difference. Despite the high potential difference, 80Sn/20Zn coatings are also tolerated, because of their high polarization resistance (Skar and Albright, 2002). No galvanic corrosion of magnesium is caused by anodized $AlMg_3$ (Reinhold, 2000), which is a result of low potential difference and high resistance. Completely incompatible with magnesium alloy are steels, stainless steels, copper, nickel and copper-containing aluminum alloys, such as the

A380 alloy (Skar and Albright, 2002). However, in spite of coating the Mg part, one should always consider coating the counterpart as well, especially if there are no alternatives for a more compatible material selection. Detailed studies of the galvanic corrosion between AZ91 (steering gear from cast magnesium) and typical attachments (e.g., retainer, threaded sockets, plugs, screws) with various types of coatings, revealed that in this case the best protection was achieved by a zinc coating plus cathodic e-coat (KTL, 15 microns) (Boese *et al.*, 2001). Conventional galvanized-steel screws can be used with magnesium silicate sealers. The silicate sealer was used successfully for galvanized-steel bolts for the B80 magnesium gearbox of Audi and VW in the series [4]. Insulating polymer coatings are also used (nylon) or plastic caps for the screw heads (Skar and Albright, 2002) (Magnesium).

Another important aspect is the fact that magnesium alloys will not have good corrosion resistance unless they are free from surface contamination by flux or metallic impurities introduced during forging, rolling, casting or other production processes (Allsebrook, 1955; Lafront *et al.*, 2008; Blawert *et al.*, 2005; Nwaogu *et al.*, 2009; Nwaogu *et al.*, 2010). This is true, even if magnesium is coated. On the one hand, the build-up of coating systems can be severely affected by surface contaminations (growth, uniformity and adhesion) and on the other hand most of the coating systems are not completely dense. Under such circumstances moisture can penetrate the coating, leading to reactions in the interface, causing degradation followed by final failure of coating systems.

7.5 Conclusions

In summary, it can be said that there are satisfactory solutions for the corrosion protection of magnesium available. Still difficult to protect are the mixed-material combinations produced by welding, riveting and clinching. After joining the materials cannot be separated anymore and require the same surface treatment. Most of the available processes (especially pre-treatments) are not suitable for both materials, due to their different electrochemical responses. Some solution might be offered by PEO processing, at least for Mg/Al and Mg/Ti combinations.

Another concern of the last decade, the development of chrome-free conversion coatings is at least partly solved (Schreckenberger, 2001). New alternatives with similar properties are available on the industrial scale (Kurze, 2008), but the pre-treatment prior to conversion coating is more important now to guarantee a uniform formation of the conversion coating. The new alternatives are more demanding regarding the surface conditions for example, a better cleaning removing impurities and processing lubricants is required compared to chromate treatments. The ultimate goal for corrosion protection of magnesium – self-healing coatings – is still not solved.

A number of concepts have been developed, and partly tested successfully in the laboratory scale, but a transfer into industrial application has not yet been achieved.

7.6 References

AIMT. 2011. *Homepage* (Online). Available: http://www.ahc-surface.com/en/surface-treatment/processes/magpass-coat (Accessed).

Allsebrook, W. E. 1955. The coating of magnesium alloys. *Corrosion Technology*, **2**, 113–116.

Amini, R. and Sarabi, A. 2011. The corrosion properties of phosphate coating on AZ31 magnesium alloy: The effect of sodium dodecyl sulfate (SDS) as an eco-friendly accelerating agent. *Applied Surface Science*, **257**, 7134–7139.

Ardelean, H., Frateur, I. and Marcus, P. 2008. Corrosion protection of magnesium alloys by cerium, zirconium and niobium-based conversion coatings. *Corrosion Science*, **50**, 1907–1918.

Arrabal, R., Matykina, E., Skeldon, P. and Thompson, G. 2008a. Incorporation of zirconia particles into coatings formed on magnesium by plasma electrolytic oxidation. *Journal of Materials Science*, **43**, 1532–1538.

Arrabal, R., Matykina, E., Viejo, F., Skeldon, P. and Thompson, G. 2008. Corrosion resistance of WE43 and AZ91D magnesium alloys with phosphate PEO coatings. *Corrosion Science*, **50**, 1744–1752.

Atrens, A. and Dietzel, W. 2007. The negative difference effect and unipositive Mg^+. *Advanced engineering materials*, **9**, 292–297.

Bala Srinivasan, P., Blawert, C. and Dietzel, W. 2009a. Dry sliding wear behaviour of plasma electrolytic oxidation coated AZ91 cast magnesium alloy. *Wear*, **266**, 1241–1247.

Bala Srinivasan, P., Blawert, C., Störmer, M. and Dietzel, W. 2010. Characterisation of tribological and corrosion behaviour of plasma electrolytic oxidation coated AM50 magnesium alloy. *Surface Engineering*, **26**, 340–346.

Bala Srinivasan, P., Liang, J., Blawert, C., Störmer, M. and Dietzel, W. 2009. Effect of current density on the microstructure and corrosion behaviour of plasma electrolytic oxidation treated AM50 magnesium alloy. *Applied Surface Science*, **255**, 4212–4218.

Bender, S. 2010. *Eine neue Theorie zum negativen Differenzeffekt bei der Magnesiumkorrosion.* Dissertation, Otto-von-Guericke-Universität Magdeburg.

Bender, S., Goellner, J., Heyn, A. and Boese, E. 2007a. Corrosion and corrosion testing of magnesium alloys. *Materials and Corrosion*, **58**, 977–982.

Bender, S., Goellner, J., Heyn, A. and Schultze, S. 2007b. Application of material specific testing on magnesium alloys using electrochemical noise. *Corrosion*, NACE Paper No. 07372, 1–12.

Bender, S., Goellner, J., Heyn, A. and Schultze, S. 2008. Investigations on defined pretreated magnesium alloys by means of electrochemical noise. *Corrosion*, NACE Paper No. 08400, 08400/1–17.

Bender, S., Goellner, J. and Atrens, A. 2008. Corrosion of AZ91 in 1N NaCl and the mechanism of magnesium corrosion. *Advanced Engineering Materials*, **10**, 583–587.

Blawert, C., Dietzel, W., Ghali, E. and Song, G. 2006. Anodizing treatments for magnesium alloys and their effect on corrosion resistance in various environments. *Advanced Engineering Materials*, **8**, 511–533.

Blawert, C., Heitmann, V., Morales, E., Dietzel, W., Jin, S. and Ghal, E. 2005. Corrosion properties of the skin and bulk of semisolid processed and high pressure die cast AZ91 alloy. *Canadian Metallurgical Quarterly*, **44**, 137–146.

Blawert, C., Manova, D., Störmer, M., Gerlach, J. W., Dietzel, W. and Mändl, S. 2008. Correlation between texture and corrosion properties of magnesium coatings produced by PVD. *Surface and Coatings Technology*, **202**, 2236–2240.

Boese, E., Göllner, J., Heyn, A., Strunz, J., Baierl, Chr. and Schreckenberger, H. 2001. Kontaktkorrosion einer Magnesiumlegierung mit beschichteten Bauteilen [Galvanic corrosion behaviour of magnesium alloy in contact with coated components]. *Materials and Corrosion*, **52**, 247–256.

Bohne, Y., Manova, D., Blawert, C., Störmer, M., Dietzel, W. and Mändl, S. 2007. Influence of ion energy on morphology and corrosion properties of Mg alloys formed by energetic PVD processes. *Nuclear Instruments and Methods in Physics Research Section B: Beam Interactions with Materials and Atoms*, **257**, 392–396.

Bohne, Y., Manova, D., Blawert, C., Störmer, M., Dietzel, W. and Mändl, S. 2007. Deposition and properties of novel microcrystalline Mg alloy coatings. *Surface Engineering*, **23**, 339–343.

Brunelli, K., Dabala, M., Calliari, I. and Magrini, M. 2005. Effect of HCl pre-treatment on corrosion resistance of cerium-based conversion coatings on magnesium and magnesium alloys. *Corrosion Science*, **47**, 989–1000.

Cai, Q., Wang, L., Wei, B. and Liu, Q. 2006. Electrochemical performance of micro-arc oxidation films formed on AZ91D magnesium alloy in silicate and phosphate electrolytes. *Surface and Coatings Technology*, **200**, 3727–3733.

Chemetall. 2011. *Homepage* (Online). Available: www.chemetall.com (Accessed).

Chen, H., Lv, G., Zhang, G., Pang, H., Wang, X., Lee, H. and Yang, S. 2010. Corrosion performance of plasma electrolytic oxidized AZ31 magnesium alloy in silicate solutions with different additives. *Surface and Coatings Technology*, **205**, S32–S35.

Conceicao, T. F., Scharnagl, N., Blawert, C., Dietzel, W. and Kainer, K. U. 2010. Corrosion protection of magnesium alloy AZ31 sheets by spin coating process with poly(ether imide) [PEI]. *Corrosion Science*, **52**, 2066–2079.

Dabala, M., Brunelli, K., Napolitani, E. and Magrini, M. 2003. Cerium-based chemical conversion coating on AZ63 magnesium alloy. *Surface and Coatings Technology*, **172**, 227–232.

Delong, H. K. and Gross, W. H. 1964. Magnesium. *In:* H.W. Dettner, J. Elze (ed.) *Handbuch der Galvanotechnik*. München: Carl Hanser Verlag.

DIN-88 1988. DIN 50021, Sprühnebelprüfungen mit verschiedenen Natriumchlorid-Lösungen, 1988.

Ding, J., Liang, J., Li, T., Hao, J. and Xue, Q. 2007. Effects of sodium tungstate on characteristics of microarc oxidation coatings formed on magnesium alloy in silicate-KOH electrolyte. *Transactions of Metals Society of China*, **17**, 244–249.

Duan, H., Du, K., Yan, C. and Wang, F. 2006. Electrochemical corrosion behavior of composite coatings of sealed MAO film on magnesium alloy AZ91D. *Electrochemica Acta*, **51**, 2898–2908.

Duan, H., Yan, C. and Wang, F. 2007. Effect of electrolyte additives on performance of plasma electrolytic oxidation films formed on magnesium alloy AZ91D. *Electrochimica Acta*, **52**, 3785–3793.

El Mahallawy, N., Bakkar, A., Shoeib, M., Palkowski, H. and Neubert, V. 2008. Electroless Ni-P coating of different magnesium alloys. *Surface and Coatings Technology*, **202**, 5151–5157.

Elektron, M. 2006. Surface treatments for magnesium alloys in aerospace and defence. *In:* REPORT, Elekton Magnesium – *Datasheet 256.*

Elektron, M. 2011. *Surface treatments for magnesium alloys in aerospace and defence* (Online). Available: www.magnesium-elektron.com (Accessed).

Elsentriecy, H., Azumi, K. and Konno, H. 2007. Effect of surface pre-treatment by acid pickling on the density of stannate conversion coatings formed on AZ91 D magnesium alloy. *Surface and Coatings Technology*, **202**, 532–537.

Elsentriecy, H., Azumi, K. and Konno, H. 2008. Effects of pH and temperature on the deposition properties of stannate chemical conversion coatings formed by the potentiostatic technique on AZ91 D magnesium alloy. *Electrochimica Acta*, **53**, 4267–4275.

Fracassi, F., D'Agostino, R., Palumbo, F., Angelini, E., Grassini, S. and Rosalbino, F. 2003. Application of plasma deposited organosilicon thin films for the corrosion protection of metals. *Surface and Coatings Technology*, **174–175**, 107–111.

Gao, Y., Wang, C., Lin, Q., Liu, H. and Yao, M. 2006. Broad-beam laser cladding of Al-Si alloy coating on AZ91HP magnesium alloy. *Surface and Coatings Technology*, **201**, 2701–2706.

Ghali, E. 2000. Magnesium and magnesium alloys. *Uhlig's Corrosion Handbook*, 793–830.

Goretta, K., Cunningham, A., Chen, N., Singh, D., Routbort, J. and Rateick, J. R. 2007. Solid-particle erosion of an anodized Mg alloy. *Wear*, **262**, 1056–1060.

Gray, J. E. and Luan, B. 2002. Protective coatings on magnesium and its alloys – a critical review. *Journal of Alloys and Compounds*, **336**, 88–113.

Grigull, U. 1963. Das Prinzip von Le Chatelier und Braun. *International Journal of Heat and Mass Transfer*, **7**, 23–31.

Gu, C., Lian, J. and Jiang, Z. 2005. Multilayer Ni-P coating for improving the corrosion resistance of AZ91D magnesium alloy. *Advanced Engineering Materials*, **7**, 1032–1036.

Gulbrandsen, E. 1992. Anodic behaviour of Mg in HCO-3/CO2–3 buffer solutions. Quasi-steady measurements. *Electrochimica Acta*, **37**, 1403–1412.

Hanawalt, J. D., Nelson, C. E. and Peloubet, J. A. 1942. Corrosion studies of magnesium and its alloys. *Transactions of the American Institute Mining Metals Engineering*, **147**, 273–299.

Hao, S., Gao, B., Wu, A., Zou, J., Qin, Y., Dong, C., An, J. and Guan, Q. 2005. Surface modification of steels and magnesium alloy by high current pulsed electron beam. *Nuclear Instruments and Methods in Physics Research Section B: Beam Interactions with Materials and Atoms*, **240**, 646–652.

Haßenpflug, C. 2001. Galvanisieren von Magnesiumdruckguss – Eigenschaften und Anwendungen. *In:* Leichtmetallanwendungen – Neue Entwicklungen in der Oberflächentechnik, 2001 Münster. DFO, 144–152.

Heller, F. 1960. Method of forming a chromate conversion coating on magnesium.

Henkel. 2011. *Homepage* (Online). Available: www.henkelna.com (Accessed).

Hoche, H., Blawert, C., Broszeit, E. and Berger, C. 2005. Galvanic corrosion properties of differently PVD-treated magnesium die cast alloy AZ91. *Surface and Coatings Technology*, **193**, 223–229.

Hydro Magnesium, Korrosionsschutz und Oberflächenbehandlung von Magnesiumlegierungen, Hydro Magnesium, 34358 Hydro Media Porsgrunn 02.02.

Ignat, S., Sallamand, P., Grevey, D. and Lambertin, M. 2004. Magnesium alloys laser (Nd:YAG) cladding and alloying with side injection of aluminium powder. *Applied Surface Science*, **225**, 124–134.

Jones, D. A. 1996. *Principles and prevention of corrosion,* New York, Macmillan Publishing Company.

Keronite International Ltd. 2011. Available: http://www.keronite.com (Accessed).

Khramov, A. N., Balbyshev, V. N., Kasten, L. S. and Mantz, R. A. 2006. Sol-gel coatings with phosphonate functionalities for surface modification of magnesium alloys. *Thin Solid Films*, **514**, 174–181.

King, P. F. 1966. The role of the anion in the anodic dissolution of magnesium. *Journal of Electrochemical Society*, **113**, 536–539.

Kisza, A., Kazmierczak, J., Borresen, B., Haarberg, G. M. and Tunold, R. 1995. Kinetics and mechanism of the magnesium electrode reaction in molten magnesium chloride. *Journal of Applied Electrochemistry*, **25**, 940–946.

Knoll, E. 2008. Beschichtung von magnesiumbauteilen für die automobilindustrie – praxisbericht aus sicht des lackierers-. *16. Magnesium Abnehmer-und Automotive Seminar.* Aalen.

Kurze, P. and Leyendecker, F. 2008. Surface treatment of magnesium materials. In: 16. Magnesium Automotive and User Seminar 2008. European Research Association for Magnesium.

Lafront, A. M., Dube, D., Tremblay, R., Ghali, E., Blawert, C. and Dietzel, W. 2008. Corrosion resistance of the skin and bulk of die cast and thixocast Az91d alloy in Cl- solution using electrochemical techniques. *Canadian Metallurgical Quarterly*, **47**, 459–468.

Lee, K., Shin, K., Namgung, S., Yoo, B. and Shin, D. 2011. Electrochemical response of ZrO2-incorporated oxide layer on AZ91 Mg alloy processed by plasma electrolytic oxidation. *Surface and Coatings Technology*, **205**, 3779–3784.

Lee, M. H., Bae, I. Y., Kim, K. J., Moon, K. M. and Oki, T. 2003. Formation mechanism of new corrosion resistance magnesium thin films by PVD method. *Surface and Coatings Technology*, **169**–170, 670–674.

Li, G., Lian, J., Niu, L., Jiang, Z. and Jiang, Q. 2006. Growth of zinc phosphate coatings on AZ91D magnesium alloy. *Surface and Coatings Technology*, **201**, 1814–1820.

Liang, J., Bala Srinivasan, P., Blawert, C. and Dietzel, W. 2009a. Comparison of electrochemical corrosion behaviour of MgO and ZrO2 coatings on AM50 magnesium alloy formed by plasma electrolytic oxidation. *Corrosion Science*, **51**, 2483–2492.

Liang, J., Bala Srinivasan, P., Blawert, C. and Dietzel, W. 2010. Influence of chloride ion concentration on the electrochemical corrosion behaviour of plasma electrolytic oxidation coated AM50 magnesium alloy. *Electrochimica Acta*, **55**, 6802–6811.

Liang, J., Bala Srinivasan, P., Blawert, C., Störmer, M. and Dietzel, W. 2009. Electrochemical corrosion behaviour of plasma electrolytic oxidation coatings on AM50 magnesium alloy formed in silicate and phosphate based electrolytes. *Electrochimica Acta*, **54**, 3842–3850.

Liang, J., Guo, B., Tian, J., Liu, H., Zhou, J. and Xu, T. 2005. Effect of potassium fluoride in electrolytic solution on the structure and properties of microarc oxidation coatings on magnesium alloy. *Applied Surface Science*, **252**, 345–351.

Liang, J., Hu, L. and Hao, J. 2007. Characterization of micro-arc oxidation coatings formed on AM60B magnesium alloy in silicate and phosphate electrolytes. *Applied Surface Science*, **253**, 4490–4496.

Lin, C., Lin, H., Lin, K. and Lai, W. 2006. Formation and properties of stannate conversion coatings on AZ61 magnesium alloys. *Corrosion Science*, **48**, 93–109.

Luo, H., Cai, Q., Wei, B., Yu, B., He, J. and Li, D. 2009. Study on the microstructure and corrosion resistance of ZrO2-containing ceramic coatings formed on magnesium alloy by plasma electrolytic oxidation. *Journal of Alloys and Compounds*, **474**, 551–556.

Makar, G. L. and Kruger, J. 1990. Corrosion studies of rapidly solidified magnesium alloys. *Journal of the Electrochemical Society*, **137**, 412–421.

Makar, G. L., Kruger, J. and Joshi, A. 1988. The effect of alloying elements on the corrosion resistance of rapidly solidified magnesium alloys. In: *Proceedings Annual Meeting of the Minerals, Metals and Materials Society, Advances in Magnesium Alloys and Composites, 26.01.1988*, Phoenix, Arizona. 105–121.

Makoto, D., Mitsuo, S. and Susumu, T. 1994. Coating pre-treatment and coating method for magnesium alloy product, *JP6116739A2*.

Malayoglu, U., Tekin, K. and Shrestha, S. 2010. Influence of post-treatment on the corrosion resistance of PEO coated AM50B and AM60B Mg alloys. *Surface and Coatings Technology*, **205**, 1793–1798.

Mondal, A. K., Kumar, S. Blawert, C. and Dahotre, N. B. 2006. Effect of laser surface treatment on microstructure and properties of MRI 230D Mg alloy. *In:* T. Chandra, K. T., M. Militzer, C. Ravindran, (eds). THERMEC 2006, 2006 Vancouver. TTP, 1153–1158.

Montemor, M., Simoes, A. and Carmezim, M. 2007. Characterization of rare-earth conversion films formed on the AZ31 magnesium alloy and its relation with corrosion protection. *Applied Surface Science*, **253**, 6922–6931.

Mu, W. and Han, Y. 2008. Characterization and properties of the MgF_2/ZrO_2 composite coatings on magnesium prepared by micro-arc oxidation. *Surface and Coatings Technology*, **202**, 4278–4284.

Murray, R. W. and Hillis, J. E. 1990. Magnesium finishing: chemical treatment and coating practices. In: *SAE International Congress and Exposition*, Detroit, US. SAE.

Naohiro, U., Yoshiaki, K., Yukio, N., Yoshihiko, N., Yoshinori, S. and Takeshi, F. 1999. Surface treated magnesium or magnesium alloy product, primary treatment for coating and coating method, JP11323571A2.

Niu, L., Jiang, Z., Li, G., Gu, C. and Lian, J. 2006. A study and application of zinc phosphate coating on AZ91D magnesium alloy. *Surface and Coatings Technology*, **200**, 3021–3026.

Nordlien, J. H., Kemal, N., Ono, S. and Masuko, N. 1996. Morphology and structure of oxide films formed on MgAl alloys by exposure to air and water. *Journal of the Electrochemical Society*, **143**, 2564–2572.

Nwaogu, U. C., Blawert, C., Scharnagl, N., Dietzel, W. and Kainer, K. U. 2009. Influence of inorganic acid pickling on the corrosion resistance of magnesium alloy AZ31 sheet. *Corrosion Science*, **51**, 2544–2556.

Nwaogu, U. C., Blawert, C., Scharnagl, N., Dietzel, W. and Kainer, K. U. 2010. Effects of organic acid pickling on the corrosion resistance of magnesium alloy AZ31 sheet. *Corrosion Science*, **52**, 2143–2154.

Petrova, L. M. and Krasnoyarskii, V. V. 1990. Corrosion-electrochemical behavior of binary magnesium yttrium alloys in neutral solutions. *Protection of Metals*, **26**, 633–635.

Pieper, C. 2005. *Korrosion und Oxidation von Magnesium-Legierungen.* Dissertation, Technische Universität Dortmund.

Pourbaix, M. 1974. *Atlas of Electrochemical Equilibria in Aqueous Solutions,* Houston, Texas, USA, National association of corrosion engineers.

Reinhold B. and Brettmann, M. 2000. Korrosionsschutz für magnesium im automobilbau. *Metalloberfläche*, **54**, 26–31.

Robinson, J. L. and King, P. F. 1961. Electrochemical behavior of the magnesium anode. *Journal of Electrochemical Society*, **108**, 36–41.

Rudd, A., Breslin, C. and Mansfeld, F. 2000. The corrosion protection afforded by rare earth conversion coatings applied to magnesium. *Corrosion Science*, **42**, 275–288.

Scharnagl, N., Blawert, C. and Dietzel, W. 2009. Corrosion protection of magnesium alloy AZ31 by coating with poly(ether imides) (PEI). *Surface and Coatings Technology*, **203**, 1423–1428.

Schreckenberger, H. 2001. *Korrosion und Korrosionsschutz von Magnesiumwerkstoffen für den Automobilbau – Problematik der Kontaktkorrosion,* Düsseldorf, VDI Verlag.

Schreckenberger, H., Izquierdo, P., Klose, S. G., Blawert, C., Heitmann, V., Höche, D. and Kainer, K. U. 2010. Vermeidung von Bimetallkorrosion – Systematische Entwicklung eines Magnesium Karosseriebauteils. Preventing galvanic corrosion – Systematic development of a magnesium car body component. *Materialwissenschaft und Werkstofftechnik*, **41**, 853–860.

Schreckenberger, H. and Laudien, G. 2000. Das Korrosionsschutzkonzept der A luminum-Magnesium-Hybridheckklappe des VW Lupo. *In:* Fortschritte mit Magnesium im Automobilbau, 2000 Bad Nauheim. 41–50.

Shao, Y., Huang, H., Zhang, T., Meng, G. and Wang, F. 2009. Corrosion protection of Mg-5Li alloy with epoxy coatings containing polyaniline. *Corrosion Science*, **51**, 2906–2915.

Simaranov, A., Marshakov, S. and Mikhailovskii, Y. 1992. The composition and protective properties of chromate conversion coatings on magnesium. *Protection of metals*, **28**, 576–580.

Skar, J. I. and Albright, D. 2002. Emerging trends in corrosion protection of magnesium die castings. *In:* Kaplan, H. I. (ed.) *TMS Annual Meeting.* Seattle: TMS.

Song, G. 2005. Recent progress in corrosion and protection of magnesium alloys. *Advanced Engineering Materials*, **7**, 563–586.

Song, G., Atrens, A. and St John, D. 2001. A hydrogen evolution method for the estimation of corrosion rate of magnesium alloys. *TMS*, 255–262.

Song, G. L. and Atrens, A. 1999. Corrosion mechanisms of magnesium alloys. *Advanced Engineering Materials*, **1**, 11–33.

Song, G. L. and Atrens, A. 2003. Understanding magnesium corrosion: A framework for improved alloy performance. *Advanced Engineering Materials*, **5**, 837–858.

Song, Y. W., Shan, D. Y. and Han, E. H. 2007. Comparative study on corrosion protection properties of electroless Ni-P-ZrO2 and Ni-P coatings on AZ91D magnesium alloy. *Materials and Corrosion*, **58**, 506–510.

Swiatowska, J., Volovitch, P. and Ogle, K. 2010. The anodic dissolution of Mg in NaCl and Na2SO4 electrolytes by atomic emission spectroelectrochemistry. *Corrosion Science*, **52**, 2372–2378.

Tan, A., Soutar, A., Annergren, I. and Liu, Y. 2005. Multilayer sol-gel coatings for corrosion protection of magnesium. *Surface and Coatings Technology*, **198**, 478–482.

Tunold, R., Holtan, H., Hägg Berge, M.-B., Lasson, A. and Steen-Hansen, R. 1977. The corrosion of magnesium in aqueous solution containing chloride ions. *Corrosion Science*, **17**, 353–365.

Umehara, H., Takaya, M. and Ito, T. 1999. Corrosion resistance of the die casting AZ91D magnesium alloys with paint finishing. *Aluminium*, **75**, 634–641.

Williams, G. and McMurray, H. N. 2008. Localized corrosion of magnesium in chloride-containing electrolyte studied by a scanning vibrating electrode technique. *Journal of the Electrochemical Society*, **155**, C340–C349.

Wray, R. I. 1941. Painting magnesium alloys. *Industrial and Engineering Chemistry*, **33**, 932–937.

Yamauchi, N., Ueda, N., Okamoto, A., Sone, T., Tsujikawa, M. and Oki, S. 2007. DLC coating on Mg-Li alloy. *Surface and Coatings Technology*, **201**, 4913–4918.

Yang, K., Ger, M., Hwu, W., Sung, Y. and Liu, Y. 2007. Study of vanadium-based chemical conversion coating on the corrosion resistance of magnesium alloy. *Materials Chemistry and Physics*, **101**, 480–485.

Yang, L., Luan, B. and Nagata, J. 2006. Novel copper immersion coating on magnesium alloy AZ91D in an alkaline bath. *Journal of Coatings Technology and Research*, **3**, 241–246.

Yerokhin, A. L., Nie, X., Leyland, A., Matthews, A. and Doweyc, S. J. 1999. Plasma electrolysis for surface engineering. *Surface Engineering*, **122**, 73–93.

Yfantis, A., Paloumpa, I., Schmeißer, D. and Yfantis, D. 2002. Novel corrosion-resistant films for Mg alloys. *Surface and Coatings Technology*, **151–152**, 400–404.

Zaharescu, M., Predoana, L., Barau, A., Raps, D., Gammel, F., Rosero-Navarro, N. C., Castro, Y., Durán, A. and Aparicio, M. 2009. SiO2 based hybrid inorganic-organic films doped with TiO2-CeO2 nanoparticles for corrosion protection of AA2024 and Mg-AZ31B alloys. *Corrosion Science*, **51**, 1998–2005.

Zhang, J. and Chaoyun, W. 2010. Corrosion and protection of magnesium alloys – a review of the patent literature. *Recent Patents on Corrosion Science*, **2**, 55–68.

Zhang, W. X., Huang, N., He, J. G., Jiang, Z. H., Jiang, Q. and Lian, J. S. 2007. Electroless deposition of Ni-W-P coating on AZ91D magnesium alloy. *Applied Surface Science*, **253**, 5116–5121.

Zhang, W. X., Jiang, Z. H., Li, G. Y., Jiang, Q. and Lian, J. S. 2008. Electroless Ni-Sn-P coating on AZ91D magnesium alloy and its corrosion resistance. *Surface and Coatings Technology*, **202**, 2570–2576.

Zhao, H. and Cui, J. 2007. Electroless plating of silver on AZ31 magnesium alloy substrate. *Surface and Coatings Technology*, **201**, 4512–4517.

Zhao, H., Huang, Z. and Cui, J. 2007. A new method for electroless Ni-P plating on AZ31 magnesium alloy. *Surface and Coatings Technology*, **202**, 133–139.

Zhao, M., Wu, S., Luo, J., Fukuda, Y. and Nakae, H. 2006. A chromium-free conversion coating of magnesium alloy by a phosphate-permanganate solution. *Surface and Coatings Technology*, **200**, 5407–5412.

Zhong, S. W., Liufa, L. and Wen, J. D. 2004. Al Arc Spray Coating on AZ31 Mg Alloy and Its Corrosion Behavior *In:* W. Ke, E. H. Han, Y.F. Han , K. Kainer and A.A.

Luo, (eds) *International Conference on Magnesium – Science, Technology and Applications*, 2004 Beijing, China 685–688.
Zucchi, F., Frignani, A., Grassi, V., Trabanelli, G. and Monticelli, C. 2007. Stannate and permanganate conversion coatings on AZ31 magnesium alloy. *Corrosion Science*, **49**, 4542–4552.

8
Applications: aerospace, automotive and other structural applications of magnesium

A. A. LUO, General Motors Global Research
& Development, USA

DOI: 10.1533/9780857097293.266

Abstract: This chapter discusses the material properties and mass saving potential of magnesium alloys in comparison with major structural materials: mild steel, advanced high-strength steel (AHSS), aluminum, polymers, and polymer composites. The alloy development and manufacturing processes of cast and wrought magnesium alloys are summarized. Structural applications of magnesium alloys in automotive, aerospace, and power tools industries are reviewed in this chapter. The opportunities and challenges of magnesium alloys for structural applications are discussed.

Key words: magnesium alloy development, magnesium casting, magnesium extrusion, magnesium sheet, structural applications.

8.1 Introduction

Magnesium, an alkaline earth metal, is the eighth most abundant element in the earth crust and the third most abundant element dissolved in seawater. The metal was first produced in England by Sir Humphry Davy in 1808 using electrolysis of a mixture of magnesia and mercury oxide.[1] Today, China produces about 80% of the world's magnesium output (about 800 000 metric tons in 2010), using the Pidgeon process, a silicothermic reduction process for extracting high purity magnesium crowns from dolomite and ferrosilicon.

According to the US Geological Survey,[2] the leading use (41%) of primary magnesium is as an alloying element in aluminum-based alloys for packaging, transportation and other applications. Structural uses of magnesium (castings and wrought products) accounted for about 32% of primary metal consumption in 2010. Desulfurization of iron and steel accounted for 13% of US consumption of primary metal, and other uses were 14%. Figure 8.1 shows the magnesium consumption by end use in the last decade and the forecast until 2015.[3]

Magnesium is the third most commonly used structural metal, following steel and aluminum. With its density about one-fourth that of steel and two-thirds

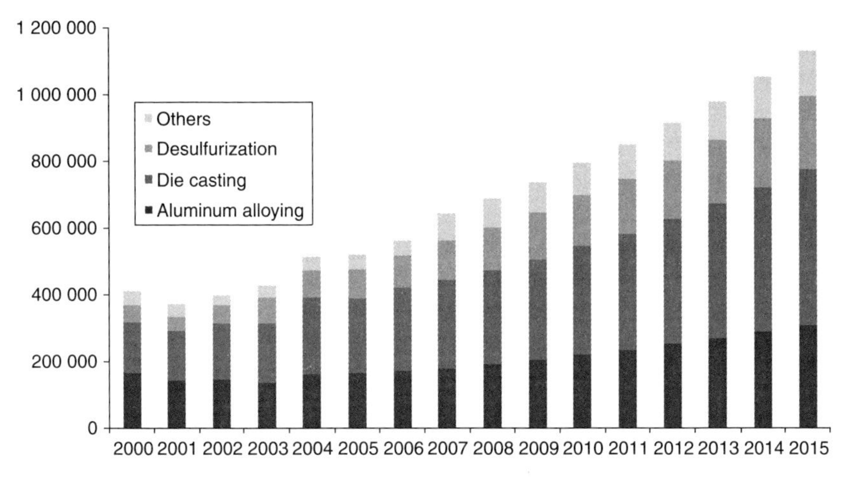

8.1 Magnesium consumption by end use in 2000–2015.[3]

that of aluminum, magnesium is the lightest structural metal, which offers significant opportunities for lightweight applications in automotive, aerospace, power tools, and 3C (computer, communication and consumer products) industries. For example, magnesium components are increasingly being used by major automotive companies including General Motors (GM), Ford, Volkswagen and Toyota.[4–11] Current major automotive magnesium applications include instrument-panel beam, transfer case, steering components and radiator support. However, the magnesium content in a typical family sedan built in North America is only about 0.3% of the total vehicle weight.[4]

In this chapter, the mechanical properties, structural performance and mass saving potential of cast and wrought magnesium alloys are compared with those of several major automotive materials: mild steel, AHSS, aluminum, polymers and polymer composites. Manufacturing processes, including welding and joining of magnesium castings and wrought products, are critically reviewed. This chapter will also review the historical, current and potential structural uses of magnesium, with a focus on automotive applications. The technical challenges of magnesium structural applications are also discussed.

8.2 Material properties

8.2.1 Property comparison with other automotive materials

Table 8.1 summarizes the mechanical and physical properties of typical cast and wrought magnesium alloys in comparison with other materials for automotive applications.[12–15] Being the lightest structural metal, magnesium has a density less than one-fourth that of ferrous alloys (cast iron, mild steel and

Table 8.1 Comparison of mechanical and physical properties of various automotive materials[12–15]

Material	Cast Mg		Wrought Mg		Cast Iron	Steel		Cast Al
Alloy/Grade	AZ91	AM50	AZ80-T5	AZ31-H2	Class 40	Mild steel Grade 4	AHSS[1] DP340/600	380
Process/Product	Die-cast	Die-cast	Extrusion	Sheet	Sand cast	Sheet	Sheet	Die-cast
Density (d, g/cm^3)	1.81	1.77	1.80	1.77	7.15	7.80	7.80	2.68
Elastic Modulus (E, GPa)	45	45	45	45	100	210	210	71
Yield Strength (YS, MPa)	160	125	275	220	N/A	180	340	159
Ultimate Tensile Strength (S_t, MPa)	240	210	380	290	293	320	600	324
Elongation (e_f, %)	3	10	7	15	0	45	23	3
Fatigue Strength (S_f, MPa)	85	85	180	120	128	125	228	138
Thermal Cond. (l W/m.K)	51	65	78	77	41	46		96
Thermal Exp. Coefficient (d, μm/m.K)	26	26	26	26	10.5	11.70		22
Melting Temp. (T_m, °C)	598	620	610	630	1175	1515		595

1. AHSS: advanced high-strength steel;
2. GFRP: glass fiber reinforced polymer;
3. CFRP: carbon fiber reinforced polymer;
4. P/M: permanent mold; and
5. T8X: simulated paint-bake (2% strain plus 30 min. at 177°C).

AHSS) and offers similar mechanical and physical properties to aluminum alloys, but with about one-third the mass saving. Extrusion and sheet alloys, for example alloys AZ80 and AZ31, respectively, provide tensile strength comparable to the aluminum extrusion alloy 6061 and the commonly used 5XXX and 6XXX sheet alloys, but they are less ductile. The wrought magnesium alloys are much less formable than steel or aluminum at room temperature due to their hcp (hexagonal close-packed) crystal structure, although their tensile ductility often appears reasonable.

While magnesium components are slightly heavier than polymers or polymer composites, they are generally stiffer, due to their higher elastic modulus. PC/ABS, a plastic blend of polycarbonate and acrylonitrile-butadiene-styrene, for example, has a modulus of about 1/20 that of magnesium and has been used in a limited way in instrument-panel beams where the design requirements

Table 8.1 continued

Wrought Al			Polymers (PC/ABS)	GFRP[2] (glass/polyester)		CFRP[3] (carbon/epoxy)	
A356-T6	6061-T6	5182-H24/ 6111-T8X[5]	Dow Pulse 2000	Structural (50% uniaxial)	Exterior (27%)	Structural (58% uniaxial)	Exterior (60%)
P/M[4] cast	Extrusion	Sheet	Injection molding	Liquid molding	Compression molding	Liquid/ compression molding	Autoclave molding
2.76	2.70	2.70	1.13	2.0	1.6	1.5	1.5
72	69	70	2.3	48	9	189	56
186	275	235/230	53				
262	310	310/320	55	1240	160	1050	712
5	12	8/20	5 at yield and 125 at break	< 1	2	< 1	< 1
90	95	120/186		N/A	N/A	N/A	N/A
159	167	123		0.6	0.3	0.5	0.5
21.5	23.6	24.1/23.4	74	12	14	2	4
615	652	638/585	143 (softening temperature)	130–160 (molding temperature)		175 (maximum service temperature)	

are met through the geometry of the closed sections. Fiber reinforcements can be used to increase the modulus of polymers, as seen in many polymer-based composites such as the conventional glass fiber reinforced polymer (GFRP) and advanced carbon fiber reinforced polymers (CFRP). Depending on applications, polymer composites can be classified into two groups:

- Structural composites (Structural-GFRP and Structural-CFRP), in which glass or carbon fibers are uniaxially orientated to provide unidirectional strength/stiffness for structural applications; and
- Exterior composites (Exterior-GFRP and Exterior-CFRP), in which glass or carbon fibers are appropriately aligned to provide 'quasi-isotropic' strength/stiffness in planar directions but not in the thickness direction, for exterior panel applications.

Exterior-GFRP is used in many low/medium-volume body panel applications (e.g., Corvette and Saturn). CFRP, on the other hand, is much more expensive and is, today, generally used in aerospace parts and high-performance cars; a recent GM application of exterior-CFRP is the hood outer for the Commemorative Corvette Z06. Recently, however, there has been a greater push to use CFRP in automotive body and closure applications, but an extensive discussion of this is outside the scope of this chapter. Polymers and composites are prone to creep, and thus not suitable for elevated-temperature applications due to their low service temperatures (e.g., 143°C for PC/ABS and 175°C for CFRP). Compared to aluminum, magnesium has a higher thermal expansion coefficient and lower thermal conductivity, which needs to be considered when substituting for aluminum in elevated-temperature applications.

8.2.2 Structural performance and mass saving potential

Materials selection for automotive structural applications is an extremely complex process in which component geometries, loading conditions, material properties, manufacturing processes and costs provide conflicting requirements.[6,15,16] As the bending mode is often the primary loading condition in many automotive structures, such as instrument-panel beams and frame rails, the following analyses on structural performance and mass saving potential of magnesium over mild steel (presently the dominant automotive material) are based on bending stiffness and strength calculations.

For a panel (plate) under bending loads, the minimum thickness (t) and mass (m) can be calculated using the 'materials performance index' concept.[16] Designating steel and magnesium properties with subscripts S and Mg, the thickness ratios and mass ratios of components made of the two materials for an equal stiffness design may be expressed as:

$$\frac{t_{\mathrm{Mg}}}{t_s} = \left(\frac{E_{\mathrm{s}}}{E_{\mathrm{Mg}}}\right)^{1/3} \qquad [8.1]$$

$$\frac{m_{\mathrm{Mg}}}{m_s} = \left(\frac{d_{\mathrm{Mg}}}{d_{\mathrm{s}}}\right)\left(\frac{E_{\mathrm{S}}}{E_{\mathrm{Mg}}}\right)^{1/3} \qquad [8.2]$$

where E and d are the elastic modulus and density of the materials, respectively. Using the property data as shown in Table 8.1, the thickness and mass ratios of magnesium (AZ91 alloy) compared with a mild steel beam can be calculated:

$$\frac{t_{\mathrm{Mg}}}{t_s} = 1.67 \qquad [8.3]$$

$$\frac{m_{\mathrm{Mg}}}{m_s} = 0.39 \qquad [8.4]$$

Therefore, in order to achieve the same bending stiffness, a magnesium panel will be required to have 1.67 times the thickness of a steel one, with a mass saving of 61%. For bending strength-limited design (same bending strength at minimum mass), such ratios for AZ91 magnesium alloy compared with mild steel become:

$$\frac{t_{\mathrm{Mg}}}{t_s} = \left(\frac{YS_{\mathrm{S}}}{YS_{\mathrm{Mg}}}\right)^{1/2} = 1.06 \qquad [8.5]$$

$$\frac{m_{\mathrm{Mg}}}{m_s} = \left(\frac{d_{\mathrm{Mg}}}{d_{\mathrm{s}}}\right)\left(\frac{YS_{\mathrm{S}}}{YS_{\mathrm{Mg}}}\right)^{1/2} = 0.25 \qquad [8.6]$$

where YS is the yield strength of the materials.

For a beam in an equal stiffness design, Equations [8.1] and [8.2] become:

$$\frac{t_{\mathrm{Mg}}}{t_s} = \left(\frac{E_{\mathrm{S}}}{E_{\mathrm{Mg}}}\right)^{1/2} \qquad [8.7]$$

$$\frac{m_{\mathrm{Mg}}}{m_s} = \left(\frac{d_{\mathrm{Mg}}}{d_{\mathrm{s}}}\right)\left(\frac{E_{\mathrm{S}}}{E_{\mathrm{Mg}}}\right)^{1/2} \qquad [8.8]$$

Similarly, for a solid beam in an equal strength design, Equations [8.5] and [8.6] become:

$$\frac{t_{\mathrm{Mg}}}{t_s} = \left(\frac{YS_{\mathrm{S}}}{YS_{\mathrm{Mg}}}\right)^{2/3} \qquad [8.9]$$

$$\frac{m_{\mathrm{Mg}}}{m_s} = \left(\frac{d_{\mathrm{Mg}}}{d_{\mathrm{s}}}\right)\left(\frac{YS_{\mathrm{S}}}{YS_{\mathrm{Mg}}}\right)^{2/3} \qquad [8.10]$$

Based on the above nomenclature and the property data provided in Table 8.1, Table 8.2 summarizes the thickness and mass ratios of various materials

Table 8.2 Thickness and mass ratios of various materials compared with mild steel for equal bending stiffness- and strength-limited design

Material		AHSS	Al (Cast)	Al (Wrought)	Mg (Cast)	Mg (Wrought)	PC/ABS	GFRP (Exterior)	CFRP (Exterior)
Equal stiffness panel	Thickness ratio	1.00	1.44	1.45	1.67	1.67	4.50	2.86	1.55
	Mass ratio	1.00	0.49	0.50	0.39	0.39	0.65	0.59	0.30
Equal strength panel	Thickness ratio	0.73	1.06	0.81	1.06	0.81	1.84	1.06	0.50
	Mass ratio	0.73	0.37	0.28	0.25	0.19	0.27	0.22	0.10
Equal stiffness beam	Thickness ratio	1.00	1.72	1.74	2.16	2.16	9.56	4.83	1.94
	Mass ratio	1.00	0.59	0.60	0.50	0.50	1.38	0.99	0.37
Equal strength beam	Thickness ratio	0.65	1.09	0.75	1.08	0.75	2.26	1.08	0.40
	Mass ratio	0.65	0.37	0.26	0.25	0.17	0.33	0.22	0.08

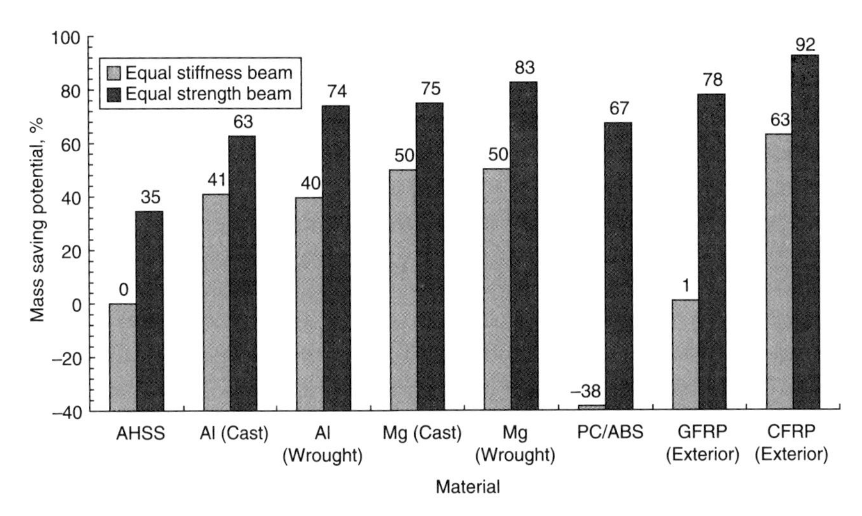

8.2 Percentage mass savings of various materials *vs.* mild steel for designing a structural panel with equivalent bending stiffness or bending strength.

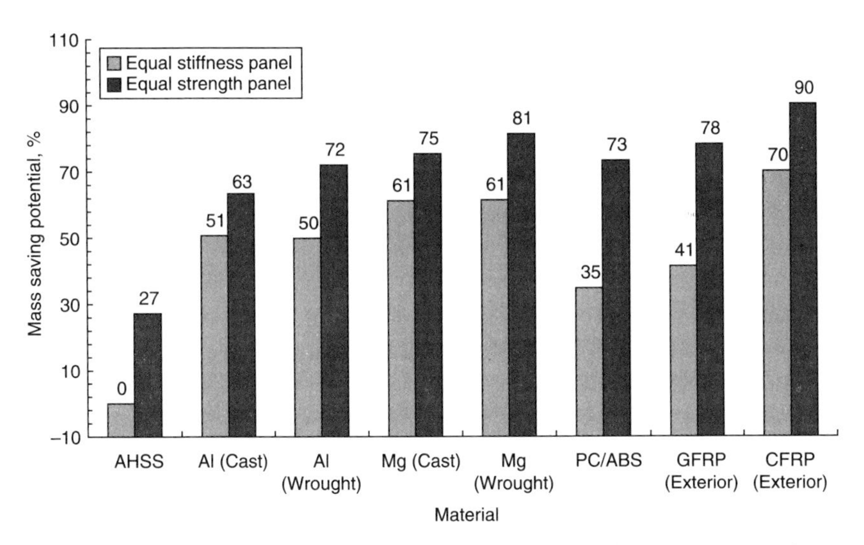

8.3 Percentage mass savings of various materials *vs.* mild steel for designing a structural beam with equivalent bending stiffness or bending strength.

compared with mild steel for solid panel and beam designs, respectively. For a generic comparison, the uniaxial properties of structural composites are not included in the analysis. Instead, the 'quasi-isotropic' properties of exterior-GFRP and CFRP composites are used to compare with the isotropic properties of other materials for mass saving analysis. Figures 8.2 and 8.3 highlight the thickness ratios and the resultant percentage mass savings of the materials when replacing a mild steel component. These results show that magnesium alloys have higher mass saving potential than AHSS, aluminum and polymers, in substituting for mild steel structures for equal stiffness or strength. To overcome its much lower elastic modulus, a polymer part will have to be reinforced with fibers or a metal–polymer hybrid structure has to be used. Compared to GFRP composites, magnesium alloys have higher mass saving potential for equal stiffness and similar savings for equal strength. While CFRP has the highest mass saving potential, wrought magnesium alloys provide slightly less mass savings, but at a lower cost.

8.3 Alloy development

8.3.1 Commercial extrusion alloys

Table 8.3 lists the nominal composition and typical room-temperature tensile properties of extruded magnesium alloy tubes.[12–14] Of the commercial extrusion alloys, AZ31 is most widely used in non-automotive applications. With higher aluminum content, AZ61 and AZ80 offer higher strength than AZ31 alloy, but with much lower extrudability. The high-strength Zr-containing ZK60 was designed for application in racing cars and bicycles, such as wheels and stems.[13] The extrusion speed for making ZK60 tubes is extremely low, rendering it uneconomical for automotive applications. The maximum extrusion speed of AZ31 alloy, the most extrudable commercial magnesium alloy, is only about a half of that of aluminum extrusion alloy 6063, which makes magnesium extrusions much more expensive due to the higher material and processing costs. New magnesium alloys are being developed to improve the extrusion speed while maintaining good mechanical properties.

8.3.2 New extrusion alloys

This section introduces two experimental extrusion alloys developed at GM for structural application: AM30 (Mg–3%Al–0.3%Mn) alloy for high-strength applications, and ZE20 (Mg–2%Zn–0.2%Ce) alloy for high-ductility applications.

AM30 Alloy

A new experimental extrusion alloy, AM30 (Mg–3%Al–0.3%Mn), was recently developed with improved extrudability and formability.[14] Aluminum also improves strength, hardness and corrosion resistance of the alloying elements considered[12] but reduces ductility. An aluminum content of about 5–6% yields the optimum combination of strength and ductility for structural applications. Increasing aluminum content widens the freezing

Table 8.3 Nominal composition and typical room-temperature tensile properties of extruded magnesium alloys[12–15]

Alloy	Temper	Composition (wt%)				Tensile properties		
		Al	Zn	Mn	Zr	Yield strength (Mpa)	Tensile strength (Mpa)	Elongation (%)
AZ31	F[1]	3.0	1.0	0.20	–	165	245	12
AZ61	F	6.5	1.0	0.15	–	165	280	14
AZ80	T5	8.0	0.6	0.30	–	275	380	7
ZK60	F	–	5.5	–	0.45	240	325	13
ZK60	T5[2]	–	5.5	–	0.45	268	330	12
AM50	*F*	*5.0*	–	*0.30*	–	*168*	*268*	*18*

[1] F signifies as extruded.
[2] T5 signifies artificially aged after extrusion.

range and makes the alloy easier to cast, but more difficult to extrude due to increased hardness. For example, alloys containing less than 3% Al can be extruded at higher extrusion speeds compared with high-strength alloys such as AZ61 (Mg–6%Al–1%Zn) and ZK60 (Mg–6%Zn–0.5%Zr).[13] To maximize the ductility and extrudability, while maintaining reasonable strength and castability (for billet casting prior to extrusion), an aluminum content of 3% was selected for the new alloy.

Zinc is next to aluminum in effectiveness as an alloying ingredient to strengthen magnesium.[12] However, it reduces ductility and increases hot-shortness of Mg–Al based alloys. Zinc-containing magnesium alloys are prone to microporosity.[17] Zinc was also reported to have mild to moderate accelerating effects on corrosion rates of magnesium as determined by alternate immersion in 3% NaCl solution.[12] Therefore, unlike most commercial magnesium alloys, Zn was not selected in this experimental alloy. Manganese does not have much effect on tensile strength, but it does slightly increase the yield strength of magnesium alloys. Its most important function is to improve the corrosion resistance of Mg–Al based alloys by removing iron and other heavy-metal elements into relatively harmless intermetallic compounds, some of which separate out during melting. For this purpose, Mn is added at about 0.4% as recommended by the ASTM Specification B93–94a.

Based on these analyses and experiments, a new magnesium alloy, AM30 (Mg–3%Al–0.4%Mn), was formulated. The extrudability of an alloy billet is defined by the maximum extrusion speed at which the billet can be pushed through an extrusion die at a given temperature without causing visible surface defects, such as cracking or fracture.[14] Compared with the current workhorse commercial magnesium wrought alloy AZ31 (Mg–3%Al–1%Zn), the new AM30 alloy can be extruded 20% faster (Fig. 8.4[14]), has a 50% increase

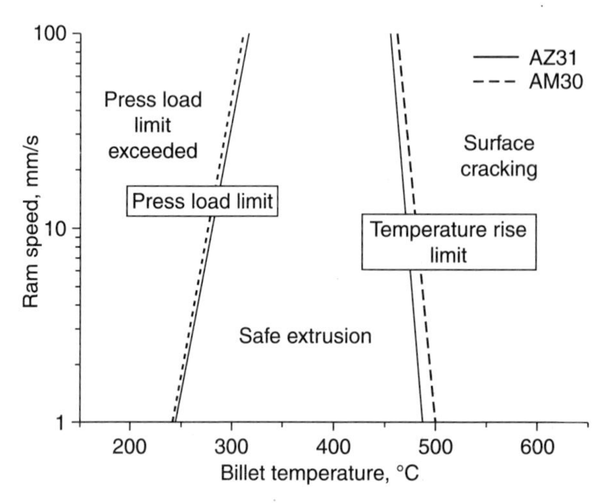

8.4 Extrusion limit diagram for AM30 and AZ31 alloys.[14]

in room-temperature ductility with similar strength, and has up to a 30% improvement in ductility at elevated temperatures up to 200°C (Fig. 8.5[14]).

Mg–Zn–Ce (ZE) alloys

It was recently found that a small addition of only 0.2% Ce to pure Mg significantly improved the ductility of extruded bars, Fig. 8.6,[18] due most likely to the reduced grain size and weakening of the texture. However, the strength of the Mg–0.2%Ce alloy remained too low for automotive structural applications. A follow-on investigation[19] examined the addition of aluminum (Al) to Mg–0.2%Ce binary alloy, which improved its strength but decreased its ductility considerably. This is due to the fact that Ce has a higher affinity for Al and forms $Al_{11}Ce_3$ in the Mg–Al–Ce ternary alloys, offsetting the beneficial effect of the Ce addition.

Zinc additions of 2–6%, on the other hand, significantly improved the strength of the Mg–0.2%Ce alloy, due to solid solution strengthening by Zn, while retaining the beneficial effect of Ce on randomizing the texture and the associated high ductility.[20] Clearly, the ZE20 alloy (Mg–2%Zn–0.2%Ce) has significantly higher strength compared with Mg–0.2%Ce alloy (135 MPa *vs.* 69 MPa yield strength and 225 MPa *vs.* 170 MPa ultimate tensile strength). Although the ZE20 alloy has slightly lower elongation (27.4% *vs.* 31%), compared with the binary Mg–0.2%Ce alloy, it has significantly higher ductility compared with 16.9% for commercial AZ31 alloy; a 62% increase. This increase is obtained with only a minor reduction of about 16% in tensile strength. As also shown in Fig. 8.7, increasing the Zn content from 2% to 8% increased the ultimate tensile strength of the Mg–Zn–Ce

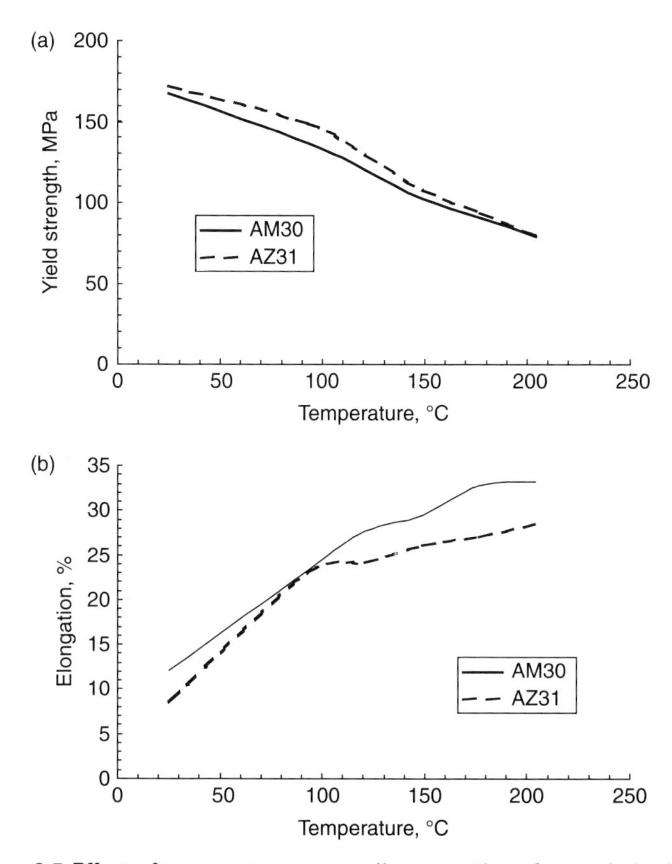

8.5 Effect of temperature on tensile properties of extruded tubes of AM30 and AZ31 alloys[14]: (a) Tensile yield strength; and (b) tensile elongation.

(ZE) alloy, but the elongation was reduced considerably. The ZE20 alloy is considered most promising and has a somewhat lower strength but much higher ductility compared with AM30 (170 MPa yield strength; 240 MPa ultimate tensile strength and 12% elongation).

The extrusion experiments on ZE alloys show that ZE20 alloy has excellent extrudability, about 25% higher maximum extrusion speed compared with the AZ31 alloy (0.08 m/s) at an extrusion temperature of 400°C, and is similar to that of AM30 which also showed about 20% better extrudability than AZ31.[14] When the Zn content is increased to 5% (ZE50 alloy), the extrudability is reduced to the level for AZ31. A further increase in Zn content to 8% (ZE80 alloy) results in a significant decrease in extrudability (0.02 m/s). Higher Zn contents have been reported to increase surface cracking and oxidation during extrusion and, hence, to lower the extrusion speed limits.[21]

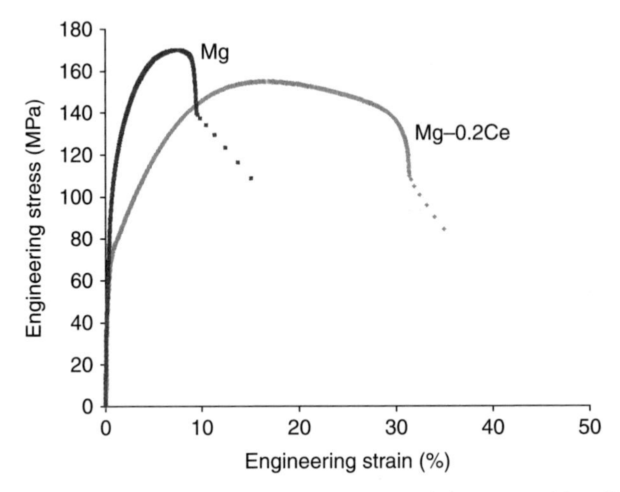

8.6 Tensile curves for (a) pure Mg; and (b) Mg–0.2%Ce alloy, after extruding bars with a ratio of 25:1.[18]

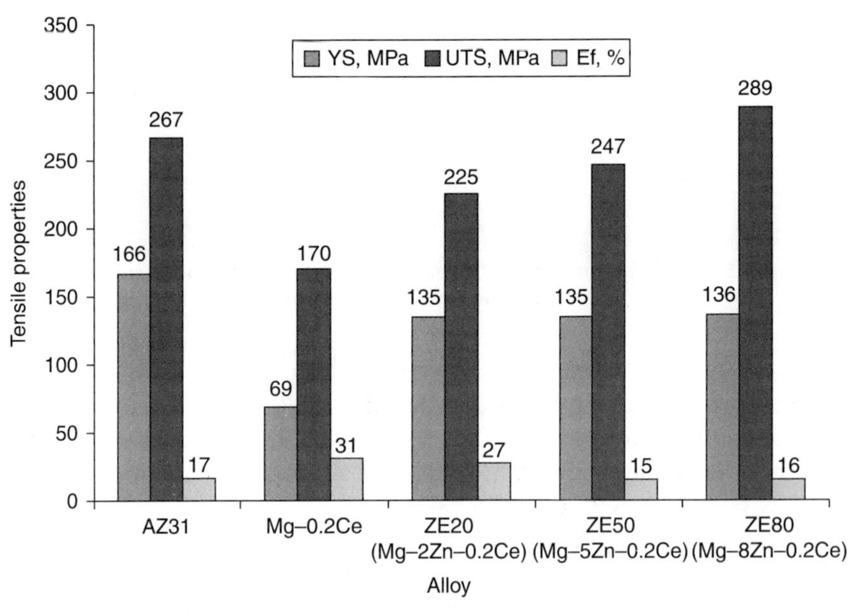

8.7 Tensile properties of Mg–Zn–Ce alloy extruded tubes compared with AZ31 and Mg–0.2%Ce alloy tubes processed in similar conditions.[20]

8.3.3 Commercial sheet alloys

There are three types of commercial sheet magnesium alloys: Mg–Al–Zn (AZ), Mg–Th (HK and HM) and Mg–Li–Al (LA) alloys, which are

Table 8.4 Nominal composition and typical room-temperature tensile properties of magnesium sheet alloys[12,22]

Alloy	Temper	Composition (wt%)					Tensile properties		
		Al	Zn	Mn	Zr	Other	Yield strength (Mpa)	Tensile strength (Mpa)	Elongation (%)
AZ31	H24[1]	3.0	1.0	0.20	–		180	325	15
HK31	H24	3.0	0.3	–	0.7	3.25 Th	160	285	9
HM21	T8[2]	–	–	0.45	–	2.0Th	130	270	11
LA141	T7[3]	1				14 Li	105–160	132–165	11–24

[1] H24 signifies strain hardened and partially annealed.
[2] T8 signifies solution heat-treated, cold worked, and artificially aged.
[3] T7 signifies solution heat-treated and stabilized.

summarized in Table 8.4. Alloy AZ31 is the most widely used alloy for sheet and plate and is available in several grades and tempers. It can be used at temperatures up to 100°C. The HK31 and HM21 alloys are suitable for use at temperatures up to 315°C and 345°C, respectively. However, HM21 has superior strength and creep resistance; but it has been banned due to the radioactivity of thorium.

Alloys with very low density and good formability have been developed by adding lithium to magnesium. For example, LA141A, which contains 14% Li, has a density of 1.35 g/cm^3, or only 78% that of pure magnesium. This alloy has found limited use in missile and aerospace applications because of its ultra-low density and excellent formability at room temperature or slightly elevated temperatures.[22]

8.3.4 New sheet alloys

As demonstrated in the commercial LA141A alloy, Mg–Li alloy system is very promising for high formability and can be potentially formed at room temperature. Similar to the extrusion alloys developed, the Mg–Zn–Ce alloy system has also been shown to provide significant improvements in tensile ductility and sheet formability. Microalloying has been found to be effective in refining Mg–Al-based sheet alloys for formability enhancement and property improvement.

Mg–Li based alloys

According to the Mg–Li phase diagram, Fig. 8.8, the alloys exhibit two-phase structures of α (hcp) Mg-rich and β (bcc) Li-rich phases at room temperature when 5–11% Li is added to magnesium. Further additions of Li (more

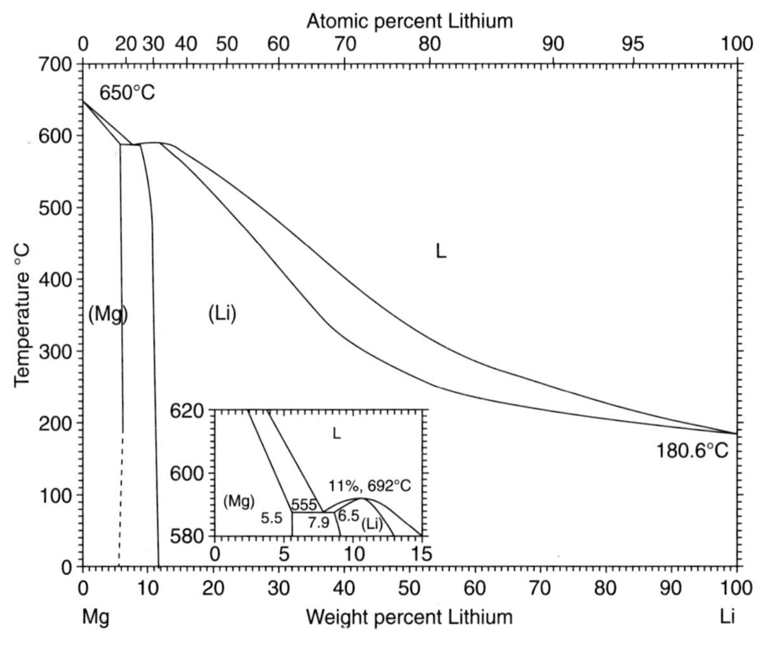

8.8 Mg–Li phase diagram.[12]

than 11%) can transform the hcp α-Mg solid solution into highly workable, body-centered cubic alloys.[23] It has been shown that the rolling textures of Mg–14%Li alloy sheet with a bcc structure are very similar to those of other bcc metals, such as interstitial-free (IF) steel, ferritic steel and Fe–3%Si when they are similarly processed.[24] However, the binary Mg–Li alloys offer poor corrosion resistance, and limited mechanical properties (especially low fatigue strength) at room and elevated temperatures. Alloy development efforts have been on the addition of third alloying elements (Al, Zn, Si, RE, etc.) on the mechanical properties of Mg–Li alloys.[25–29] While aluminum improves the creep resistance and mechanical properties at elevated temperatures, Mg–Li–Al alloys have poor corrosion resistance due to the existence of the chemically active fcc LiAl phase in the microstructure, in addition to the β (bcc) phase solid solution. Rare earth additions can improve the formability and chemical and thermal stability of Mg–Li–Al alloys, especially when the Li content is less than 6% and the alloys are in the α-Mg single phase region, due to the formation of Al_2RE and Al_4RE phases in the microstructure.

The microstructure and formability of Mg–Li–Zn alloy sheets were studied and the results are summarized in Table 8.5.[30] The Mg–6%Li–1%Zn alloy with dominant α-Mg (hcp) phase has limited ductility and formability, while the Mg–12%Li–1%Zn alloy with dominant β-Li (bcc) microstructure has high ductility at comparatively low strain rates. On the other hand, the

Table 8.5 Microstructure and tensile properties of Mg–Li–Zn alloys obtained from uniaxial tension tests for an initial stain rate of $8.3 \times 10^{-4}\ s^{-1}$ [30]

Alloy	Mg–6%Li–1%Zn	Mg–9.5%Li–1%Zn	Mg–12%Li–1%Zn
Microstructure	α-Mg (hcp)	α-Mg + β-Li	β-Li (bcc)
Yield strength (Mpa)	112	121	124
Ultimate tensile strength (Mpa)	155	134	125
Elongation (%)	32.2	71.4	56.0
Work-hardening exponent (*n*)	0.15	0.06	0.00
Normal anisotropy parameter (*r*)	5.98	0.87	0.70

Mg–9.5%Li–1%Zn alloy sheet with an (α + β) two-phase microstructure has a better combination of tensile properties and formability compared with the alloys with higher or lower Li contents.

Mg–Zn–Ce (ZE) alloys

Similar to ZE20 extrusion alloy (Section 16.3.2.2), a Mg–1.5%Zn–0.2%Ce sheet alloy was reported[31] to provide high tensile elongation (around 30%) and improved formability, attributed to the TD (transverse direction)-split texture produced when the alloy was rolled at 450°C and annealed at 350°C. However, the strength of this alloy is still considered low (yield strength < 120 MPa and ultimate tensile strength < 210 MPa) for many automotive structural applications.

The conventional Mg–Al based alloy sheets generally show strong basal texture, in which the basal plane is parallel to the rolled-sheet surface. The basal and prismatic slip can operate parallel to the rolling direction and width direction, but not in the thickness direction. Therefore, the rolled sheets hardly deform in the thickness direction, resulting in premature fracture at an initial stage of forming at room temperature, and thus the poor formability.

A Mg–Zn alloy with a small addition of RE (such as Ce) processed by hot rolling has significantly different texture, with the basal poles at 35° from the thickness direction.[31] As a result, the alloy sheet can readily deform in the thickness direction, thus a significant improvement in room temperature formability. The unique texture in the Mg–Zn–Ce alloy is attributed to the activation of prismatic slip by the Ce addition.

Microalloying of Mg–Al based alloys

Various alloying elements, such as Sr, Ti, Ce, Ca, Sb, Sn and Y, have been studied aimed at improving the mechanical properties and formability of

Mg–Al–Zn alloys. These alloying elements, either individually or in combination, form additional phases and/or refine the grain structure of the alloys.

Small additions of Sr (0.01–0.1%) and Ti (0.01–0.03%) can effectively decrease the grain size of the AZ31 alloy billets,[21] promising improved formability and mechanical properties in the resultant sheet or extrusion products. Additions of up to 1%Ce or about 0.1%Ca can refine the grains of AZ31 alloy, and small additions of Sb and Sn can improve the strength of AZ alloys due to the formation of precipitation phases Mg_3Sb_2 and Mg_2Sn.[32] An addition of 0.7%Y to AZ31 was reported to modify the $Mg_{17}Al_{12}$ intermetallic phase to more rounded particles, leading to better properties in the sheet samples.[32]

8.4 Manufacturing process development

This section briefly reviews some of the key manufacturing processes that are utilized in current or potential applications of magnesium alloys, especially in the automotive industry. In addition to the conventional processes, certain other methods are being currently developed for producing and shaping automotive components. These are important for wrought alloys for automotive applications.

8.4.1 Extrusion and forging processes

Conventional extrusion processes

Magnesium alloys can be warm or hot extruded in hydraulic presses to form bars, tubes and a wide variety of profiles.[12] It is well known that a tubular section is significantly stiffer than a solid beam for the same mass. While hollow magnesium extrusions can be made with a mandrel and a drilled or pierced billet, it is generally preferable to use a bridge die where the metal stream is split into several branches which recombine before the die exit.

Hydrostatic extrusion process

The hydrostatic extrusion process, typically used for copper-tubing fabrication, is a much faster extrusion process compared with conventional direct extrusion. It was reported that seamless magnesium tubes were extruded using the hydrostatic process at speeds up to 100 m/min, due to the absence of friction between the billet and container since the billet is suspended in hydraulic oil.[33] Although the process is capable of extrusion ratios up to 700, the outer diameter of tubes produced by this process is limited to about 45 mm, even with a large 4000-ton press.[33]

Forging processes

Commercial extrusion alloys listed in Table 8.3 (AZ31, AZ61, AZ80 and ZK60), can also be forged into high-integrity components using hydraulic presses or slow-action mechanical presses. Forging is normally done within 55°C of the solidus temperature of the alloy. Corner radii of 1.6 mm, fillet radii of 4.8 mm, and panels or webs 3.2 mm thick can be achieved by forging. The draft angles required for extraction of the forgings from the dies can be held to 3° or less.[12]

8.4.2 Tube bending and hydroforming processes

Tube hydroforming

Tube hydroforming is a metal forming process which uses pressurized fluids such as water to make various perimeter shapes from tubes. Compared with stampings and castings, hydroformed tubular sections provide further mass savings for structural components. While hydroformed steel and aluminum tubes are currently used in many structural applications, including frame rail, engine cradle, radiator support and instrument-panel beams,[34] the hydroforming of magnesium tubes is not yet developed.

Tube bending

Tube bending is generally needed as a pre-form step for hydroforming automotive parts, but the bendability of magnesium extrusions at room temperature is very limited.[35] A moderate temperature (150–200°C) bending process has been developed at GM for magnesium alloy extrusions.[36] A bend radius of twice the tube outer diameter (2D) was achieved on magnesium alloy extrusions at 150°C, Fig. 8.9.[36] The mechanical properties and microstructure of magnesium alloys at elevated temperatures indicate that

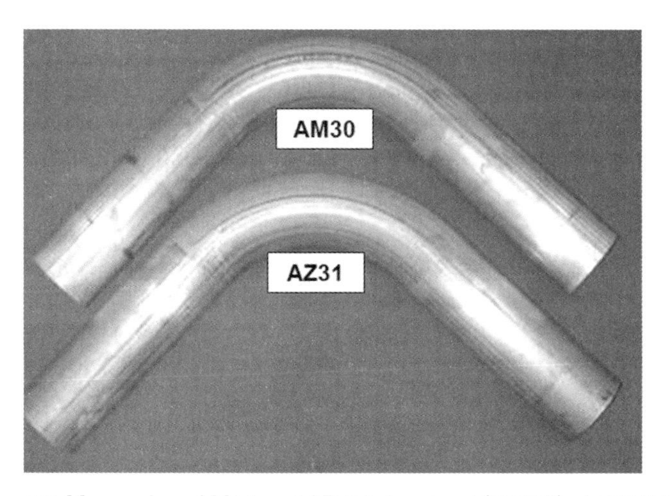

8.9 Magnesium AM30 and AZ31 tubes bent (2D/90°) at 150°C.[36]

moderate temperature hydroforming and other forming processes at this temperature range (150–200°C) should be explored.[37]

8.4.3 Sheet production and forming processes

Sheet production processes

The direct-chill (DC) casting process is generally used to cast magnesium slabs (about 50 mm thick), which are then hot-rolled at 315–370°C to produce magnesium sheet and plates.[12] Unlike aluminum, for which cold-roll is usually the final step in sheet production, warm finish roll is applied to magnesium sheet products. Large grains of 200–300 μm in the slab can be reduced by warm rolling to fine recrystallized grains between 7 and 22 μm for a sheet gage of 1.3–2.6 mm.[38,39] Recently, twin-roll continuous casting (CC) process has been investigated for the production of low-cost magnesium sheet. Pilot plants have been established in Germany, Austria, Australia and Turkey,[40–43] and production plants are being built in China and Korea. In this process, molten magnesium is poured into a gap between two rolls to produce a continuous strip about 2.5–10 mm thick, which is then warm-rolled to a final gage. Currently, magnesium sheet using this process is available from Salzgitter (Germany) with a maximum width of 1850 mm and a minimum gage of 1.0 mm.[44]

Sheet forming processes

The majority of the processes used to convert sheet metal into automobile components occur at room temperature including: stamping, flanging, bending, hemming, trimming, etc. The processes are very robust for high formability materials such as mild steel, but with some concessions on draw depth and corner radii, have been successfully used with less formable materials such as aluminum and high-strength steel. Unfortunately, the limited formability of magnesium, due to its HCP structure (discussed earlier), makes the use of these processes very difficult. An example of the problem is shown in Fig. 8.10 where the results of a forming trial using a simple rectangular pan for a mild steel and AZ31B magnesium sheet are compared. The pan could be formed to a 125 mm depth with the steel, Fig. 8.10a, but split after only about 12 mm with the magnesium, Fig. 8.10b.[45]

Warm stamping (200–400°C) of magnesium sheet has been used to make complex products for aerospace and luggage components, and the optimum temperature for warm stamping was reported as approximately 350°C.[39] Figure 8.10c shows that magnesium alloy sheet can be stamped at reasonably moderate temperatures of 150–175°C,[46,47] similar to that for tube-bending temperatures reported earlier.[36] The sheet hydroforming process differs from the conventional stamping process in that the solid punch

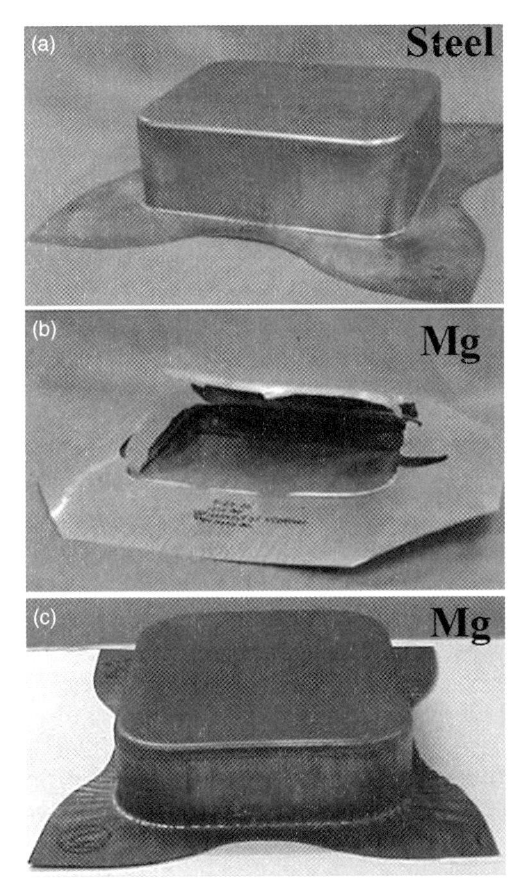

8.10 Comparison of a forming trial on a 125 mm deep pan with (a) mild steel formed at room temperature, (b) AZ31B magnesium formed at room temperature, and (c) AZ31B magnesium formed at 350°C.[45]

(upper die) or female (bottom) die is replaced with a forming medium. When the female die is replaced with a fluid or a pad of flexible polyurethane, it is also called hydro-mechanical drawing.[48,49] Another variant, called active sheet hydroforming, is to use a forming medium instead of a solid punch to directly press the blank against a die contour.[50] The optimum conditions for hydro-mechanical drawing magnesium sheet were reported to be temperatures of 180–220°C and pressures of 600–800 bar.[47] Obviously, forming at lower temperatures would be more economical and yield much better dimensional accuracy, due to less thermal contraction and distortion during cooling to room temperature.

Superplastic forming (SPF) is a gas forming process using one-sided tools (like sheet hydroforming) based on the superplasticity of many materials (aluminum, titanium, magnesium, etc.) at elevated temperatures and under controlled strain rates. SPF is often used to fabricate large and complex aluminum parts in the aerospace industry. Extensive R&D at GM on aluminum SPF has shortened the process cycle time to an acceptable level for automotive panels, leading to the development of the quick plastic forming (QPF) process.[51] QPF has been implemented to produce aluminum decklids for the Cadillac STS. Recently, the QPF process has been used to produce many prototype magnesium inner closure panels at a forming temperature of about 475°C.[52] Examples of QPF prototype products are shown in a later section.

8.4.4 Welding and joining techniques

Welding processes

Various welding processes can be used for joining magnesium to magnesium. Most magnesium alloys are readily fusion-welded with higher speeds than aluminum due to the lower thermal conductivity and latent heat of magnesium.[12,53–55] While gas metal arc welding (MIG or GMAW) is used for joining magnesium sections ranging from 0.6 to 25 mm, gas tungsten arc welding (TIG or GTAW) is more suited for thin sections up to 12.7 mm.[12] It should be recognized that hot-shortness may produce cracks in welding magnesium alloys containing more than 1% zinc, which can often be overcome by using proper filler wires: ER AZ61A is the preferred filler wire for welding wrought alloys containing aluminum, while ER AZ91A has been found to lower crack sensitivity in AZ and AM cast alloys.[54] Magnesium sheet and extrusions ranging from 0.5 to 3.3 mm can be joined by all types of resistance welding, including seam, projection and flash, but the most common type is spot welding.[12] A higher welding speed for thin magnesium sheet can be achieved with laser or electron beam welding, due to its narrow heat affected zones and less weld distortion. Welding speeds of 2.5 to 9 m/min can be achieved with a 4 kW solid-state laser for welding AZ31 sheet 1.0 to 3.2 mm thick.[55] Non-vacuum electron beam welding of AZ31 alloy can reach as high as 12–15 m/min.[55]

Other joining processes

Fusion welding of magnesium die castings can be challenging due to the presence of porosity and the formation of brittle intermetallic phase ($Mg_{17}Al_{12}$) in the welds.[53] Solid-state welding techniques, such as friction-stir welding and magnetic pulse welding, can be used to improve the weld quality of magnesium die castings.[56–58] While these solid-state welding techniques can potentially be used to join magnesium to dissimilar materials such as aluminum,[56,59] mechanical joining (self-piercing rivets, clinching and hemming) and adhesive bonding are preferred for dissimilar material joining involving magnesium to aluminum or steel. Adhesive bonding of magnesium parts (with or without dissimilar materials) requires proper pre-treatment of the joint surfaces, which includes cleaning, etching and wet chemical passivation. Many adhesives including epoxy and polyurethane can be used as long as they are chemically stable and have aging stability.[60]

8.5 Aerospace applications

8.5.1 Historical applications

Historically, magnesium was one of the main aerospace construction metals, and was used for German military aircraft as early as World War I and extensively for German aircraft in World War II.[3] The United States Air Forces' long-range bombers B-36 and B-52 also contained a large amount of magnesium sheet, castings, forgings and extrusions.[3] The B-36 was reported[3] to use 12 200 lbs of magnesium sheet components (Fig. 8.11); 1500 lbs of magnesium forgings; and 660 lbs of magnesium castings in the 1950s. The 1832 of Boeing 727 airplanes built from 1962 to 1984 contained 1200 magnesium part numbers including leading and trailing edge flaps, control surfaces, actuators, door frames, wheels, engine gear boxes, power generation components, structural items (not primary) and others.[61] Magnesium was also used intensively in the former Soviet aircraft industry: for example, TU-95MS plane had 1550 kg of magnesium, and TU-134 had 780 kg of magnesium components in various locations of airplanes.[62] Unfortunately, many of these applications were reduced in modern aviation, due to perceived hazards with magnesium parts

8.11 B 36 assembly line in 1950s (magnesium sheets appearing dark in the picture).[3] With permission from Robert Brown.

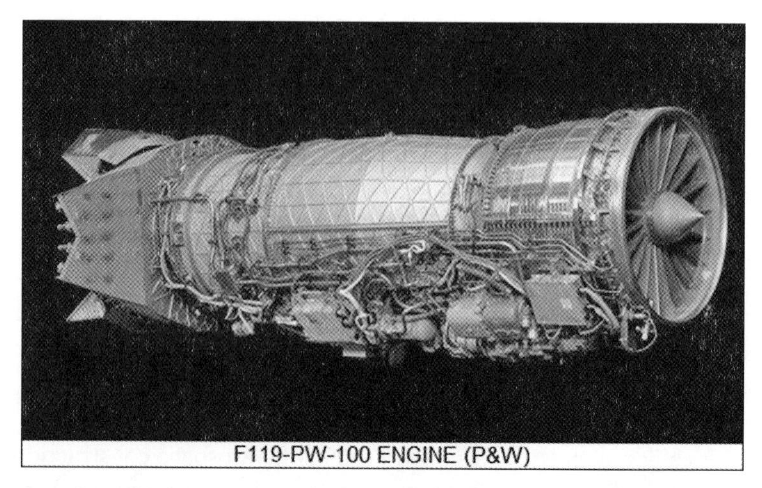

8.12 Pratt & Whitney F119 auxiliary casing in ELEKTRON WE43 alloy[61] (Source: http://www.vectorsite.net/avf22.html).

in the event of fire and the International Air Transport Association (IATA) legislation limiting magnesium alloys to non-structural parts due to corrosion problems reported in the 1950s and 1960s.[61]

8.5.2 Current and future applications

Today, despite the remarkable improvement in corrosion resistance of modern magnesium alloys reviewed in this book and this chapter, the application of magnesium in the commercial aerospace industry is generally restricted to engine- and transmission-related castings and landing gears. Magnesium is not used in structural application by major aircraft manufacturers such as Airbus, Boeing and Embraer,[62] but has found many applications in the helicopter industry, such as cast gearboxes and some other non-structural components. Some of the notable aerospace applications include:[61]

i. Sikorsky UH60 Family (Blackhawk) main transmission in ZE41 alloy;
ii. Sikorsky S92 main transmission in WE43A alloy;
iii. Thrust reverser cascade casting in AZ92A alloy found on Boeing 737, 747, 757 and 767;
iv. Pratt & Whitney F119 auxiliary casing in WE43 alloy (Fig. 8.12);
v. Pratt & Whitney Canada PW305 turbofan in ZE41 alloy; and
vi. Rolls-Royce tray in ELEKTRON ZRE1 alloy (Fig. 8.13).

There has been renewed interest from the aerospace industry around the world (Europe, North America and China) in resolving the regulatory

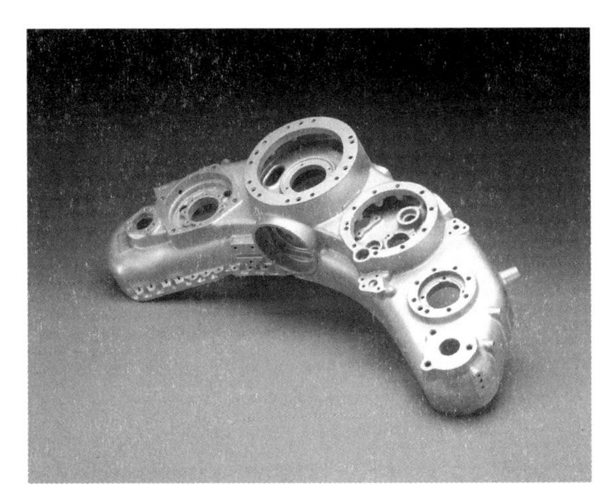

8.13 Rolls-Royce tray in ELEKTRON ZRE1 alloy.[61] With permissions from Magnesium elektron.

barriers and technical challenges for magnesium applications. A high-profile European research program, Magnesium for Aerospace Applications (FP6 AEROMAG), has assembled a number of magnesium alloy and component producers to work with the aerospace industry to develop new magnesium alloys and manufacturing processes for aerospace applications. It is expected that, with efforts like this around the world, magnesium will become a major structural material in the future aerospace industry.

8.6 Automotive applications

8.6.1 Historical applications

Magnesium has a long history of automotive use, and a pictorial summary of many of these historic and current applications is provided in Fig. 8.14.[45] The first automotive magnesium application was the racing engine pistons for Indy 500 in 1918. Another early application of magnesium as an automotive material was a sand cast crankcase on the 1931 Chevrolair. Commercial applications of magnesium sand castings were also reported in England including lower crankcases for city buses and transmission housings for tractors[63] in 1930's. Crankcases and housings were also produced in Germany by high-pressure die casting process.[64] Magnesium usage grew throughout the 1930's and then grew exponentially during World War II. With the introduction of the Volkswagen Beetle, automotive magnesium consumption again accelerated and reached a peak in 1971, the major applications being the air-cooled engine and gearbox, which together weighed about 20 kg.[65]

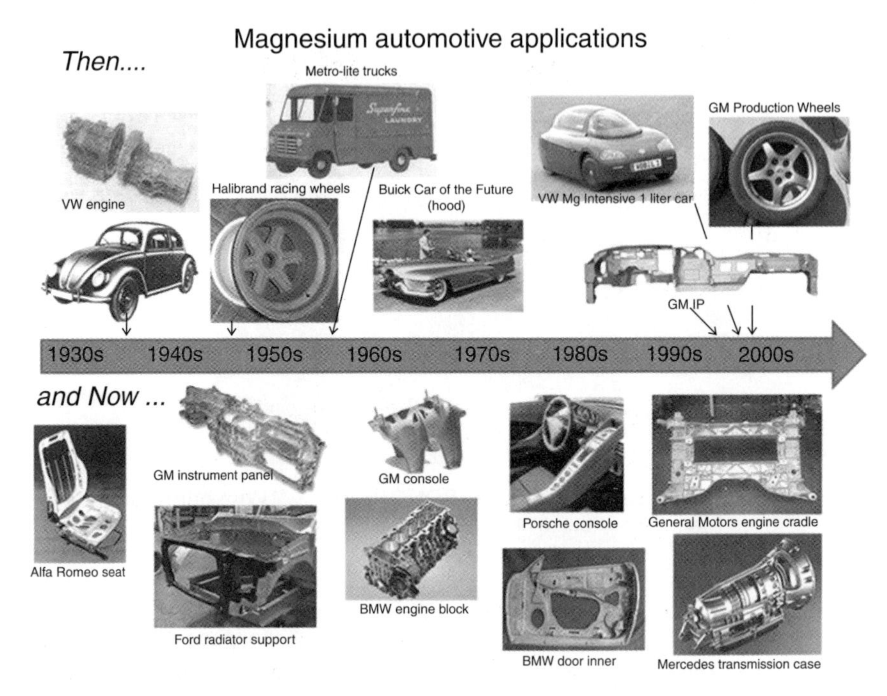

8.14 Pictorial summary of past and current magnesium automotive applications.[45]

However, several factors emerged and combined to cause the reduction and eventual elimination of magnesium as a structural powertrain component after the 1970s.[45] These factors included greater power requirements for the engine, which increased both its operating temperature and load, ultimately resulting in the conversion of the engine from air cooling to water cooling as the AZ81 alloy, and later the AS41 or the AS21 alloys, could not keep up with the required operating environment.[66] The use of water cooling put magnesium at a disadvantage compared with other engine materials because of its poor corrosion resistance. At that time, the effect of iron, copper and nickel impurities (in ppm amounts) on promoting the corrosion of magnesium had not been recognized.[67] By the time the 'high purity' alloys AZ91D and AM60B, which replaced AZ91C and AM60A, respectively, were developed in the 1980s, the cost of magnesium alloys had begun to increase, and the use of magnesium in automotive applications decreased, although some applications remained.

While the majority of historical magnesium applications have been cast components, which take advantage of the excellent fluidity of magnesium and the ability to cast very complex and thin walled shapes, there were many automotive applications of magnesium sheet and extruded components. The

8.15 (a) 1951 Buick LeSabre concept car with magnesium and aluminum body panels, (b) 1961 Chevrolet Corvette with prototype hood made from magnesium sheet, and (c) 1957 Chevrolet Corvette SS Race Car with 'featherweight magnesium body'.[45]

first commercial ground transportation applications were developed in the early 1950s. MetroLite trucks were manufactured between 1955 and 1965 and featured magnesium sheet panels as well as structures made from magnesium plate and extrusions.[68] These trucks had increased payload capacity and were excellent applications for magnesium because they did not require extreme formability. GM made prototype hoods for the Buick LeSabre in 1951, various body panels for the Chevrolet Corvette SS Race Car in 1957, and hoods for the Chevrolet Corvette in 1961 – see Fig. 8.15.[45] However, sheet magnesium has not been used in high-volume production in the mainstream automobile industry.

8.6.2 Current and future applications

Table 8.6 is a summary of the current major magnesium applications in automotive industry to the best of the Author's knowledge. It shows that magnesium has made significant gains in world-wide interior applications, replacing mostly steel stampings in instrumental panels, steering wheels and steering column components. In the powertrain area, North America is leading in the application of magnesium 4WD (four-wheel-drive) transfer cases

Table 8.6 Global magnesium applications in automobiles

System	Component	North America	Europe	Asia
Interior	Instrument panel	Yes	Yes	Yes
	Knee bolster retainer	Yes		
	Seat frame	Yes	Yes	Yes
	Seat riser	Yes	Yes	Yes
	Seat pan	Yes	Yes	
	Console bracket	Yes		
	Airbag housing	Yes		
	Center console cover		Yes	
	Steering wheel	Yes	Yes	Yes
	Keylock housing	Yes		
	Steering column parts	Yes	Yes	Yes
	Radio housing	Yes		
	Glove box door	Yes		
	Window motor housing	Yes	Yes	
Body	Door frame		Yes	
	Liftgate	Yes	Yes	
	Roof frame	Yes	Yes	
	Sunroof panel	Yes	Yes	
	Mirror bracket	Yes	Yes	
	Fuel filler lid		Yes	
	Door handle		Yes	Yes
	Spare tire carrier	Yes		
Chassis	Wheel (racing)	Yes	Yes	Yes
	ABS mounting bracket	Yes		
	Brake pedal bracket	Yes		Yes
	Brake/accelerator bracket	Yes		
	Brake/clutch bracket	Yes		
	Brake pedal arm	Yes		
Powertrain	Engine block		Yes	
	Valve cover/cam cover	Yes	Yes	Yes
	4WD transfer case	Yes		
	Transmission case		Yes	Yes
	Clutch housing and piston	Yes		
	Intake manifold	Yes	Yes	
	Engine oil pan		Yes	Yes
	Alternator/AC bracket	Yes		
	Transmission stator	Yes		
	Oil filter adapter	Yes		Yes
	Electric motor housing	Yes		

in high-volume truck production; while Europe is aggressively expanding the use of magnesium in engine blocks and transmission cases using recently developed creep-resistant magnesium alloys. Only a limited number of body and chassis components are currently made of magnesium, which presents a great opportunity for magnesium to expand its applications in lightweight

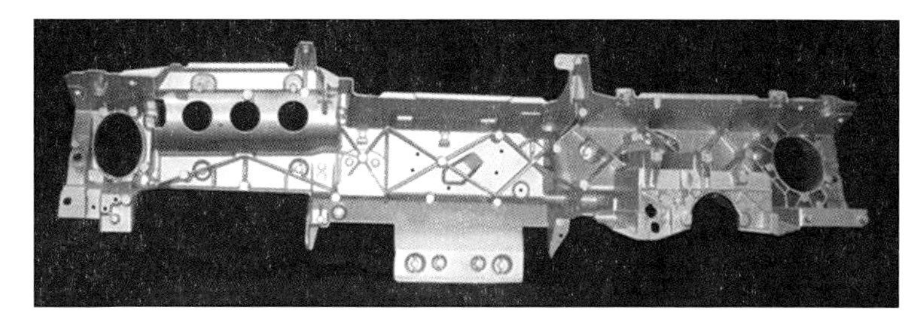

8.16 Magnesium die cast instrument-panel beam for Buick LaCrosse (6.9 kg).

vehicle construction. This section discusses the current and potential magnesium applications in the vehicle subsystems.

Interior

Since corrosion is of less concern in the interior, this area has seen the most magnesium applications, with the biggest growth in the instrument panels (IP) and steering structures. The first magnesium IP beam was die-cast by GM in 1961 with a mass saving of 4 kg over the same part cast-in zinc. The design and die casting of magnesium IP beams have advanced dramatically in the last decade. For example, the current IPs normally have a thickness of 2–2.5 mm (compared with 4–5 mm for the earlier IP beam applications) with more part consolidation and mass savings. Figure 8.16 shows a magnesium die-cast instrument-panel beam (6.9 kg) in current GM production of Buick LaCrosse. However, the use of cast magnesium IP beams is recently facing strong competition. IP beams made of aluminum extrusions are used by Mercedes in Europe. IP designs using bent steel tubes (with or without hydroforming) are slightly heavier than magnesium die casting, but significantly less expensive. To maintain and grow its use in IP production, magnesium design and thin-wall casting technology must continue to improve in further reducing weight and cost. Tubular designs using magnesium extrusions and sheet components should also be explored.

The use of magnesium seat structures began in Germany, where Mercedes used magnesium die castings in its integrated seat structure with a three-point safety belt in the SL Roadster.[69] This seat structure consisted of five parts (two parts for the seat back frame and three parts for the cushion frame) with a total weight of 8.5 kg and varying wall thickness of 2–20 mm.[69] In the material selection for the seat program, magnesium was chosen over plastic, steel sheet and aluminum gravity casting designs. Magnesium die castings made of the high-ductility alloys of AM50 and AM20 offered the best combination of high strength, extreme rigidity, low weight and cost.[69] Similar to the IP development, magnesium seat design and manufacturing have gone through significant improvements in recent years. The latest example is the two-piece (backrest and cushion) design used in the Alfa Romeo 156. The die castings used today for seating structures are as thin as 2 mm, providing even greater weight savings. In North America, Chrysler recently introduced the 'Stow-n-Go' seating and storage system for its minivans, where the folding mechanisms require light weight for easy operation, thus some aluminum is used in the second-row seats and the back frame of the third-row folding seats is magnesium casting.[70] While other materials, such as AHSS and aluminum, are also being used for these applications, it is expected that magnesium will make significant inroads into seat components as a lightweight and cost-effective solution.

Compared to magnesium castings, wrought magnesium provides further mass saving opportunities for many interior applications, such as seat and IP beams. The only current production application of magnesium sheet is the center console in the low-volume Porsche Carrera, as shown in Fig. 8.17.[45] Research is needed to reduce the cost of wrought magnesium and its forming processes.

Body

The use of magnesium in automotive body applications is limited but recently expanding. GM has been using a one-piece die-cast roof frame since the C-5 Corvette introduction in 1997. Magnesium is also used in the Cadillac XLR roadster's retractable hard-top convertible roof and the roof top frame. The Ford F-150 trucks and SUVs (sport utility vehicles) have coated magnesium castings for their radiator support. In Europe, Volkswagen and Mercedes have pioneered the use of thin-wall magnesium die castings in body panel applications. The one-piece die-cast door inner for the Mercedes S-Class

8.17 Sheet magnesium center console cover in Porsche Carrera GT automobile.[45]

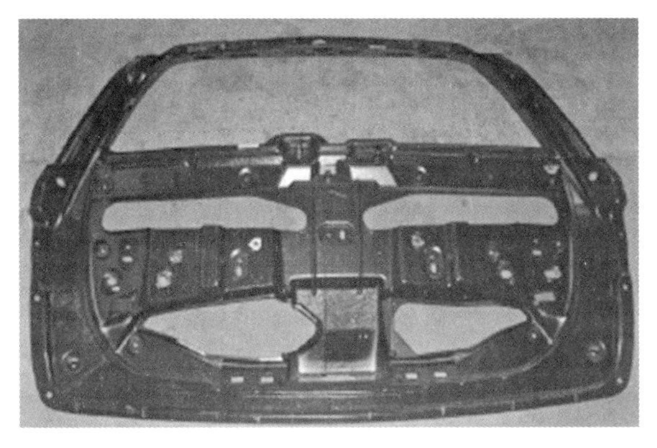

8.18 2010 Lincoln MKT magnesium liftgate inner casting.[72] Courtesy of Meridian Lightweight Technologies.

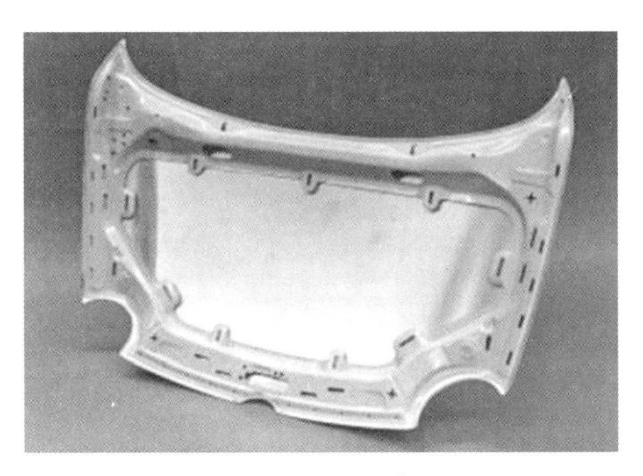

8.19 VW Lupo magnesium hood.[42]

Coupe is only 4.56 kg.[71] The 2010 Lincoln MKT magnesium liftgate inner panel is the first die-cast magnesium closure ever to satisfy 55 mph rear crash requirements.[72] As shown in Fig. 8.18, the 8-kg inner casting is perhaps the world's largest magnesium casting (1379 × 1316 mm).[72] The key to manufacturing these thin-wall castings (approximately 2 mm) lies in casting design using proper radii and ribs for smooth die filling and to stiffen the parts. These thin-wall die castings, such as closure inners, can often offset the material cost penalty of magnesium over steel sheet metal construction due to part consolidation.

In body panel applications where bending stiffness is frequently the design limit, magnesium sheet metal can offer as much as 61% mass saving (see Fig. 8.2). Volkswagen made a prototype hood for the Lupo (Fig. 8.19) a few years ago.[42] GM has made numerous panels including a hood, door inner

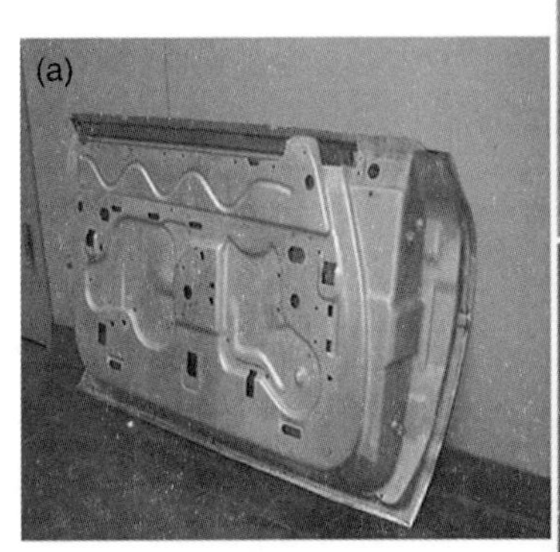

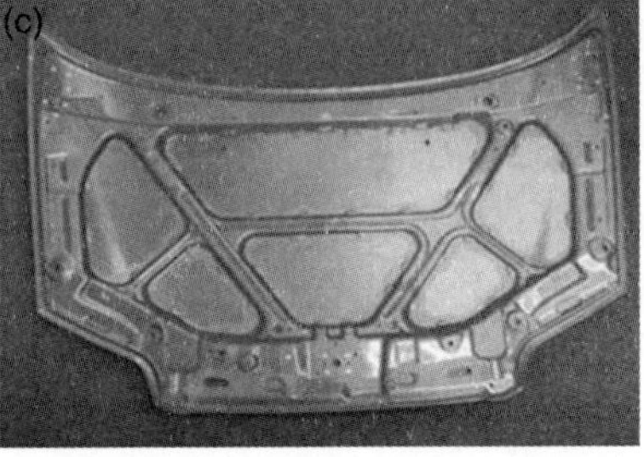

8.20 Magnesium sheet panels formed recently by General Motors: (a) door inner panel;[40] (b) decklid inner panel;[52] and (c) hood.[73]

8.21 Inner panel drawn by Chrysler LLC using magnesium sheet.[74]

panel, decklid inner, liftgate and various reinforcements, some of which are shown in Fig. 8.20.[40,52,73] Chrysler LLC has performed a number of studies using magnesium sheet including the inner panel and a magnesium intensive body structure, Fig. 8.21.[74]

The majority of the applications discussed above were 'inner' panels, which create the structure of the vehicle closures but are generally not visible on the outside of the vehicle. This is due to two factors. First, the surface quality of the currently available magnesium sheet requires

8.22 Cast magnesium alloy wheel for Chevrolet Corvette.

significant finishing compared with aluminum or steel, and second, the limited formability at room temperature makes assembly processes for outer panels, such as hemming, difficult. This likely means the first commercial applications for magnesium sheet will be inner panels. However, elevated-temperature forming of magnesium and corrosion protection coatings further imposes a cost penalty for magnesium sheet applications. The development of new sheet alloys for near-room temperature forming is needed, along with low-cost and robust coatings for corrosion resistance.

Similar to magnesium sheet, magnesium extrusions for automotive applications are still in the development stage, with many prototype parts, such as bumper beams and frame rail for VW 1-Liter Car.[46] However, the use of magnesium tubes/extrusions in body applications would require more development to meet all structural and cost targets, as well as a supply base for high-volume automotive production.

Chassis

Cast or forged magnesium wheels have been used in many high-priced race cars or high-performance roadsters, including GM's Corvette (Fig. 8.22). However, the relatively high cost and potential corrosion problems of magnesium wheels prevented their use in high-volume vehicle production. The first-in-industry one-piece magnesium die-cast cradle for the Chevrolet Corvette Z06 weighs only 10.5 kg (Fig. 8.23) and demonstrates a 35% mass savings over the aluminum cradle it replaced.[75] This cradle uses a new AE44 (Mg–4Al–4RE) alloy, which offers high strength and ductility at room and elevated temperatures. The production of lightweight and low-cost magnesium chassis components, such as wheels, engine cradles and control arms, depends on the improvement of magnesium casting processes. Various casting processes have been developed for the production of aluminum wheels and chassis parts. These processes include permanent mold casting, low-pressure casting, squeeze casting and semi-solid metal (SSM) casting. The successful adaptation of these processes to magnesium alloys will make magnesium castings more competitive to aluminum in the chassis area. The development of low-cost, corrosion-resistant coatings and new magnesium alloys with improved fatigue and impact strength will also accelerate the further penetration of magnesium in chassis applications.

Hollow structures of aluminum castings and extrusions are presently used in high-volume cradle production, such as the welded structures for

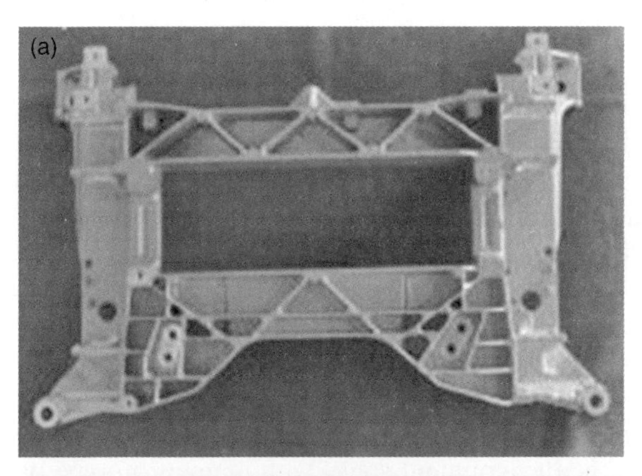

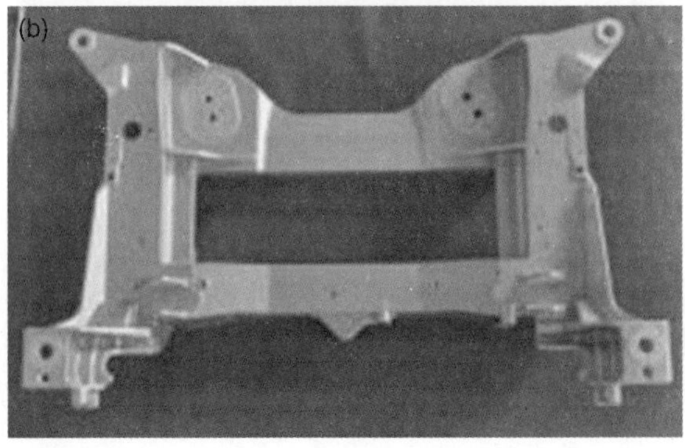

8.23 Cast magnesium engine cradle for Chevrolet Corvette Z06.

GM's mid-size cars and the hydroformed tubular subframe for BMW 5 and 7 series. Hollow designs are generally more mass-efficient than solid castings. Cradles and subframes using magnesium tubes and hollow castings would offer more mass savings. The development of hollow magnesium casting and low-cost extrusion processes are needed for these lightweight and efficient chassis applications.

Powertrain

The majority of powertrain castings (such as engine block, cylinder head, transmission case and oil pan) are made of aluminum alloys, which represent the most significant opportunity for lightweighting with magnesium due to the excellent castability of magnesium alloys. At present, millions of pickup trucks and SUVs produced in North America have magnesium transfer cases. VW and Audi have high-volume production of manual transmission cases of magnesium in Europe and China. Magnesium valve covers are used by many vehicles in North America, Europe and Asia. The operating temperatures for these applications are below 120°C, and AZ91 is the alloy of choice due to its excellent combination of mechanical properties, corrosion resistance and castability.

Higher-temperature applications such as automatic transmissions and engine blocks require creep-resistant magnesium alloys. The Mercedes 7-speed automatic transmission case (Fig. 8.24) uses AS31 alloy with marginally better creep resistance than AZ91 alloy.[76] Honda introduced a new alloy referred as ACM522 (Mg–5%Al–2%Ca–2%RE) in the production of Honda Insight (a low-volume hybrid gas/electric car) oil pans, achieved a 35% weight saving over the aluminum design.[77] Another significant development is the BMW Mg/Al composite engine block, Fig. 8.25.[10] The composite block consists of an aluminum insert (about 2/3 of the total block weight) surrounded by magnesium alloy AJ62 (Mg–6%Al–2%Sr) (1/3 of

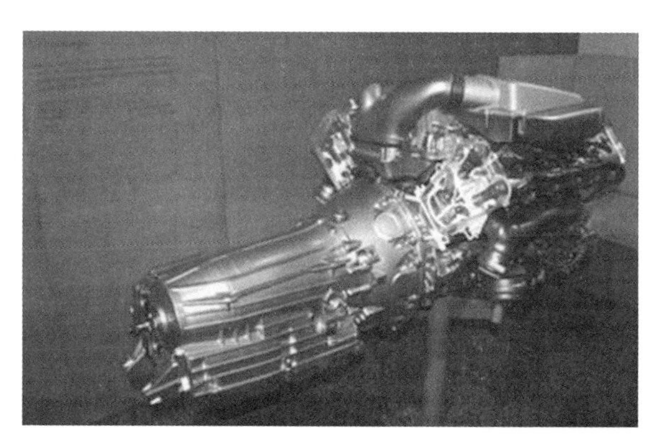

8.24 The Mercedes 7-speed automatic transmission case.[76]

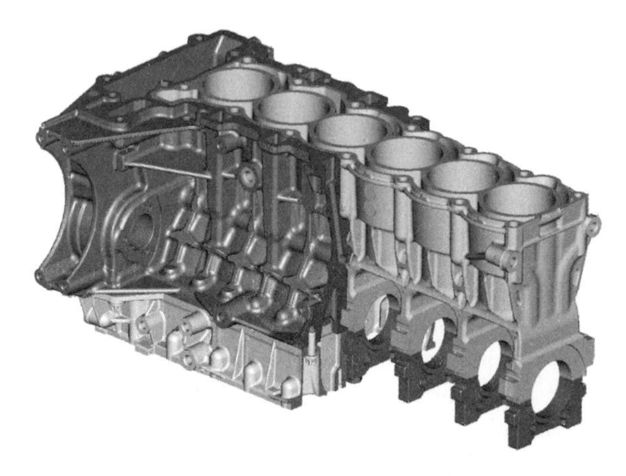

8.25 The BMW composite engine showing a cutaway of the magnesium exterior revealing the aluminum interior.[10]

the total weight) in the upper section of the cylinder liners and water-cooling jacket. The hypereutectic aluminum alloy insert avoids the use of additional liner technology and facilitates the highly loaded bolt joints for both cylinder head and crankshaft bearings. The insert also includes the water jacket, avoiding the potential problem of coolant corrosion with magnesium.[10] The magnesium housing surrounding the aluminum insert, in turn, primarily serves the oil ducts and the connection of ancillary units. The gearbox cover and the mounts for both the alternator and vacuum pump are integrated in the housing. The lower section of the crankcase, in turn, also made of die-cast magnesium AJ62 alloy, comprises cast-in sintered steel inserts for the crankshaft mounts. This composite design was reported to be 10 kg lighter than the aluminum block, and has since been in BMW production since 2005.

In the United States, much of the development is jointly sponsored by the US Department of Energy (DOE) and the US Council for Automotive Research (USCAR), and led by the three OEMs (GM, Ford and Chrysler). The USCAR Magnesium Powertrain Cast Components (MPCC) Project had the objective of demonstrating the readiness of magnesium alloys for completely replacing the major aluminum components of a V block engine.[78] The MPCC cylinder block achieved a mass reduction of 25% (29% of the cast aluminum components were replaced by magnesium). A prototype engine made with a low-pressure sand cast cylinder block, a Thixomolded rear-seal carrier, and a high-pressure die-cast oil pan and front cover, with all other parts carried over from the baseline aluminum engine, has been completed,[79] see Fig. 8.26. Significant learning has been generated in the dynamometer test of the prototype Mg-intensive engine, promising more magnesium powertrain applications in the future.

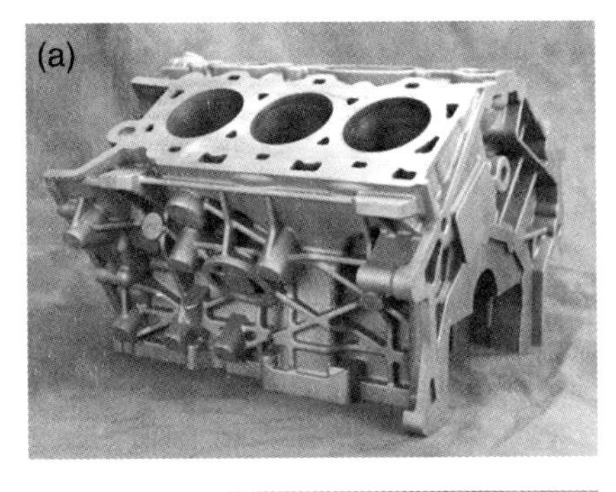

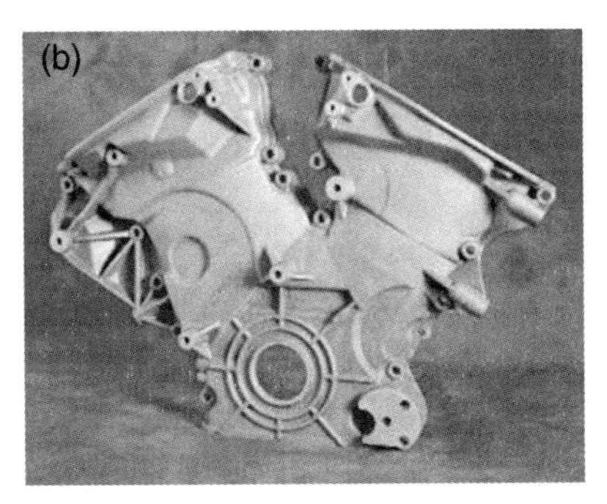

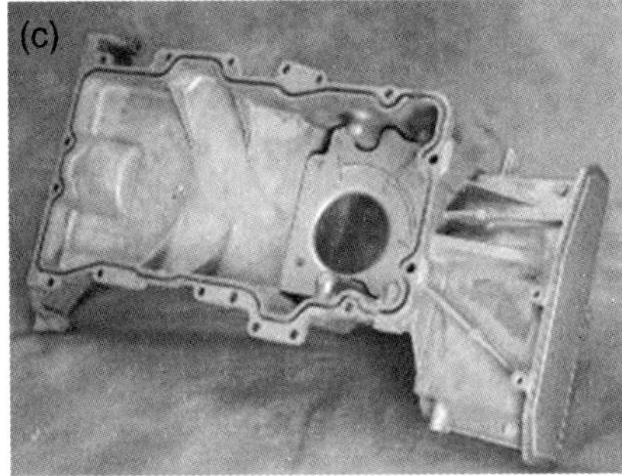

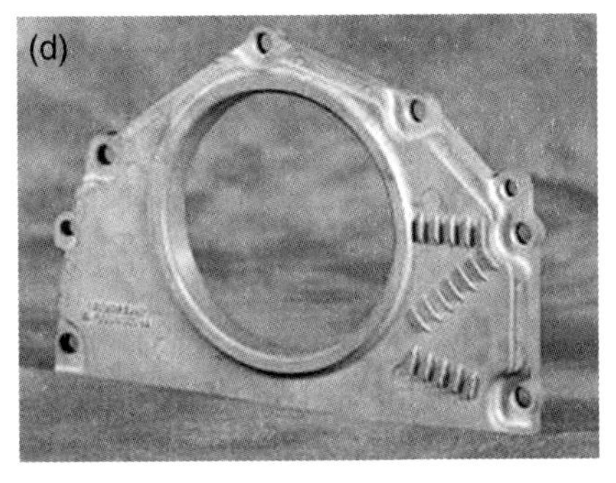

8.26 Magnesium powertrain components from the USCAR Magnesium Powertrain Cast Components Project; (a) cylinder block, (b) front engine cover, (c) oil pan and (d) rear seal carrier.[79]

8.7 Other applications

The light weight of magnesium alloys has attracted many other applications beyond the transportation (automotive and aerospace) industries, most notably electronics and power tools. In addition to its low density, magnesium offers 100 times better heat dissipation than plastics, the best vibration dampening of any metal, ease of machining, electromagnetic shielding, and the major environmental advantage of being recyclable.[80]

8.7.1 Applications in electronics

Driven by environmental programs across the consumer electronics industry, portable electronics product manufacturers are opting for light, yet tough, magnesium for everything from flash audio/video players to digital cameras, mobile phones, computer notebooks, radar detectors and more. Magnesium meets the design challenges that are instrumental to consumer electronics becoming lighter, thinner and more mobile. Components that house and protect highly sensitive technology inside these entertainment and communications devices must exhibit strength and durability to withstand daily

8.27 The Philips GoGear SA52 portable digital audio/video player with magnesium housing lets consumers experience high-quality music and video content. © Photo courtesy of Royal Philips Electronics. Used with permission.[80]

8.28 Canon's EOS 50D Digital SLR digital camera protected by strong, durable magnesium casing. © Photos courtesy of Canon, U.S.A., Inc. Used with permission.[80]

abuse from being dropped, stepped on, bumped, banged around in transit, and survive even the ultimate test – teenagers. Figures 8.27–8.30 show magnesium applications in casings for audio/video players, cameras, cell phones and laptop computers.[80]

8.7.2 Applications in power tools

The power tool industry increasingly relies on die-cast magnesium components to offer durable, lighter weight designs that are easier to handle and manage over long work shifts – an important feature, especially for framing and construction crews on the job site.[81] Figures 8.31 and 8.32 shows magnesium usage in housings of hand-held power nailers and saws, with better ergonomics when weight is reduced, resulting in less user fatigue issues during work shifts.[81]

8.29 Thixomolded magnesium alloy AZ91D cell phone frames. Photo courtesy of Japan Steel Works. Used with permission.[80]

8.30 The HP EliteBook 2530p technology and LCD screen are well-protected by the magnesium base and aluminum-clad magnesium case and display enclosure. Photo courtesy of HP.[80]

8.8 Future trends

While magnesium is the lightest structural metal and the third most commonly used metallic material in automobiles following steel and aluminum, many challenges remain in various aspects of alloy development and manufacturing processes to exploit its high strength-to-mass ratio for widespread lightweight applications in the transportation industry.

8.8.1 Material challenges

Compared with the numerous aluminum alloys and steel grades, there are only a limited number of low-cost wrought magnesium alloys available for automotive applications. The conventional Mg–Al based alloys

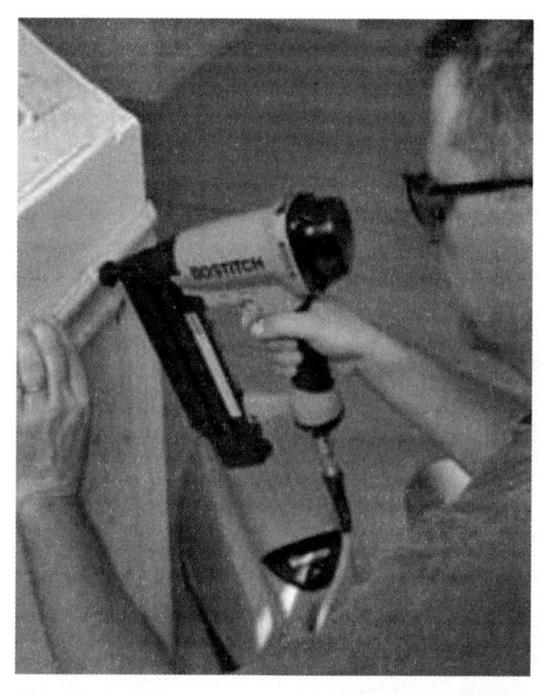

8.31 Magnesium cast housings provide user-preferred lightness, balance, and durability for the Bostitch F-Series of pneumatic framing nailers. Photos courtesy of Stanley-Bostitch. Used with permission.[81]

8.32 The 7-1/4-inch Worm Drive power SKILSAW® SHD77M features magnesium alloy AZ91D for both the gear and motor housings, shaving a full two pounds off the unit's weight. Photo courtesy of SKIL-Robert Bosch Tool Corporation. Used with permission.[81]

offer moderate mechanical properties due to the limited age-hardening response of this alloy system. Since the development of vacuum die casting and other high-integrity casting processes, magnesium castings can be heat-treated with no blisters. Alloy systems with significant precipitation hardening, such as Mg–Sn[82,83] and Mg–RE,[84] should be developed with improved mechanical properties. New alloys with improved ductility, fatigue strength, creep resistance and corrosion resistance should also be explored.

For wrought magnesium alloys, the strong plastic anisotropy and tension–compression asymmetry due to texture remain obstacles for many structural applications. Microalloying with elements such as Ce[18] to improve the plasticity of magnesium alloys has been proven an effective approach in wrought magnesium alloy development. Alloys systems such as Mg–Li, Mg–RE and Mg–Zn–RE have shown more 'isotropic' mechanical properties. Computational thermodynamics and kinetics[85] will be used to design and optimize these new alloys.

The properties of magnesium alloys can be significantly enhanced if micro- and nano-particles are introduced to form metal matrix composites (MMC). Micro- and nano-sized particles offer strengthening mechanisms in different length scales and provide a tremendous opportunity for a new class of engineering materials with tailored properties and functionalities for automotive applications.

8.8.2 Process challenges

Although the success of magnesium is primarily attributed to its superior die-castability compared with aluminum alloys, these castings cannot generally be heat-treated due to the porosity intrinsic to die casting that is present. Several recent developments show promise, including super vacuum die casting and squeeze casting, that drive porosity to minimal levels to enable their heat treatment without blistering. Combined with advanced low-cost alloys, these processes could provide competitive advantages for increased use of magnesium die castings. Other casting processes, such as gravity, semi-permanent mold, low-pressure, etc., have also been considered for magnesium, although casting rules developed for aluminum need to be modified to compensate for the larger shrinkage with magnesium. These processes are still, nevertheless, important for magnesium due to the need for large hollow castings for structural subsystems, such as engine, cradles that provide the highest mass efficiency. Melt handling, molten metal transfer with minimal turbulence, grain-refinement, die coating as well as casting parameters need to be developed specifically for magnesium alloys to fully utilize their intrinsic properties in these casting processes.

Various forming processes need to be optimized for magnesium alloys. Elevated-temperature forming is needed for most extrusion and sheet components. Research efforts have been directed to lowering the forming temperatures and reducing the cycle times. New forming processes should be developed to utilize the dramatically improved formability of magnesium alloys at certain ranges of temperature and strain rate. Room temperature (RT) or near-RT forming techniques are also being explored for new magnesium alloys such as Mg–Zn–Ce alloys.[86]

8.8.3 Performance challenges

There are several performance-related challenges that need significant research efforts. Some of them are highlighted in the current Canada–China–USA 'Magnesium Front End Research & Development' project.[87]

Crashworthiness

Magnesium castings have been used in many automotive components such as the instrument-panel beams and radiator support structures. High-ductility AM50 or AM60 alloys are used in these applications and have performed well in crash simulation and tests; and many vehicles, with these magnesium components, achieved a five-star crash rating. However, there is limited material performance data available for component design and crash simulation. A recent study shows that magnesium alloys can absorb significantly more energy than either aluminum or steel on an equivalent mass basis.[88] While steel and aluminum tubes fail by progressive folding in crash loading (more desirable situation), magnesium alloy (AZ31 and AM30) tubes tend to fail by sharding or segment fracture.[88,89] However, the precise fracture mechanisms for magnesium under crash loading are still not clear, and material models for magnesium fracture are needed for crash simulation involving magnesium components. Additionally, new magnesium alloys need to be developed to have progressive folding deformation in crash loading.

Noise, vibration and harshness (NVH)

It is well known that magnesium has high damping capability, but this can be translated into better NHV performance only for mid-range sound frequency; 100–1000 Hz.[74] The low-frequency (<100 Hz) structure-borne noise can be controlled by the component stiffness between the source and receiver of the sound. The lower modulus of magnesium, compared with steel, is often compensated by thicker gages and/or ribbing designs. For high-frequency (>1000 Hz) airborne noise, a lightweight panel, regardless of material, would transmit significantly more road and engine noise into the occupant compartment unless the acoustic frequencies could be broken up and damped. Magnesium, with its low density, is disadvantaged for this type of applications unless new materials with laminated structures are developed for sound isolation.

Fatigue and durability

Fatigue and durability are critical in magnesium structural applications and there is limited data in the literature, especially for wrought alloys.[90–92] The effect of alloy chemistry, processing and microstructure on the fatigue characteristics of magnesium alloys need to be studied. Extrusion and sheet products need to be characterized sufficiently to establish links between microstructural features and fatigue behavior. Multi-scale simulation tools can be used to predict the fatigue life of magnesium components and subsystems, which can be validated for automotive applications.

Corrosion and surface finishing

Pure magnesium has the highest standard reduction potential of the structural automotive metals.[45] As noted earlier, while pure magnesium (at least with very low levels of iron, nickel and copper) has atmospheric corrosion rates that are similar to that of aluminum, magnesium's high reduction potential makes it very susceptible to galvanic corrosion when it is in electrical contact with other metals below it in the reduction potential table. The impact of this susceptibility to galvanic corrosion on the application of magnesium in exposed environments is severe in both the macro-environment and the micro-environment. In the macro-environment, magnesium alloys must be electrically isolated from other metals to prevent the creation of galvanic couples; for example, steel bolts cannot be in direct contact with magnesium. Isolation can be achieved by replacing the bolt with a less reactive metal, as has been done in the Mercedes automotive transmission case where steel bolts have been replaced with aluminum bolts.[76] Isolation can also be achieved by coating the 'other' metal. Finally, isolation can be achieved by the use of shims or spacers of compatible materials of sufficient geometry and size to prevent electrical contact in the presence of salt water, as shown, for example, for the Corvette cradle, Fig. 8.33.[76] While the component cost can be competitive with aluminum, the isolation strategies required can often make the application more expensive and thus restrictive in its use.

A major challenge in magnesium automotive applications is to establish the surface finishing and corrosion protection processes. The challenge is two-fold, since surface treatments for magnesium play roles in both

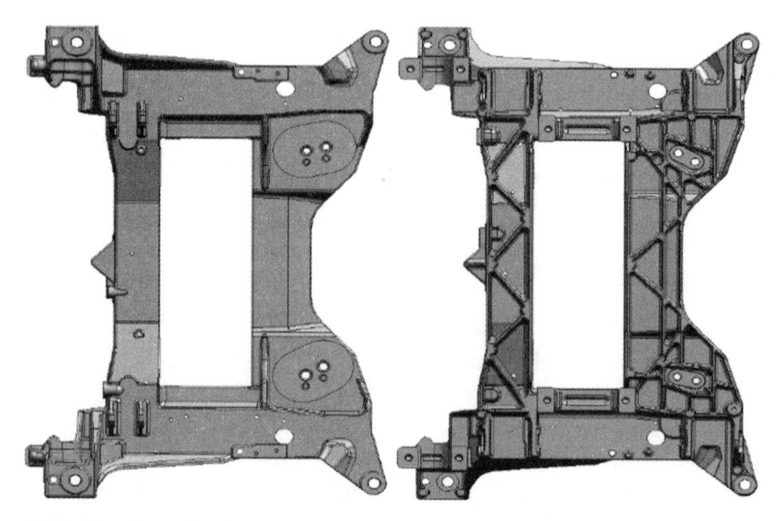

8.33 Aluminum isolator locations for Chevrolet Corvette Z06 magnesium cradle (bottom and top views).[75]

manufacturing processes (e.g., adhesive bonding) as well as the product life cycle that demands corrosion resistance. Furthermore, the current manufacturing paradigm for steel-intensive body structures employs chemistries in the paint shop that are corrosive to magnesium and are additionally aggravated by galvanic couples primarily steel fasteners. Future research will explore novel coating and surface treatment technologies including pretreatments such as micro-arc anodizing, non-chromated conversion coatings, and 'cold' metal spraying of aluminum onto aluminum surfaces. Since most studies of corrosion protection and pre-treatment of magnesium have focused on die castings, the behavior of sheet, extrusion and high-integrity castings will be explored for process compatibility.

8.8.4 Future development

The future success of magnesium as a major automotive material will depend on how these technical challenges are addressed. These challenges are huge and global, and would require significant collaboration among industries, governments and academia from many counties. One example is the current Canada–China–USA 'Magnesium Front End Research & Development' project funded by the three governments.[87] This project has brought together a unique team of international scope, from the United States, China and Canada, and has developed some key enabling technologies and knowledge base for automotive magnesium applications. The technologies and knowledge base developed in this project not only benefit the

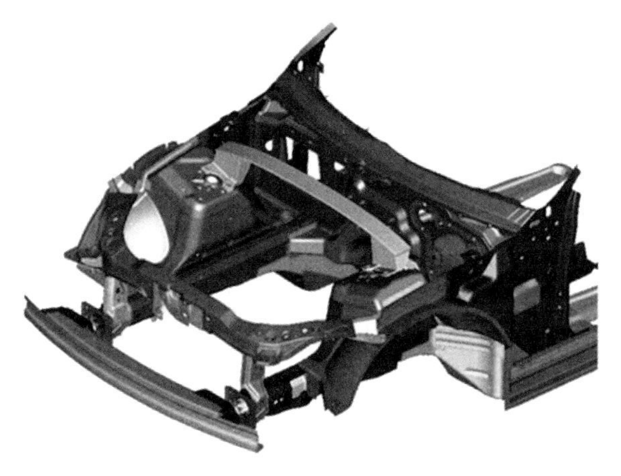

8.34 A schematic of the front end structure of a production sedan.[87]

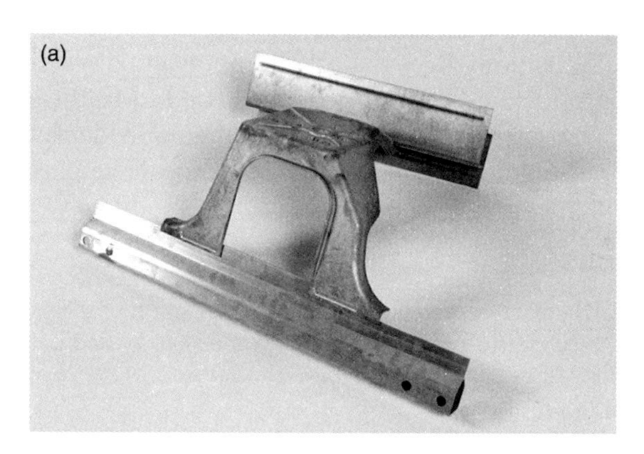

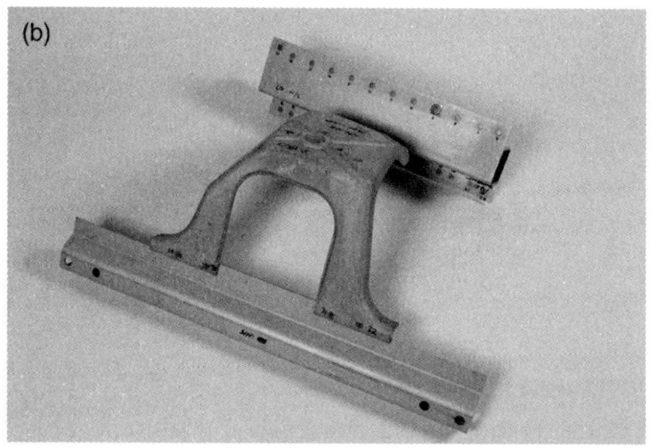

8.35 USCAR demo structure build using (a) FSLW (friction stir linear welding); and (b) LSPR (laser-assisted self pierce rivet joining processes:[93] (a) USCAR demo structure build using FSLW joining process; and (b) USCAR demo structure build using LSPR joining process.

automotive magnesium applications using front end structure as a test bed (see Fig. 8.34), they also promote primary magnesium production, component manufacturing, and fundamental research to advanced computational tools like Integrated Computational Materials Engineering (ICME). Such technologies have been demonstrated in a 'demo' structure designed and built by the USCAR team,[93] employing friction-stir linear lap welding (FSLW) and laser-assisted self-pierce rivets (LSPR), both with and without adhesive bonding, see Fig. 8.35. It is very encouraging that many of these international and interdisciplinary collaborations are being nurtured for magnesium. It is expected that future developments exploiting the new characterization tools available will provide the much needed breakthroughs to design new wrought magnesium alloys and low-cost corrosion mitigating solutions to increase the use of magnesium, the lightest structural metal!

8.9 Acknowledgements

The author gratefully acknowledges the collaboration and discussions with Drs. Anil Sachdev, Paul Krajewski, Bob R. Powell Raj K. Mishra and Mark Verbrugge of GM, Warren, MI, USA, and many other individuals who provided support and information on magnesium research and applications.

8.10 References

1. H. Davy, 'Electro-chemical researches on the decomposition of the earths; with observations on the metals obtained from the alkaline earths, and on the amalgam procured from ammonia', *Philosophical Transactions of the Royal Society of London*, 1808, **98**, 333–370.
2. U.S. Geological Survey, *Mineral Commodity Summaries – Magnesium Metal*, January 2011, http://minerals.usgs.gov/minerals/pubs/commodity/magnesium/mcs-2011-mgmet.pdf, 98–99.
3. R.E. Brown, '*Future of Magnesium Developments in 21st Century'*, Presentation at *Materials Science & Technology Conference*, Pittsburgh, PA, USA, October 5–9, 2008.
4. A.I. Taub, P.E. Krajewski, A.A. Luo, and J.N. Owens, 'The evolution of technology for materials processing over the last 50 years: the automotive example', *JOM (Journal of Metals)*, 2007, **59**(2), 48–57.
5. K.U. Kainer, *Magnesium – Alloys and Technologies*, Wiley-VCH, Weinheim, Germany, 2003.
6. A.A. Luo, 'Magnesium: current and potential automotive applications', *JOM (Journal of Metals)*, 2002, **54**(2), 42–48.
7. H.-H. Becker, ' Status, potential and challenges for automotive magnesium applications from the point of view of an OEM', presentation at *65th Annual World Magnesium Conference*, May 18–20, 2008, Warsaw, Poland.
8. Y. Michiura, 'Current magnesium research and application in automotive industry in Japan', presentation at *International Automotive Body Congress*, November 7–8, 2007, Troy, MI, USA.

9. J.S. Balzer, P.K. Dellock, M.H. Maj, G.S. Cole, D. Reed, T. Davis, T. Lawson and G. Simonds, *Structural Magnesium Front End Support*, SAE Technical Paper 2003–01–0185, SAE International, Warrendale, PA, 2003.
10. M. Hoeschl, W. Wagener and J. Wolf, *BMW's Magnesium-Aluminium Composite Crankcase, State-of-the-Art Light Metal Casting and Manufacturing*, SAE Technical Paper 2006–01–0069, SAE International, Warrendale, PA, 2006.
11. C.J. Duke and S. Logan, 'Lightweight magnesium spare tire carrier', *Proceedings of the 64th Annual World Magnesium Conference, International Magnesium Association*, 2007, Wauconda, IL 60084, USA, 75–80.
12. M.M. Avedesian and H. Baker, *ASM Specialty Handbook, Magnesium and Magnesium Alloys*, ASM International, Materials Park, OH, 1999.
13. Timminco Magnesium Wrought Products, Timminco Corporation Brochure, Aurora, CO, 1998.
14. A.A. Luo and A.K. Sachdev, 'Development of a new wrought magnesium-aluminum-manganese alloy AM30', *Metallurgical and Materials Transactions A*, 2007, **38A**, 1184–1192.
15. ASM, *Metals Handbook, Desk Edition*, ASM International, Materials Park, OH, 1998.
16. M.F. Ashby, 'Performance indices', *Metals Handbook*, Vol. **20**, ASM International, Materials Park, OH, 1998, 281–290.
17. R.S. Busk, *Magnesium Product Design*, Marcel Dekker, Inc., New York and Basel, 1987.
18. R.K. Mishra, A.K. Gupta, P.R. Rao, A.K. Sachdev, A.M. Kumar and A.A. Luo, 'Influence of cerium on the texture and ductility of magnesium extrusions', *Scripta Materialia*, 2008, **59**, 562–565.
19. A.A. Luo, W. Wu, L. Jin, R.K. Mishra, A.K. Sachdev, W. Ding, 'Microstructure and mechanical properties of extruded magnesium-aluminum-cerium alloy tubes', *Metallurgical and Materials Transactions A*, 2010, **41A**, 2662–2674.
20. A.A. Luo, R.K. Mishra and A.K. Sachdev, 'High-ductility magnesium-zinc-cerium extrusion alloy', *Scripta Materialia*, 2011, **64**, 410–413.
21. X. Zeng, Y. Wang, W. Ding, A.A. Luo and A.K. Sachdev, 'Effect of strontium on the microstructure, mechanical properties, and fracture behavior of AZ31 magnesium alloy', *Metallurgical and Materials Transactions A*, 2006, **37A**, 1333–1341.
22. Alloy Digest, 'Magnesium – LA141A', Filing Code M-56 Magnesium Alloy, February 1964.
23. J. H. Jackson, P.D. Frost, A.C. Loonam, L.W. Eastwood and C. H. Lorig, 'Mg-Li Base Alloys — Preparation, Fabrication, and General Characteristics', *Transactions of AIME*, 1949, **185**, 149.
24. G.J. Shen and B.J. Duggan, 'Texture development in a cold-rolled and annealed body-centered-cubic Mg-Li alloy', *Metallurgical and Materials Trans. A*, 2007, **38A**, 2593–2601.
25. G.V. Raynor and J.R. Kench, 'The theta phase in magnesium-lithium-silver alloys, with reference to instability after ageing', *Journal of the Institute of Metals*, 1959–60, **88**, 209.
26. J.C. McDonald, *Journal of the Institute of Metals*, 1969, **97**, 353.
27. Y. Kojima, M. Inoue and O. Tanno, 'Superplasticity in Mg-Li alloy', *Journal of the Japanese Institute of Metals*, 1990, **54**, 354.
28. S. Hori and W. Fujitani, 'Cold Workability of Mg-Li and Mg-Li-Zr Alloys', *Journal of Japanese Institute of Light Metals*, 1990, **40**, 285.

29. K. Matsuzawa, T. Koshihara, S. Ochiai and Y. Kojima, *Journal of Japanese Institute of Light Metals*, 1990, **40**, 659.
30. H. Takuda, H. Matsusaka, S. Kikuchi and K. Kubota, 'Tensile properties of a few Mg-Li-Zn alloy thin sheets', *Journal of Materials Science*, 2002, **37**, 51–57.
31. Y. Chino, K. Sassa and M. Mabuchi, 'Texture and stretch formability of Mg-1.5mass%Zn-0.2mass%Ce alloy rolled at different rolling temperatures', *Materials Transactions*, 2008, **49**, 2916–2918.
32. F. Pan and E. Han, *High-Performance Wrought Magnesium Alloys and Processing Technologies*, Science Press, Beijing, 2007.
33. K. Savage and J.F. King, 'Hydrostatic extrusion of magnesium', in *Magnesium Alloys and Their Applications*, ed. K.U. Kainer, Wiley-VCH, Weinheim, Germany, 2000, 609–614.
34. A.A. Luo and A.K. Sachdev, 'Bending and hydroforming of aluminum and magnesium alloy tubes', *Hydroforming for Advanced Manufacturing*, Woodhead Publishing Ltd, Cambridge, UK, 2008, 238–266.
35. J. Becker, G. Fischer and K. Schemme, 'Light weight construction using extruded and forged semi-finished products made of magnesium alloys', in *Magnesium Alloys and Their Applications*, eds. B.L. Mordike and K.U. Kainer, Werkstoff-Informationsgesellschaft Frankfurt, Germany, 1998, 15–29.
36. A.A. Luo and A.K. Sachdev, 'Development of a moderate temperature bending process for magnesium alloy extrusions', *Materials Science Forum*, 2005, **488–489**, 447–482.
37. A.A. Luo and A.K. Sachdev, 'Mechanical properties and microstructure of AZ31 magnesium alloy tubes', in *Magnesium Technology 2004*, ed. A.A. Luo, TMS, Warrendale, PA, 2004, 79–85.
38. J. Enss, T. Everetz, T. Reier and P. Juchmann, 'Properties of magnesium rolled products', in *Magnesium Alloys and Their Applications*, ed. K.U. Kainer, Wiley-VCH, Weinheim, Germany, 2000, 591–595.
39. P.E. Krajewski, 'Elevated temperature forming of sheet magnesium alloys', in *Magnesium Technology 2002*, ed. H.I. Kaplan, TMS, Warrendale, PA, 2002, 175–179.
40. P.E. Krajewski, *Elevated Temperature Behavior of Sheet Magnesium Alloys*, SAE Technical Paper 2001–01–3104, SAE International, Warrendale, PA, 2001.
41. H. Palkowski and L. Wondraczek, 'Thin slab casting as a new possibility for economic magnesium sheet production', in *Magnesium – Proceedings of the 6th International Conference Magnesium Alloys and Their Applications*, ed. K.U. Kainer, Wiley-VCH, Weinheim, Germany, 2003, 774–782.
42. A. Moll, M. Mekkaoui, S. Schumann and H. Friedrich, 'Application of Mg sheets in car body structures', in *Magnesium – Proceedings of the 6th International Conference Magnesium Alloys and Their Applications*, ed. K.U. Kainer, Wiley-VCH, Weinheim, Germany, 2003, 935–942.
43. O. Duygulu, S. Ucuncuoglu, G. Oktay, D. Sultan Temur, O. Yucel and A. Arslan Kaya, 'Development of 1500mm wide wrought magnesium alloys by twin roll casting technique in Turkey', in *Magnesium Technology 2009*, eds. E.A. Nyberg, S.R. Agnew, N.R. Neelameggham and M.O. Pekguleryuz, TMS, Warrendale, PA, 2009, 379–384.
44. S. Braunig, M. During, H. Hartmann and B. Viehweger, 'Magnesium sheets for industrial applications', in *Magnesium – Proceedings of the 6th International*

Conference Magnesium Alloys and Their Applications, ed. K.U. Kainer, Wiley-VCH, Weinheim, Germany, 2003, 955–961.

45. B.R. Powell, P.E. Krajewski and A.A. Luo, 'Magnesium alloys', *Materials Design and Manufacturing for Lightweight Vehicles*, Woodhead Publishing Ltd, Cambridge, UK, 2010, 114–168.
46. V. Kaese, L. Greve, S. Juttner, M. Goede, S. Schumann, H. Friedrich, W. Holl and W. Ritter, 'Approaches to use magnesium as structural material in car body', in *Magnesium – Proceedings of the 6th International Conference Magnesium Alloys and Their Applications*, ed. K.U. Kainer, Wiley-VCH, Weinheim, Germany, 2003, 949–954.
47. P. Juchmann and S. Wolff, 'New perspectives with magnesium sheet', in *Magnesium – Proceedings of the 6th International Conference Magnesium Alloys and Their Applications*, ed. K.U. Kainer, Wiley-VCH, Weinheim, Germany, 2003, 1006–1012.
48. R. Neugebauer and M. Seifert, 'Results of tempered hydroforming of magnesium sheets with hydroforming fluids', in *Magnesium – Proceedings of the 6th International Conference Magnesium Alloys and Their Applications*, ed. K.U. Kainer, Wiley-VCH, Weinheim, Germany, 2003, 306–312.
49. G. Kurz, 'Heated hydro-mechanical deep drawing of magnesium sheet metal', in *Magnesium Technology 2004*, ed. A.A. Luo, TMS, Warrendale, PA, 2004, 67–71.
50. K. Siegert and S. Jaeger, 'Pneumatic bulging of magnesium AZ31 sheet metal at elevated temperatures', in *Magnesium Technology 2004*, ed. A.A. Luo, TMS, Warrendale, PA, 2004, 87–90.
51. P.E. Krajewski and J.G. Schroth, 'Overview of quick plastic forming technology', *Materials Science Forum*, 2007, **551/552**, 3–12.
52. J.T. Carter, P.E Krajewski and R. Verma, 'The hot blow forming of AZ31 Mg sheet: formability assessment and application development', *Journal of Minerals, Metals, and Materials*, 2008, **60**(11), 77–81.
53. A. Stern, A. Munitz and G. Kohn, 'Application of welding technologies for joining of Mg alloys: microstructure and mechanical properties', in *Magnesium Technology 2003*, ed. H.I. Kaplan, TMS, Warrendale, PA, 2003, 163–168.
54. S. Lathabai, K.J. Barton, D. Harris, P.G. Lloyd, D.M. Viano and A. McLean, 'Welding and weldability of AZ31B by gas tungsten arc and laser beam welding processes', in *Magnesium Technology 2003*, ed. H.I. Kaplan, TMS, Warrendale, PA, 2003, 157–162.
55. K.G. Watkins, 'Laser welding of magnesium alloys', in *Magnesium Technology 2003*, ed. H.I. Kaplan, TMS, Warrendale, PA, 2003, 153–156.
56. R. Johnson and P. Threadgill, 'Friction stir welding of magnesium alloys', in *Magnesium Technology 2003*, ed. H.I. Kaplan, TMS, Warrendale, PA, 2003, 147–152.
57. N. Li, T.-Y. Pan, R. Cooper and D.Q. Houston, 'Friction stir welding of magnesium AM60 alloy', in *Magnesium Technology 2004*, ed. A.A. Luo, TMS, Warrendale, PA, 2004, 19–23.
58. J.I. Skar, H. Gjestland, L.D. Oosterkamp and D.L. Albright, 'Friction stir welding of magnesium die castings', in *Magnesium Technology 2004*, ed. A.A. Luo, TMS, Warrendale, PA, 2004, 25–30.
59. A.C. Somasekharan and L.E. Murr, 'Fundamental studies of the friction stir welding of magnesium alloys to 6061-T6 aluminum', in *Magnesium Technology 2004*, ed. A.A. Luo, TMS, Warrendale, PA, 2004, 31–36.

60. L. Budde, J. Bischoff and T. Widder, 'Low-temperature joining of magnesium-materials in vehicle constructions', in *Magnesium Alloys and Their Applications*, eds. B.L. Mordike and K.U. Kainer, Werkstoff-Informationsgesells chaft Frankfurt, Germany, 1998, 613–618.
61. G.D. Wardlow, 'A changing world with different rules – new opportunities for magnesium alloys?', presentation at the *64th Annual World Magnesium Conference*, Vancouver, BC, Canada, May 13th–15th, 2007.
62. O. Pashkova, I. Ostrovsky and Y. Henn, 'Present state and future of magnesium application in aerospace industry', presentation at *New Challenges in Aeronautics*, Moscow, 2007.
63. E.F. Emley, *Principles of Magnesium Technology*, Pergamon Press, 1966.
64. S. Schumann and H. Friedrich, 'The use of magnesium in cars – today and in the future', in *Magnesium Alloys and their Applications*, eds. B. Mordike and K. Kainer, Werkstoff-Informationsgesellschaft, Frankfurt, Germany, 3–13, 1998.
65. H. Friedrich and S. Schumann, 'Research for a 'new age of magnesium' in the automotive industry', *Journal of Materials Processing Technology*, 2001, **117**, 276–281.
66. F. Hollrigl-Rosta, 'Magnesium in Volkswagen', *Light Metal Age*, 1980, 22–29.
67. J. Hillis, *The Effects of Heavy Metal Contamination on Magnesium Corrosion Performance*, SAE Technical Paper No. 830523, SAE International, Warrendale, PA, USA, 1983.
68. L.T. Barnes, 'Rolled magnesium products, what goes around, comes around', *Proceedings of the International Magnesium Association*, Chicago, IL, 1992, 29–43.
69. A. Hector and W. Heiss, *Magnesium Die-Castings as Structural Members in the Integral Seat of the New Mercedes-Benz Roadster*, SAE Technical Paper No. 900798, SAE, Warrendale, PA, 1990.
70. D. Alexander, *Intier Seats are Customer Driven*, Automotive Engineering International, May 2004, 22–24.
71. L. Riopelle, 'Magnesium applications', *International Magnesium Association (IMA) Annual Magnesium in Automotive Seminar*, Livonia, MI, April 20, 2004.
72. S. Gibbs, 'Magnesium structural part parts with Myth', *Metal Casting Design and Purchasing*, July/August 2010, 29–33.
73. R. Verma and J.T. Carter, *Quick Plastic Forming of a Decklid Inner Panel with Commercial AZ31 Magnesium Sheet*, SAE International Technical Paper No. 2006–01–0525, 2006, SAE International, Warrendale, Pennsylvania.
74. S. Logan, A. Kizyma, C. Patterson and S. Rama, *Lightweight Magnesium-Intensive Body Structure*, SAE International Technical Paper No. 2006–01–0523, 2006, SAE International, Warrendale, Pennsylvania.
75. J. Aragones, K. Goundan, S. Kolp, R. Osborne, L. Ouimet and W. Pinch, *Development of the 2006 Corvette Z06 Structural Cast Magnesium Crossmember*, SAE International Technical Paper No. 2005–01–0340, 2005, SAE International, Warrendale, Pennsylvania.
76. J. Greiner, C. Doerr, H. Nauerz and M. Graeve, *The New '7G-TRONIC' of Mercedes-Benz: Innovative Transmission Technology for Better Driving Performance, Comfort, and Fuel Economy*, SAE Technical Paper No. 2004–01–0649, 2004, SAE International, Warrendale, PA, USA.

77. S. Koike, K. Washizu, S. Tanaka, T. Baba and K. Kikawa: SAE Technical Paper No. 2000–01–1117; 2000, Warrendale, PA, SAE International, Warrendale, PA, USA.
78. B. Powell, L.J. Ouime, J.E. Allison, J.A. Hines, R.S. Beals, L. Kopka and P.P. Ried, 'Progress toward a magnesium-intensive engine: the USAMP magnesium powertrain cast components project', *SAE 2004 Transactions, Journal of Materials and Manufacturing Paper No. 2004–1–0654*, SAE International, Warrendale, PA, USA, 250–259.
79. B.R. Powell, 'Magnesium powertrain cast components', published in *FY2008 Annual Progress Report for Automotive Lightweighting Materials,* U.S. Department of Energy, Washington, D.C., April 2009.
80. International Magnesium Association, *Magnesium's Tough Strength Endures Abuse to Protect Portable Electronic Devices*, International Magnesium Association, Wauconda, IL, USA, 2008.
81. International Magnesium Association, *Lighter Magnesium Improves Power Tool Performance*, International Magnesium Association, Wauconda, IL, USA, 2008.
82. C.L. Mendis, C.J. Bettles, M.A. Gibson and C.R. Hutchinson, 'An enhanced age hardening response in Mg–Sn based alloys containing Zn', *Materials Science and Engineering A*, 2006, **435/436**, 163–171.
83. A.A. Luo, P. Fu, L. Peng, X. Kang, Z. Li and T. Zhu, 'Solidification microstructure and mechanical properties of cast magnesium-aluminum-tin alloys', *Metallurgical and Materials Transactions A*, 2012, **43A**, 360–368.
84. P. Fu, L. Peng, H. Jiang, J. Chang and C. Zhai, 'Effects of heat treatments on the microstructures and mechanical properties of Mg-3Nd-0.2Zn-0.4Zr (wt.%) alloy', *Materials Science and Engineering A*, 2008, **486**, 183–192.
85. A.A. Luo, R.K. Mishra, B.R. Powell and A.K. Sachdev, 'Magnesium alloy development for automotive applications', *Materials Science Forum*, 2012, **706–709**, 69–82.
86. J. Bohlen, M. Nuernberg, J.W. Senn, D. Letzig and S.R. Agnew, 'The texture and anisotropy of magnesium-zinc-rare earth alloy sheets', *Acta Materialia*, 2007, **55**(6), 2101–2112.
87. A.A. Luo, W. Shi, K. Sadayappan and E.A. Nyberg, 'Magnesium front end research and development: phase I progress report of a Canada-China-USA collaboration', *Proceedings of IMA 67th Annual World Magnesium Conference*, International Magnesium Association (IMA), Wauconda, IL, USA, 2010.
88. M. Easton, A. Beer, M. Barnett, C. Davies, G. Dunlop, Y. Durandet, S. Blacket, T. Hilditch, and P. Beggs, 'Magnesium alloy applications in automotive structures', *JOM*, 2008, **60**(11), 57–62.
89. D.A. Wagner, S.D. Logan, K. Wang, T. Skszek and C.P. Salisbury, 'Test results and FEA predictions from magnesium AM30 extruded beams in bending and axial compression', published in *Magnesium Technology 2009*, eds. E.A. Nyberg, S.R. Agnew, N.R. Neelameggham and M.O. Pekguleryuz, 2009, TMS, Warrendale, PA.
90. S. Begum, D.L. Chen, S. Xu and A.A. Luo, 'Low-cycle fatigue properties of an extruded AZ31 magnesium alloy', *International Journal of Fatigue*, 2009, **31**, 726–735.

91. S. Begum, D.L. Chen, S. Xu and A.A. Luo, 'Strain-controlled low-cycle fatigue properties of a newly developed extruded magnesium alloy', *Metallurgical and Materials Transactions A*, 2008, **39A**, 3014–3026.
92. C.L. Fan, D.L. Chen and A.A. Luo, 'Dependence of the distribution of deformation twins on strain amplitudes in an extruded magnesium alloy after cyclic deformation', *Materials Science and Engineering: A*, 2009, **519**, 38–45.
93. A.A. Luo, J.F. Quinn, R. Verma, Y.-M. Wang, T.M. Lee, D.A. Wagner, J.H. Forsmark, X. Su, J. Zindel, M. Li, S.D. Logan, S. Bilkhu, R.C. McCune, 'The USAMP magnesium front end research and development project – focusing on a 'Demonstration' structure', *Light Metal Age*, **70**, 2012, in press.

9
Applications: magnesium-based metal matrix composites (MMCs)

H. DIERINGA, Helmholtz-Zentrum Geesthacht, Germany

DOI: 10.1533/9780857097293.317

Abstract: This chapter gives an introduction to different forms and production routes of magnesium-based metal matrix composites (MMCs). Firstly, the different kinds and forms of reinforcements are discussed, followed by the typical casting processes used for processing the MMCs. Powder metallurgical processes have already been described, so we omit these. Subsequently, some examples of the mechanical properties are given, especially at elevated temperatures.

Key words: ceramics, particle, fibre, hybrid, reinforcement, creep strengthening.

9.1 Introduction

The extension of applications for magnesium-based materials is limited, when only alloying is taken into account. Aluminium-free alloys containing rare earths exhibit very good high-temperature strength and creep resistance, but improving these properties is possible only by reinforcement with ceramic materials. Typically, one distinguishes between the matrix, which in this case is the magnesium alloy, and the reinforcement. Different materials and shapes have been used as reinforcement: ceramic particles and ceramic fibres or a combination of these, which is called hybrid reinforcement. Whereas during the 1980s and 1990s mainly micrometer-sized particles and fibres were employed, in recent years nanometer-sized particles have also been introduced into magnesium. This is due to significantly reduced prices for nanoparticles and their easier availability. Compared to micrometer-sized reinforcements, the additional effects of Orowan strengthening and grain refinement are higher when nanoparticles are used. Various types of carbon nanotubes have also been investigated, for their strengthening of magnesium alloys (Dieringa, 2011). The prerequisite for real strengthening, when particles or fibres are introduced in metallic matrices, is a good bonding between them. This requires a chemical reaction as a minimum. An interface has to be built during processing that is able to transfer stresses. Usually the coefficient of thermal expansion of the matrix and reinforcement is different. However, the interface has to be strong enough to withstand the resulting stresses, when the composites are heated up. In order to diversify the mechanical properties, not only the kind and shape of reinforcement can be varied, but also the amount of reinforcement and the alignment. The latter is mainly true for fibres. Generally speaking, it is possible to tailor the mechanical properties of the resulting composite by using the right reinforcement. This gives the possibility to replace other materials with a weight saving Mg–MMC solution. The reason magnesium-based composites are seldom used in practice is the high cost of production. Not only the ceramic reinforcement, but also the additional production steps make this material uncompetitive in industrial applications.

9.2 Reinforcements for magnesium metal matrix composites (MMCs)

For the magnesium-based composites, ceramic reinforcements and carbon fibres were used, in order to study their mechanical and physical properties. Metallic particles or fibres were not included in the selection, due to their poor expected corrosion properties. It is possible to tailor the mechanical properties of the composite within a certain range by choosing the kind, the geometry, the amount and, in the case of fibres or whiskers, the geometric distribution of reinforcements. If more than one kind of reinforcement shape (particles, short fibres, whiskers or long fibres) is used, the composite is called a hybrid reinforced composite. In this case it is possible to profit from the different advantages inherent in each reinforcement type by combining them. Figure 9.1 shows a schematic of different kinds of continuous and discontinuous reinforcements. The different types of reinforcements will later be described in more detail.

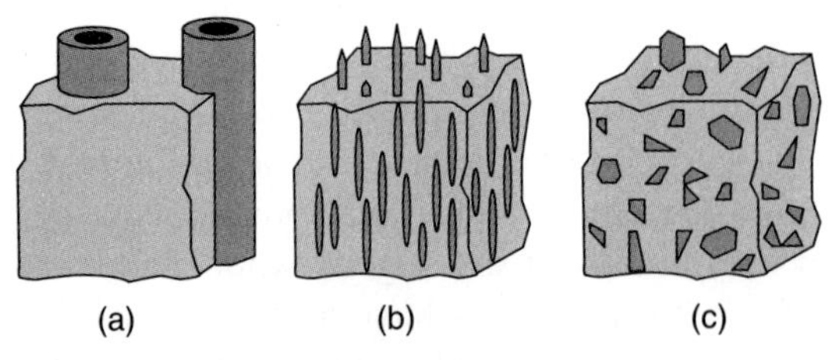

9.1 Different kinds of reinforcement types: (a) continuous long fibres, (b) discontinuous short fibres or whiskers and (c) particles in the matrix.

9.2.1 Particle reinforcement

For the reinforcement of magnesium alloys, hard ceramic particles are usually chosen that are in use for grinding and polishing applications. Typical particle materials are nitrides (BN, AlN, TiN, ZrN), carbides (B_4C, ZrC, SiC, TiC, W_2C, WC), oxides (ZrO_2, Al_2O_3, Cr_2O_3) and borides (TiB_2, ZrB_2, WB). When choosing the reinforcement type, the chemical reactivity between the matrix of the alloy and the particles has to be taken into account. To achieve high mechanical properties, a reaction layer is necessary, because external and internal (thermal) stresses have to be transferred. Not only the kind and size of reinforcement influence the properties of the composite, but also the shape of the particles. Round, blocky and platelet forms are available. Often the platelet form is preferred, because for forming a good bonding, contact with a crystal plane may be necessary, which is more possible in platelets rather than in round particles. Particles with sharp edges are usually avoided, because the edges may act as the starting points for cracks when the material is stressed. Compared to other metals, the magnesium melt is very reactive and forms reaction product layers with all these reinforcements. If the duration of contact with the melt is too long, some ceramic materials can even dissolve. In order to reach a low-density composite, the density of the reinforcement particles should be low. However, another reason for needing a good density combination is the fact that when composites are fabricated by melt-stirring processes, the differences between matrix and particle densities should be kept small, because particles would otherwise sink down or float to the top of the melt.

9.2.2 Continuous fibre reinforcement

Compared to the variety of particles, there are fewer materials available in continuous fibre form, but many more different shapes, mechanical properties and suppliers to choose from on the market. Usually, either single fibres, called monofilaments, or multifilaments are used, the latter having hundreds or thousands of thin fibres in the range 5–25 μm. For magnesium-based composites, carbon fibres are often used in R&D projects as well as in a few industrial applications. This is due to their low density, high Young's modulus and tensile strength, low coefficient of thermal expansion (CTE), good thermal and electrical conductivity, high availability and resistivity against magnesium melt. Two differently processed carbon-fibre types are available: the polyacrylonitrile (PAN) fibres (Table 9.1) and the pitch fibres (Table 9.2). Continuous long fibres based on oxides are mostly alumina-based. In their use as reinforcement for magnesium alloys, they have some advantages. Alumina-based fibres offer good processing abilities, because

Table 9.1 Properties of PAN-based C-fibres

	Standard fibres	Aircraft applications	
		Low modulus	High modulus
Young's modulus (GPa)	288	220–241	345–448
Tensile strength (MPa)	380	3450–4830	3450–5520
Elongation (%)	1.6	1.5–2.2	0.7–1.0
Electr. resistance (μΩ cm)	1650	1650	900
CTE (10^{-6} K^{-1})	–0.4	–0.55	–0.75
Manufacturer	Zoltec, SGL, Toray, Mitsubishi Rayon	Hexcel, Toray, Mitsubishi Rayon, Tenax, Soficar	

Source: ASM, 2001.

Table 9.2 Properties of pitch-based C-fibres

	Low modulus	High modulus	Ultrahigh modulus
Young's modulus (GPa)	170–241	380–620	690–965
Tensile strength (MPa)	1380–3100	1900–2750	2410
Elongation (%)	0.9	0.5	0.27–0.4
Electr. resistance (μΩ cm)	1300	900	130–220
CTE (10^{-6} K^{-1})		–0.9	–1.6
Manufacturer	BPAmoco, Mitsubishi Casei	BPAmoco	

Source: ASM, 2001.

during wetting with magnesium melt a stable spinel layer forms at the interface. All the fibres listed in Table 9.3 are cheap and stable in air and protective gases. They all have high strength, low CTE and a sufficiently low density. Depending on the content of SiO_2 and B_2O_3, different production routes can be chosen for these fibres. With higher content, fibres can be produced by spinning or melting, because the melting point is relatively low. Fibres approaching alumina-only content have a higher melting point and such processes are not economic. These fibres were produced by precursor technologies. Varying amounts of SiO_2 lead to different mechanical properties. Without silica, the fibres consist of α-alumina alone, which has a high Young's modulus, and such fibres are brittle, resulting only in reduced strength. The most recent examples of continuous long fibres are SiC multifilament fibres. These fibres can be used as reinforcement for magnesium composites that are manufactured using spun polymer precursors. Based on polycarbosilane or polytitanocarbosilane fibres, these materials are transformed to ceramic fibres at 1300°C in a protective gas atmosphere. The resulting fibres have a thin protective layer of SiO_2, which enables

Table 9.3 Properties and manufacturer of commercially available alumina-based fibres

Name	Manu-facturer	Content (wt%)	Dia (μm)	Density (g/cm^3)	Young's modulus (GPa)	UTS (MPa)	CTE (10^{-6} K^{-1})
Altex	Sumitomo	85 Al_2O_3 15 SiO_2	15	3.3	210	2000	7.9
Alcen	Nitivy	70 Al_2O_3 30 SiO_2	7–10	3.1	170	2000	
Nextel 312	3M	62 Al_2O_3 24 SiO_2 14 B_2O_3	10–12	2.7	150	1700	3.0
Nextel 440	3M	70 Al_2O_3 28 SiO_2 2 B_2O_3	10–12	3.05	190	2000	5.3
Nextel 550	3M	73 Al_2O_3 27 SiO_2	10–12	3.03	193	2000	5.3
Nextel 610	3M	>99 Al_2O_3	12	3.9	373	3100	7.9
Nextel 650	3M	89 Al_2O_3 10 ZrO_2 1 Y_2O_3	11	4.1	358	2500	8.0
Nextel 720	3M	100 Al_2O_3	12	3.4	260	2100	6.0
Almax	Mitsui	99.5 Al_2O_3	10	3.6	330	1800	8.8
Saphikon	Saphikon	100 Al_2O_3	125	3.98	460	3500	9.0
Sumica	Saphikon	85 Al_2O_3 15 SiO_2	9	3.2	250		
Saffil	Saffil	96 Al_2O_3 4 SiO_2	3.0	3.3–3.5	300–330	2000	

Source: Dieringa, 2006.

good wetting with the magnesium melt. SiC fibres have a density close to 2.5 g/cm^3 and a Young's modulus around 200 GPa. They are quite cheap and very stable, even at high temperatures and, similar to alumina-based fibres, they exhibit only small thermal expansion.

9.2.3 Whisker reinforcement

Whiskers are small, needle-like single crystals with an aspect ratio of roughly 10 or more and a diameter of 1 μm. They are processed by growing from oversaturated gases, or electrolysis from solutions or solids. Due to the manufacturing conditions, they have a very low defect density. Apart from SiC and Si_3N_4, which have been used for magnesium-based composites as reinforcement, $9Al_2O_3 \times B_2O_3$ and $K_2O \times TiO_2$ combinations are also of interest. The fact that whiskers are very thin and small has led to discussions about the health risks. They can be inhaled and do not degrade in the lung, leading to the suspicions of their being potentially carcinogenetic.

9.2.4 Short-fibre reinforcement

Concerning their reinforcing and strengthening effects, short fibres are midway between those of long fibres and particles. Whereas long-fibre reinforcement results in extremely anisotropic mechanical properties (high strength following the fibre's alignment and low strength at perpendicular vectors), particle reinforcement leads to nearly isotropic properties. By the use of short fibres it is possible to tailor the mechanical properties to determine the amount, kind and distribution of the reinforcing fibres. Different types of short fibres are described in Tables 9.1–9.3, because all the continuous fibres can be cut into short fibres.

9.2.5 Nano-sized reinforcement

Since ceramic nanoparticles or nanofibres are available and inexpensive, these types of reinforcement are also undergoing research for magnesium-based composites. The reinforcement of magnesium alloys with very small ceramic particles or carbon nanotubes offers a wider range of property modulation, compared to the micro-sized ones. This is mainly due to the fact that Orowan strengthening is barely yielded by larger particles, because the inter-particle spacing and the size of the particles is too great (Clyne and Withers, 1993). The pure Orowan strengthening effect can be described by

$$\Delta\sigma_{\mathrm{Oro}} = \frac{0.13 G_m b}{\lambda} \ln \frac{r}{b} \qquad [9.1]$$

where $\Delta\sigma_{\mathrm{Oro}}$ is the increase in yield strength by Orowan strengthening, G_m the shear modulus of the matrix, b the Burgers vector, λ the inter-particle spacing and r the particle radius. Usually λ can be estimated as:

$$\lambda \approx d_p \left[\left(\frac{1}{2V_p} \right)^{1/3} - 1 \right] \qquad [9.2]$$

where V_p is the volume fraction of reinforcement (Zhang and Chen, 2004).

When the materials are prepared by a metallurgical melting process, an additional strengthening by the grain refining induced by the nanoparticles may be achieved. Only very small particles can act as nuclei for solidification. The strengthening by grain refinement follows the Hall–Petch equation (Bata and Pereloma, 2004)

$$\sigma_y = \sigma_i + k_y d_g^{-1/2}, \qquad [9.3]$$

where σ_y is the yield strength, σ_i the lattice friction, k_y is a material dependent constant and d_g the grain diameter. For magnesium alloys, mainly SiC, Al_2O_3 and Y_2O_3 have been used as the reinforcement. Only a few studies have been performed on SiO_2 and even carbon nanotubes (Dieringa, 2011).

9.2.6 Hybrid reinforcement

With hybrid reinforcement, a mixture of at least two different kinds of reinforcement is denoted. It does not mean a mixture of two different ceramic particles, but a combination of fibres and particles, for instance. The goal is to attain composite properties that could not have been obtained by using only one type of reinforcement. An example is the use of a framework of short fibres in which hard ceramic particles are embedded, in order to distribute them homogeneously. A fibre framework can also be used to keep particles on their place, in order to let them react during an infiltration process. The reaction product may strengthen the composite after this so called *in situ* reaction process.

9.3 Processing of magnesium composites

A huge number of different process types are in use to produce magnesium-based composites. Depending on the reinforcement type, all the processes can be subdivided into solid state or powder metallurgical (PM) and liquid phase or ingot metallurgical (IM) processes. All these processes can also be used for the fabrication of composites, whereby at the beginning of the processes a mixture or blending of magnesium alloy powders is performed. Since all the PM processes have already been described, in this chapter we will focus on ingot metallurgy processes. These liquid state processes result quite often in a very good interface of reinforcement with the magnesium matrix. The liquid processes can be further subdivided into infiltration techniques, casting processes and spray deposition. Those are the most inexpensive processing technologies for discontinuous, reinforced magnesium-based composites. In the following, some of the most common processes are described.

9.3.1 Stir casting

The stir-casting process is the easiest and cheapest processing route to produce discontinuous, reinforced composites (Luo, 1995; Saravanan and Surappa, 2000). Particles, short fibres or whiskers can be used as reinforcing components. They are introduced into the molten matrix. The force produced by stirring is needed to overcome the usually bad wetting

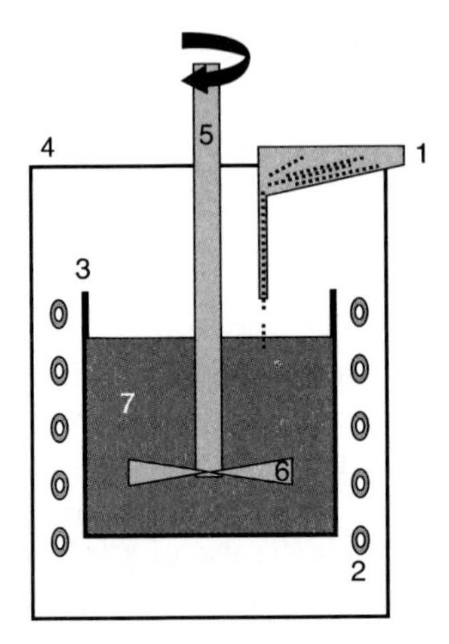

9.2 Sketch of the stir-casting process.

between the matrix and reinforcement. The duration of stirring and the melt temperature strongly influences the development of an interface between the components. A heated crucible maintains the temperature during stirring. A sketch of the process is given in Fig. 9.2. The reinforcement (1) is added into the melt (7). The crucible (3) is embedded in a vacuum chamber (4). In the heated crucible (2) the impeller (6) rotates and both a horizontal and a vertical vortex are created that deaglomerate the reinforcement and distribute it homogeneously. The use of a slight vacuum prevents the materials from forming gas entrapments or a gas boundary layer on top of the melt.

9.3.2 Compocasting

Whereas the stir-casting process is performed at temperatures above the liquidus temperature of the magnesium alloys, compocasting is done between solidus and liquidus temperature. This semi-solid material, where an already solidified fraction is surrounded by liquid, is called slurry. When the slurry is sheared at high rates the viscosity decreases. Without any introduction of reinforcement, this casting process is called rheocasting. With an addition of reinforcements the process is called compocasting. The shearing forces are larger compared to stir casting, because additional solid fractions of the magnesium alloy positively influence the rheological behaviour of the mixture. The deagglomeration, wetting and homogenisation are all improved. Three additional advantages over stir casting can be seen: (i) the temperature is lower and the semi-solid state has a much lesser tendency to burn, (ii) due to the higher viscosity of the semi-solid compared to the pure

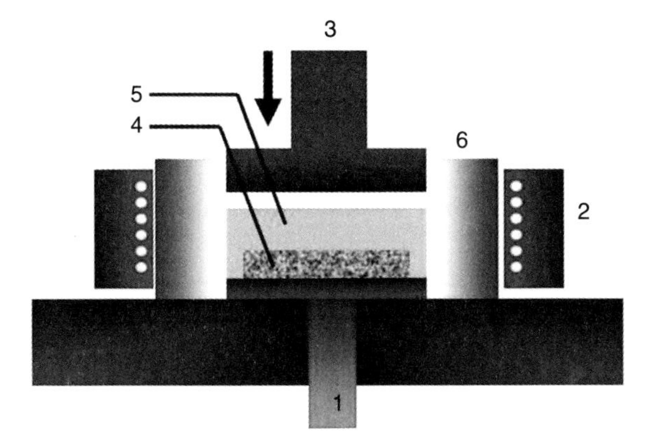

9.3 Sketch of the squeeze-casting process.

solid, the tendency of a settling of the reinforcement is reduced. Neither swimming to the top nor sinking to the bottom of the crucible takes place. Finally, (iii) due to the lower temperature, the degradation probability of the reinforcements is reduced. After distributing the reinforcement in the semi-solid, the slurry can be cast using die casting, centrifugal casting, mould casting or squeeze casting.

9.3.3 Melt infiltration casting

The melt infiltration casting or squeeze-casting process is widely used for the manufacture of fibre-reinforced magnesium composites. Prefabricated preforms containing fibres and, where required, particles are stuck together with high-temperature stable binders (silica) and have a defined volume fraction of reinforcement. This fraction is usually between 10 and 60 vol%. Most often it has a volume content of 15–25 vol% reinforcement. Due to the manufacturing process of these preforms, the fibres are frequently not randomly oriented; rather they have a planar isotropic distribution. A sketch of the squeeze-casting process is shown in Fig. 9.3. The heated preforms (4) are put in a preheated mould (6) and the superheated melt is poured over it (5). The melt is squeezed into the preform by a hydraulic ram (3). After full infiltration, the part solidifies under pressure and is taken out by an ejector (1). Due to the non-turbulent flow and solidification under pressure, the castings are free of pores and blowholes, show no shrinkage porosity, and have a fine-grained microstructure. One disadvantage is the possibility of damaging the preform during the early stages of infiltration. This process allows manufacturing of completely reinforced MMCs and parts with selectively reinforced areas.

9.3.4 Gas pressure assisted infiltration casting

Similar to the squeeze-casting process, gas pressure infiltration needs prefabricated preforms for producing composite materials. Instead of a ram, gas pressure on the melt surface is used for infiltration into the heated preform. The time taken for total infiltration in the gas pressure process is much longer compared to squeeze casting. This is important to know, because the formation of interfaces between matrix melt and reinforcement depends strongly on the time of melt contact. Often a vacuum is applied to the mould and preform, and the gas applied is usually an inert gas, such as argon.

9.3.5 Spray deposition

Spray deposition is a process in which a stream of molten melt is atomised by an inert gas, such as argon or nitrogen, and ceramic particles are added simultaneously. This process is used for manufacturing semi-finished products, which can be later extruded or forged. A substrate is located under the stream that collects the partially solidified material. Depending on the movement of the substrate, ingots, tubes or strips can be produced. Usually a density of 95% of the theoretical density is reached with this process. A subsequent consolidation step is needed to reach the full density.

9.4 Interfaces, wetting and compatibility

For improving mechanical properties, compared to unreinforced magnesium alloy, the interface between the matrix and reinforcement plays an important role. For composite materials, there has to be some chemical reaction during the manufacturing process, which leads to a more or less well-bonded interface. Without a bonding interface, no stresses can be transmitted. The two different materials (ceramic reinforcement and magnesium matrix) have moreover to fulfill some prerequisites. They should have a similar CTE. Magnesium alloys all show a CTE of approximately 27×10^{-6} K^{-1}. As can be seen in Tables 9.1–9.3, ceramic materials used for reinforcing have a thermal coefficient between 3 and 10 and C-fibres can even show negative CTE. This can lead to plastic deformation when a magnesium-based composite is heated, as shown in Fig. 9.4. The ceramic fibre with a lower CTE expands less than the metallic matrix. A compressive stress is generated in the mid section of the fibre close to it. Should this stress exceed the yield strength of the matrix, plastic deformation occurs in a way that the material shortens in the fibre's direction. A closer look at this effect is given in (Kumar *et al.*, 2003).

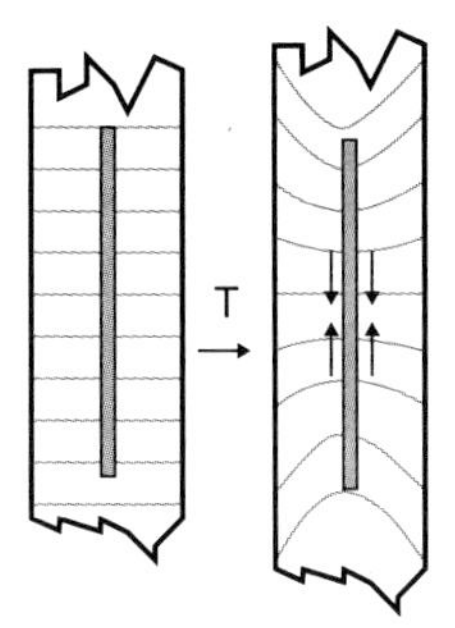

9.4 Heating of a fibre-reinforced composite material.

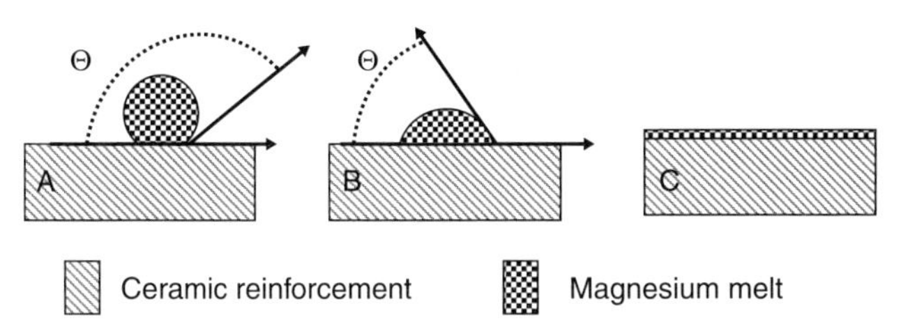

9.5 Different wetting behaviour of ceramic reinforcements. (a) $\Theta > 90°$ bad wetting; (b) $\Theta < 90°$ partial wetting and (c) $\Theta = 0$ total wetting (after Salmang and Scholze (1983)).

A prerequisite for a good matrix/reinforcement bonding is the wettability of the ceramic fibre or particle by the metallic melt. Wettability is usually determined by the ability of a liquid to spread on a solid surface. From a thermodynamic point of view, the state of the system is stable when the free energy of the complete system (liquid magnesium melt, solid ceramic reinforcement and the vapour around the liquid and the solid) reaches a minimum. By reducing the solid/liquid interface, the vapour/solid interface increases (see Fig. 9.5). Wettability is characterised by a wetting angle Θ. The smaller the Θ, the better wettability, as can be seen in Fig. 9.5.

Particularly in the case of magnesium-based MMCs, the choice of reinforcement types is limited due to severe chemical reactions that occur during the infiltration processes. The time of contact with the magnesium melt and the melt temperature influence the intensity of the chemical reaction. The chemical composition of the reinforcement and alloying elements in a magnesium alloy causes a wide spread of the interfaces that are the outcome of the chemical reactions.

It has already been mentioned that SiC, C-fibres and alumina are the best investigated reinforcements for magnesium alloys. In the case of C-fibres, magnesium carbides may form, but they are not stable in solids.

For magnesium alloys containing aluminium as alloying element (AZ31, AZ91, AE42), aluminium carbides form that are stable, but show solubility in water.

$$Mg + 2C \Rightarrow MgC_2$$

$$4Al + 3C \Rightarrow Al_4C_3$$

In case of SiC and alumina as reinforcements, the following exothermic reactions occur:

$$3Mg + Al_2O_3 \Rightarrow 3MgO + 2Al$$

$$3Mg + 4Al_2O_3 \Rightarrow 3MgAl_2O_4 + 2Al$$

$$2Mg + SiO_2 \Rightarrow 2MgO + Si$$

$$MgO + SiO_2 \Rightarrow MgSiO_3$$

$$2Mg + SiC \Rightarrow Mg_2Si + C$$

$$2Mg + Si \Rightarrow Mg_2Si$$

Concerning reactivity and enthalpies of reaction, as well as choosing temperatures and time of contact, one can tailor the interface for the application of the composite that is required.

9.5 Properties of magnesium-based MMCs

In the following section, the mechanical and physical properties of different types of magnesium-based composites are described. The section is subdivided into the different kinds of reinforcement shapes.

9.5.1 Particle reinforced magnesium

A PM route was chosen to process different SiC-particle reinforced composites based on the magnesium alloy WE54 (Garces *et al.*, 2010). Seven and 15 vol% particles of two different sizes (2 and 12 μm) were blended with WE54 powder, cold compacted and subsequently extruded at 400°C. For comparison, WE54 was processed the same way without reinforcement. After extrusion, the composites showed a smaller grain size compared to the WE54. The yield strength of the composites was a little higher and showed a drop at temperatures above 200°C, very similar to the unreinforced alloy.

The reason for the higher yield stress is assumed to be the smaller grain size and the increase in dislocation density due to a mismatch of CTE in the matrix and SiC.

Ten vol SiC particles and 3 vol Si particles were selected as reinforcement for magnesium alloy AZ91 using a squeeze-casting process for production (Trojanova *et al.*, 2009). After applying a T6 heat treatment, compression tests were performed at temperatures between RT and 300°C. Optical micrographs and analysis show a formation of Mg_2Si in coarse blocks and a Chinese script form. The composite material shows higher yield and ultimate compressive strength up to approximately 200°C. Again, the reason for the higher strength seems to be dislocation generation due to differing CTEs.

In Tiwari *et al.* (2007), pure magnesium was reinforced with 6 and 16 vol SiC particles and its corrosion behaviour was investigated and compared with unreinforced magnesium. A disintegrated melt deposition method followed by extrusion was performed for production. A polarisation experiment, galvanic corrosion tests and electrochemical impedance spectroscopy was performed on each material. In 1 molar NaCl solution, Tafel and free potential tests revealed that SiC decreases the corrosion resistance of the magnesium. However, galvanic corrosion between magnesium and SiC did not contribute significantly to the corrosion behaviour. The higher corrosion rates of the composites are assumed to depend on the defective nature of surface films on the composite.

Again, a PM route was chosen to produce AZ91 reinforced with 15 vol% SiC, AlN, TiB and Ti(C,N) particles, as well as Al_2O_3-platelets (Schröder *et al.*, 1989). AZ91 powder with a mean size of 63 μm was mixed with particles, and the strength and CTE was evaluated. An increasing strength and hardness was observed, only the AlN reinforced magnesium alloy showing a decreased strength. This phenomenon was attributed to wetting problems and the powder size distribution. In all cases, an increase in Young's modulus and wear resistance was observed, along with a decrease in CTE.

9.5.2 Short-fibre reinforced magnesium

Strength and fracture toughness was investigated in an AZ91-based composite, reinforced with 10, 15 and 20 vol% alumina fibres and compared with unreinforced AZ91 (Purazrang *et al.*, 1994). The materials were prepared by squeeze casting, where a preheated preform was infiltrated with AZ91 melt under pressure. Testing was performed on as-cast material and on T6 heat-treated material. The results are shown in Table 9.4. Hardness increases significantly when AZ91 is reinforced with alumina fibres. This effect is even higher in the T6 state. Strength and Young's modulus are

Table 9.4 Hardness and mechanical properties of AZ91 and AZ91 + 20 vol% alumina fibres in the as-cast and T6 states

	AZ91		AZ91 + 20 vol% alumina	
	as-cast	T6	as-cast	T6
Hardness HB5/125 (MPa)	461–569	637–686	745–1068	1147
Young's modulus (GPa)	41–43.5	36.2–43	63–76	64–69
Yield strength (MPa)	112–157	160–168	193–275	–
UTS (MPa)	155–211	206–216	239–295	174–194
Elongation (%)	3–4	3.2–5.4	1.6–1.8	0
K_{IC} (MPa $m^{1/2}$)	14.2–14.5	14.4–15.3	11.9–12.2	8.5–8.8

Source: Purazrang *et al.*, 1991

appreciably increased, the elongation is slightly reduced. A reduction of fracture toughness of 20% compared to unreinforced AZ91 is noteworthy.

The same authors investigated the influence of different volume fractions of alumina fibres (Purazrang *et al.*, 1994). 10, 15 and 20 vol% alumina fibres were again introduced into the AZ91 magnesium alloy via squeeze casting. The higher the proportion of fibres the higher the yield strength and ultimate tensile strength. With increasing testing temperature, yield and ultimate strength decrease and the elongation to fracture increase. Fracture toughness decrease with increasing temperature and with increasing fibre amount. The Young's modulus increases with increasing fibre content.

A creep-resistant magnesium alloy QE22 was reinforced with three different kinds of short fibres (Dieringa *et al.*, 2002). Saffil and Maftech fibres contain mainly Al_2O_3 and Supertech fibres consist of SiO_2, CaO and MgO. Composite materials were produced by a squeeze-casting process and the volume fraction of reinforcement was 20% in each case. Creep tests at 200°C and 60 MPa applied stress have shown that the already creep-resistant magnesium alloy QE22 can be further strengthened in creep resistance with Saffil fibres. However, reinforcement with Maftech and Supertech fibres weakens the alloy. The curves are shown in Fig. 9.6. A very important reason for short-fibre reinforcement is an improvement in creep resistance of the resulting composite, compared to the unreinforced matrix alloy. Usually the resistance against an applied stress at elevated temperatures is assessed by the minimum creep rate, which is the creep rate during the secondary creep stage. Unreinforced QE22 has a minimum creep rate of 6.61×10^{-9} s^{-1}. Saffil reinforced composite has 1.13×10^{-9} s^{-1}, while Maftech and Supertech composites have 3.38×10^{-8} s^{-1} and 2.46×10^{-8} s^{-1} respectively. All materials were tested for their thermal expansion behaviour. As might be expected, the unreinforced QE22 has a CTE of 26.5×10^{-6} K^{-1}. Whereas Saffil, Maftech and Supertech composites have reduced CTEs of 15.2×10^{-6} K^{-1}, 18.9×10^{-6} K^{-1}

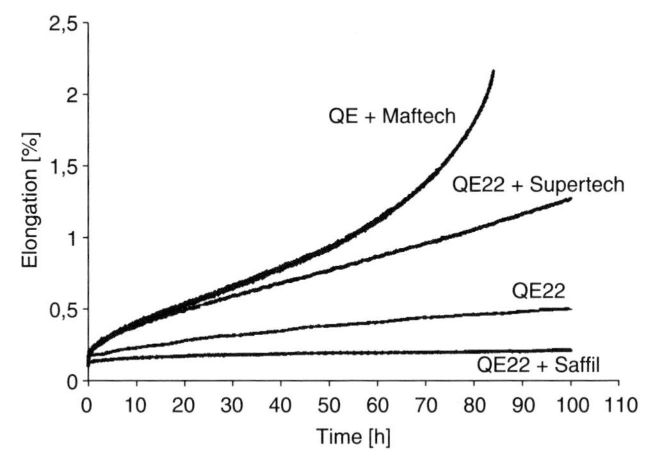

9.6 Creep curves of QE22 with different reinforcements. Tests performed at 200°C and 60 MPa.

and 19.6 × 10^{-6} K^{-1} respectively. Microstructural optical micrographs are shown in Fig. 9.7.

Minimum creep rates were evaluated in tension and compression creep over a wide range of temperatures and stresses in Dieringa *et al.* (2007). The matrix alloy AE42 and a short-fibre reinforced composite with 20 vol% Saffil fibres showed a significant increase in creep resistance in each testing direction. The minimum creep rate of the composite was two to three orders of magnitude lower, compared to unreinforced AE42. Figure 9.8 shows the creep rates of the different tests performed at 200°C. It can be seen that there is a difference in minimum creep rates even in the tension and compression creep of *one* material. Moreover, the difference between the matrix alloy and the composite is significant.

9.5.3 Nanoparticle reinforced magnesium

In Ferkel and Mordike (2001), the composite was prepared with a PM method. 3 vol% SiC particles with a mean diameter of 30 nm and pure magnesium powder with a mean diameter of 40 μm were mixed for 8 h and milled. After encapsulating and degassing, the material was extruded at 350°C. For the purposes of comparison, pure magnesium powder was treated in the same way. Light optical microscopy showed that a mean grain size of 20 μm was obtained in the pure magnesium and mixed specimens, whereas the mixed and milled specimens exhibited a grain size close to 1 μm. In TEM studies, the SiC nanoparticles were found close to the grain boundaries, forming a network. Tensile tests at RT, 100°C, 200°C and 300°C showed that the milled composite always exhibits the highest

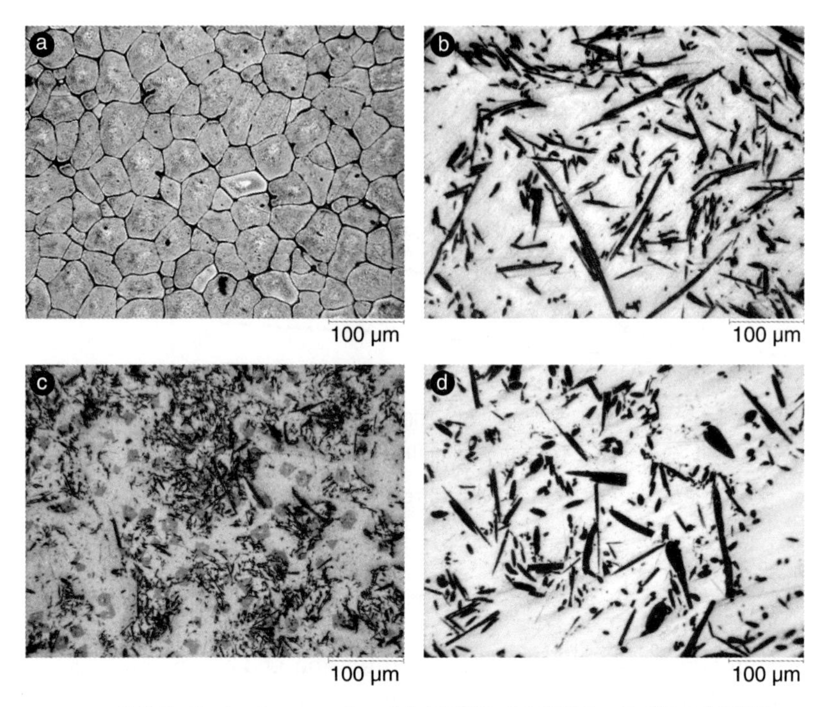

9.7 Optical micrographs of (a) QE22, (b) QE22 + Saffil, (c) QE22 + Maftech and (d) QE22 + Supertech fibres.

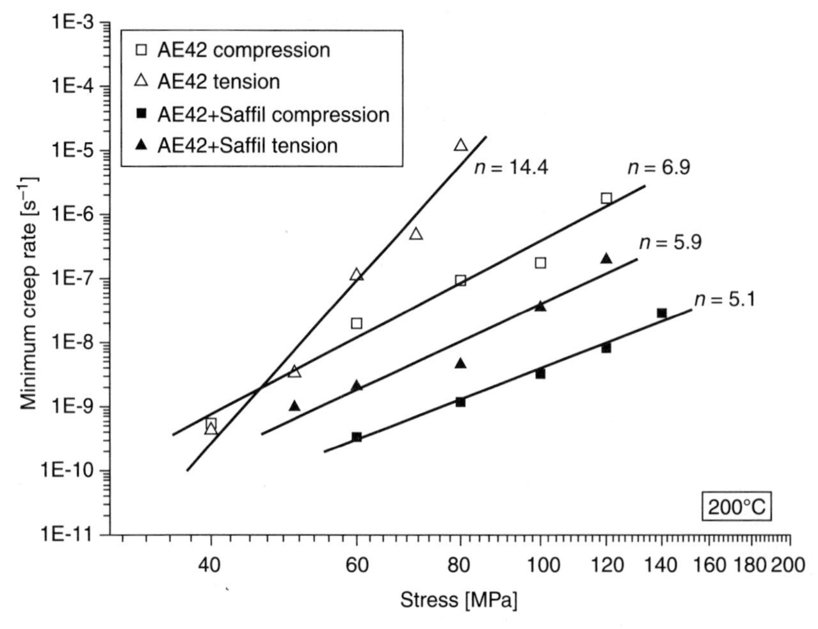

9.8 Minimum creep rates of tension and compression creep tests performed at 200°C. AE42 and its short-fibre reinforced composite (20 vol% Saffil fibres) were tested.

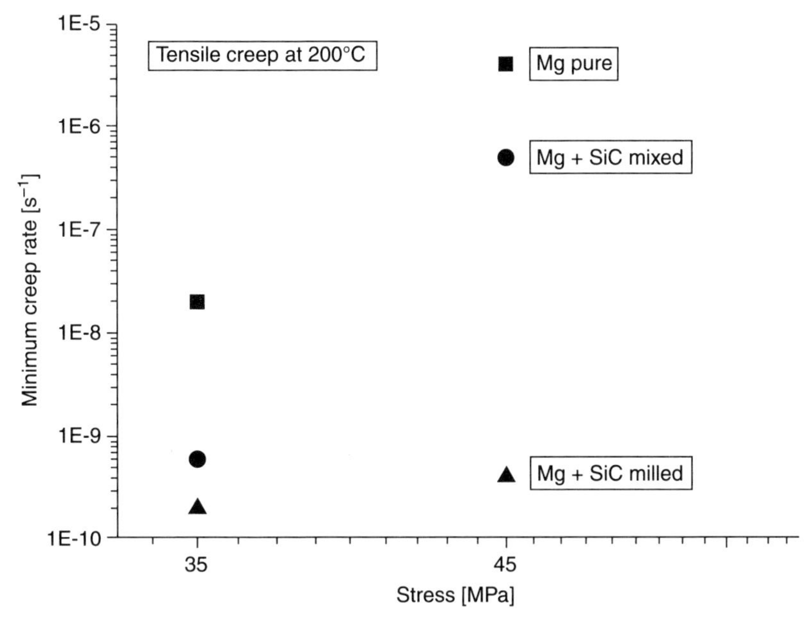

9.9 Minimum creep rates of creep tests performed at 200°C and 35 MPa and 45 MPa, respectively.

ultimate tensile stress. The smaller grain size, according to the Hall–Petch relationship, is assumed to be responsible for this. Tensile creep tests were performed at 200°C and with stresses of 35 MPa and 45 MPa. The minimum creep rates are plotted in Fig. 9.9 and it can be seen that after milling the composite shows the best creep resistance, even if the grain size is smaller compared to the others, which should enhance grain boundary sliding. SiC particles are potentially responsible for hindering this. Estimation of the stress exponent n, as a part of the power-law equation, results in values of ~10 and ~3, at temperatures of 200°C and 300°C respectively. A dislocation-movement-controlled deformation rather than diffusion creep is assumed to be the dominant process. Reducing the surface roughness of the creep specimens also significantly improves their creep resistance. Comparing the creep results with those of creep-resistant magnesium alloys (WE43, WE54 and QE22) shows that even a small amount of nano-SiC leads to similar or even better creep resistance.

Magnesium reinforced with three different amounts of 300 nm alumina particles is investigated in (Hassan, 2008). The material was produced using the disintegrated melt deposition (DMD process, followed by hot extrusion as described in (Hassan and Gupta, 2005). Three different amounts of alumina particles were introduced: 1.5, 2.5 and 5.5 wt%. The results of an investigation into the morphology, density and grain size are shown in Table 9.5. Microstructural investigations have shown that the reinforcement

Table 9.5 Density, porosity and grain size of Mg and the composites

Material	Alumina content (wt%)	Density (g/cm³)		Porosity (%)	Grain size (μm)
		Eheor.	Experim.		
Mg	–	1.7400	1.7397 ± 0.0009	0.02	49 ± 8
$Mg0.7Al_2O_3$	1.5	1.7548	1.7541 ± 0.0029	0.04	6 ± 2
$Mg1.1Al_2O_3$	2.5	1.7647	1.7636 ± 0.0013	0.06	6 ± 1
$Mg2.5Al_2O_3$	5.5	1.7954	1.7897 ± 0.0044	0.32	4 ± 1

Source: Hassan and Gupta, 2008

Table 9.6 Mechanical properties of Mg and the composites

Materials	0.2YS (MPa)	UTS (MPa)	Ductility (%)	WF (J/m³)	$\sigma_{0.2YS}/\rho$	σ_{UTS}/ρ
Mg	97 ± 2	173 ± 1	7.4 ± 0.2	11.1 ± 0.3	56	99
$Mg0.7Al_2O_3$	214 ± 4	261 ± 5	12.5 ± 1.8	28.9 ± 4.7	122	149
$Mg1.1Al_2O_3$	200 ± 1	256 ± 1	8.6 ± 1.1	20.9 ± 2.8	113	145
$Mg2.5Al_2O_3$	222 ± 2	281 ± 5	4.5 ± 0.5	10.0 ± 1.3	124	157

Source: Hassan and Gupta, 2008

was distributed randomly in the matrix and the magnesium recrystallised completely after extrusion. The finer grain size is again attributed to nucleation of magnesium grains by alumina particles and the grain growth restriction of the particles during recrystallisation. Mechanical properties are given in Table 9.6.

In a further study by the same authors, pure magnesium was reinforced with 2.5 wt% alumina of three different sizes: 50 nm, 300 nm and 1 μm (Hassan and Gupta, 2006). The materials were produced by mixing the powder, then compacting and subsequently sintering it. After hot extrusion at 250°C, microstructural investigations and mechanical testing were performed. The best 0.2 YS and UTS are found in the composite with 50 nm alumina particles. Ductility increases with increasing particle size. The work of fracture, which expresses the ability of energy absorption up to the moment of fracture, increases with increasing particle size. It is much higher in each composite compared to pure magnesium, as can be seen in Table 9.7 and Table 9.8.

Magnesium alloy AZ61 was reinforced with SiO_2 particles with a size of 20 nm using a friction stir process (FSP) in Lee *et al.* (2006). A groove was cut into a magnesium alloy bar and the nano-powder was poured into the groove. In order to close the cut, a flat FSP tool was used, followed by a normal pin and shoulder tool for mixing the reinforcement into the alloy. Two material samples were produced; containing 5 and 10 vol% SiO_2,

Table 9.7 Density, porosity, grain size and inter-particle spacing of all materials

Material	Density (g/cm^3)		Porosity (%)	Grain size (μm)	Inter-particle spacing λ (μm)
	Theor.	Experim.			
Mg	1.7400	1.7378 ± 0.0022	0.08	60 ± 10	–
Mg/Al_2O_3 (50 nm)	1.7647	1.7632 ± 0.0048	0.09	31 ± 13	0.47
Mg/Al_2O_3 (300 nm)	1.7647	1.7646 ± 0.0009	0.01	11 ± 4	2.85
Mg/Al_2O_3 (1 μm)	1.7647	1.7645 ± 0.0013	0.01	11 ± 3	9.49

Source: Hassan and Gupta, 2006

Table 9.8 Mechanical properties of Mg and the composites

Material	Hardness		0.2YS (MPa)	UTS (MPa)	Ductility (%)	Work of fracture (J/m^3)
	Macro (15HRT)	Micro (HV)				
Mg	43.5 ± 0.3	37.4 ± 0.4	132 ± 7	193 ± 2	4.2 ± 0.1	7.1 ± 0.3
Mg/Al_2O_3 (50 nm)	59.7 ± 0.5	69.5 ± 0.5	194 ± 5	250 ± 3	6.9 ± 1.0	15.5 ± 2.6
Mg/Al_2O_3 (300 nm)	56.3 ± 0.5	51.8 ± 0.3	182 ± 3	237 ± 1	12.1 ± 1.4	25.0 ± 3.3
Mg/Al_2O_3 (1 μm)	50.3 ± 0.5	51.2 ± 0.5	172 ± 1	227 ± 2	16.8 ± 0.4	34.7 ± 0.8

Source: Hassan and Gupta, 2006

respectively. The size of SiO_2 decreases with tool passes stirring the same area. The silica cluster size and average grain size are given in Table 9.9. The grain size in an unreinforced AZ61 billet was 75 μm and after 4P FSP it was 7–8 μm. The mechanical properties of AZ61 and the composites produced using FSP are given in Table 9.10. An increase in hardness and strength of unreinforced AZ61 after friction stir processing can be attributed to grain refinement during processing. An improvement of strength with a reduction of ductility is achieved by the reinforcement of silica. Tensile tests at 350°C yielded elongation of 350% at 10^{-2} s^{-1} and 420% at 10^{-1} s^{-1}, which clearly exhibits high strain-rate superplasticity.

Magnesium powder of 98.5% purity and yttria nano-powder with a particle size between 30 and 50 nm were used to produce two composites with yttria concentrations of 0.5 and 2.0 wt%, respectively (Tun and Gupta, 2007). The powders were blended for one hour and compacted to billets with a pressure of 97 bar. After a microwave sintering process, the billets were hot extruded at a temperature of 350°C. For the purposes of

Table 9.9 SiO_2 cluster size and grain size in composites with 5 vol% (1D) and 10 vol% (2D) silica after 1, 2, 3 and 4 tool passes (1P, 2P, 3P and 4P)

	1D1P	1D2P	1D3P	1D4P
SiO_2 cluster size (nm)	600	210	210	190
Average grain size (μm)	3.1	2.8	2.0	1.8
	2D1P	2D2P	2D3P	2D4P
SiO_2 cluster size (nm)	300	200	170	150
Average grain size (μm)	1.5	1.5	1.0	0.8

Source: Lee *et al.*, 2006

Table 9.10 Mechanical properties of AZ61 and the composites at room temperature

Material	Hardness (HV)	Yield strength (MPa)	UTS (MPa)	Elongation (%)
AZ61 billet	60	140	190	13
AZ61 after 4P FSP	72	147	242	11
1D2P (5 vol%, 2 passes)	91	185	219	10
1D4P (5 vol%, 4 passes)	97	214	233	8
2D2P (10 vol%, 2 passes)	94	200	246	4
2D4P (10 vol%, 4 passes)	105	225	251	4

Source: Lee *et al.*, 2006

comparison, pure magnesium was also processed. A microstructural characterisation was performed and the results are given in Table 9.11. Mechanical tests and thermal expansion measurements were performed and the results are given in Table 9.12. The reduction of CTE can be attributed to the low CTE of yttria (7.6×10^{6} K^{-1}) and a good bonding between particles and the matrix. With an increasing amount of yttria 0.2 YS, UTS and ductility also increase. The ductility increase is attributed to activation of non-basal slip systems and the tendency of yttria nanoparticles to enhance cross-slip of dislocations. The same materials were investigated to determine the extrusion ratio and its influence on the mechanical properties (Tun and Gupta, 2008). Pure magnesium and 2 wt% nano-yttria-reinforced magnesium was extruded with ratios of 12:1, 19:1 and 25:1 respectively. The density increases with increasing ratios, due to a decrease in porosity. The microhardness tests revealed an increase in hardness with increasing ratio. This is attributed to an improvement in yttria distribution, decreasing grain size and reduction of porosity. Both materials show an increase in yield strength and ultimate strength with increasing extrusion ratio. Both effects are also attributed to the grain size reduction and the decrease of porosity.

Table 9.11 Density, porosity and grain size of yttria-reinforced Mg

Material	Yttria content (vol%)	Density (g/cm^3)	Porosity (%)	Grain size (μm)
Mg	–	1.74 ± 0.01	0.13	20 ± 3
$Mg/0.5Y_2O_3$	0.17	1.73 ± 0.01	0.87	19 ± 3
$Mg/2.0Y_2O_3$	0.7	1.76 ± 0.01	0.35	18 ± 3

Source: Tun and Gupta, 2007

Table 9.12 Mechanical properties and CTE of yttria-reinforced Mg

Material	0.2YS (MPa)	UTS (MPa)	Ductility (%)	Work of fracture (MJ/m^3)	CTE (10^{-6} K^{-1})	Microhardness (HV)
Mg	134 ± 7	193 ± 1	7.5 ± 2.5	12.9 ± 4.8	28.2 ± 0.0	37 ± 2.0
$Mg/0.5Y_2O_3$	144 ± 2	214 ± 4	8.0 ± 2.8	16.6 ± 4.2	21.3 ± 0.1	38 ± 0.4
$Mg/2.0Y_2O_3$	157 ± 10	244 ± 1	8.6 ± 1.2	21.8 ± 3.1	20.8 ± 0.6	45 ± 2.0

Source: Tun and Gupta, 2007

9.5.4 Carbon nanotubes reinforced magnesium

In recent years, carbon nanotubes (CNT) or multi-walled carbon nanotubes (MWCNT) have become popular for reinforcing metals and even polymers. These nanotubes show extremely high mechanical properties and flexibility as well as a high Young's modulus. Their small size reduces the possibility of thermal mismatch-induced dislocation generation at the matrix/tube interface, which is seen as a further advantage. MWCNTs were introduced into the surface of an AZ31 alloy by FSP to modify the surface (Morisada *et al.*, 2006). The tubes have a length of 250 nm, an outer diameter of 20–50 nm and were combined with an AZ31 plate of 6 mm thickness. A groove was cut into the sheet and filled with nanotubes before the FSP was performed. It could be shown that the MWCNTs are dispersed randomly into the matrix. The best distribution was achieved at a speed of 25 mm/min and a rotation speed of 1500 rev/min. Grain refinement was promoted in the reinforced areas. A hardness of maximum 78 HV is attributed to the nanotubes and to the grain refinement. Hardness of unreinforced AZ31 in the as-received and after-FSP states was 41 HV and 55 HV respectively.

The DMD technique was used to produce composite material of pure magnesium reinforced with 0.3, 1.3, 1.6 and 2 wt% of carbon nanotubes (Goh *et al.*, 2006a). This process has already been described. For the purposes of comparison, pure magnesium was processed in the same way. Following DMD, the ingots were hot extruded at 350°C with an extrusion

Table 9.13 Density, macrohardness and mechanical properties of CNT-reinforced magnesium

Material	CNT (wt%)	Density (g/cm³)	Macro-hardness (HR15T)	0.2YS (MPa)	UTS (MPa)	Elong. (%)
Mg (99.9%)	–	1.738 ± 0.010	45 ± 1	126 ± 7	192 ± 5	8.0 ± 1.6
Mg-0.3 wt% CNT	0.3	1.731 ± 0.005	48 ± 1	128 ± 6	194 ± 9	12.7 ± 2.0
Mg-1.3 wt% CNT	1.3	1.730 ± 0.009	46 ± 1	140 ± 2	210 ± 4	13.5 ± 2.7
Mg-1.6 wt% CNT	1.6	1.731 ± 0.003	42 ± 1	121 ± 5	200 ± 3	12.2 ± 1.7
Mg-2.0 wt% CNT	2.0	1.728 ± 0.001	39 ± 1	122 ± 7	198 ± 8	7.7 ± 1.0

Source: Goh *et al.*, 2006a

Table 9.14 Density, CTE and mechanical properties of liquid (LM) and PM-processed nanocomposites

Material	Density (g/cm³)	CTE (10^{-6} K^{-1})	0.2YS (MPa)	UTS (MPa)	Elong. (%)
Mg (99.5%) LM	1.738 ± 0.010	28.73 ± 0.59	126 ± 7	192 ± 5	8.0 ± 1.6
Mg-0.25 wt% CNT LM	1.731 ± 0.005	27.82 ± 0.22	128 ± 6	194 ± 9	12.7 ± 2.0
Mg (98.5%) PM	1.738 ± 0.001	28.56 ± 0.28	127 ± 5	205 ± 4	9.0 ± 2.0
Mg-0.3 wt% CNT PM	1.736 ± 0.001	25.90 ± 0.93	146 ± 5	210 ± 6	8.0 ± 1.0

Source: Goh *et al.*, 2006b

ratio of 20.25. A slight increase in hardness is measurable with low amounts of CNT (see Table 9.13). The hardness drops for reinforcements of 1.3% and above. Yield, tensile strength and ductility show the highest values at 1.3 wt% CNT. The activation of further slip planes is attributed as responsible for this higher ductility.

A comparative study of CNT-reinforced magnesium produced using PM and liquid metallurgical (LM) techniques was performed in (Goh *et al.*, 2006b). For the liquid process, magnesium turnings of 99.9% purity and a CNT with an average diameter of 20 nm were cast using the DMD and subsequently hot extruded at 350°C. For the powder technique, magnesium powder was mixed with 0.3 wt% nanotubes for ten hours and compacted to billets. After sintering for two hours, the billets were also hot extruded at a temperature of 350°C. For comparison, both materials were processed without nanotubes, too. It can be seen in Table 9.14 that the reduction of CTE in the PM composite is more significant compared to the liquid processed. Yield and tensile strength of the PM-processed composite is higher compared to the LM material. The authors attribute this result to the higher amount of MgO in the original material for the powder route, which was found in XRD analysis. Additional MgO is

Table 9.15 Density, Young's modulus and mechanical properties of MWCNT-reinforced magnesium alloy AZ91

Material	Density (g/cm^3)	E (GPa)	0.2YS (MPa)	UTS (MPa)	Elong. (%)
AZ91D	1.80 ± 0.007	40 ± 2	232 ± 6	315 ± 5	14 ± 3
AZ91D-0.5CNT	1.82 ± 0.008	43 ± 3	281 ± 6	383 ± 7	6 ± 2
AZ91D-1CNT	1.83 ± 0.006	49 ± 3	295 ± 5	388 ± 11	5 ± 2
AZ91D-3CNT	1.84 ± 0.005	51 ± 3	284 ± 6	361 ± 9	3 ± 2
AZ91D-5CNT	1.86 ± 0.003	51 ± 4	277 ± 4	307 ± 10	1 ± 0.5

Source: Shimizu *et al.*, 2008

assumed to give a further strengthening effect to that of the CNT. In the case of the liquid processed composite, ductility was improved, compared to unreinforced magnesium, whereas it is reduced slightly in the powder version.

Again a PM method was chosen to produce an AZ91-based CNT composite in (Shimizu *et al.*, 2008). AZ91 powder with a diameter of 100 μm and MWCNT with a length of approx. 5 μm were physically blended and subsequently hot pressed. The amount of MWCNT in the composites was 0.5, 1, 3 and 5 wt% respectively. The resulting material was hot extruded at 450°C with an extrusion ratio of 9:1. Young's modulus and density increase with increasing amounts of nanotubes in the same way that the ductility reduces (Table 9.15). Yield and ultimate tensile strength show a maximum at an amount of approximately 1 wt% MWCNT addition.

9.6 Conclusions and future trends

Magnesium-based MMCs show a wide range of possibilities to fit mechanical and physical properties to the intended application. With a variation of amount, distribution and kind of ceramic reinforcement, the properties can be adjusted to the surrounding materials, which may be aluminium, steel, or even polymer. Mechanical properties are comparable to high-strength aluminium alloys, but magnesium-based composites show a lower density when compared to those. It makes them a promising material for future lightweight applications.

9.7 References

American Society for Metals (ASM) (2001), 'Composites', *ASM Handbook*, **21**, 38.

Bata V, Pereloma EV (2004), 'An alternative physical explanation of the Hall–Petch relation', *Acta Mater*, **52**, 3, 657–665.

Clyne TW, Withers PJ (1993), *An Introduction to Metal Matrix Composites*, Cambridge University Press, Cambridge, ISBN: 0–521–41808–9.

Dieringa H, Morales E, Fischer P, Kree V, Kainer KU (2002), 'Gefüge und mechanische Eigenschaften von Magnesium-Matrix Verbundwerkstoffen', *Fortschritte in der Metallographie*, Portella P (Ed.), ISBN: 3–88355–303–4, 91–96.

Dieringa H (2006), 'Particles, fibres and short fibres for the reinforcement of metal matrix composites', in: Kainer KU (Ed.) *Metal Matrix Composites*, Wiley-VCH, ISBN: 3–527–31360–5, 55–76.

Dieringa H, Hort N, Kainer KU (2007), 'Comparison of tensile and compressive creep data of AE42 magnesium alloy and its short fiber composite', *Proceedings of the PFAM* **XVI**, ISBN: 978–98105–9650–7, 248–255.

Dieringa H (2011), 'Properties of magnesium alloys reinforced with nanoparticles and carbon nanotubes: a review', *J Mat Sci*, **46**, 289–306.

Ferkel H, Mordike BL (2001): 'Magnesium strengthened by SiC nanoparticles", *Mater Sci Engin*, **A298**, 193–199.

Garces G, Rodriguez M, Perez P, Adeva P. (2010), 'Microstructural and mechanical characterisation of WE54–SiC composites', *Mat Sci Engin*, **A527**, 6511–6517.

Goh CS, Wei J, Lee LC, Gupta M (2006a) 'Simultaneous enhancement in strength and ductility by reinforcing magnesium with carbon nanotubes', *Mater Sci Engin*, **A423**, 153–156.

Goh CS, Wei J, Lee LC, Gupta M (2006b), 'Effect of fabrication techniques on the properties of carbon nanotubes reinforced magnesium', *Solid State Phenom*, **111**, 179–182.

Hassan SF, Gupta M (2005), 'Enhancing physical and mechanical properties of Mg using nanosized Al2O3 particulates as reinforcement', *Metall Mater Trans*, **36A**, 2253–2258.

Hassa SF, Gupta M (2006), 'Effect of length scale of Al2O3 particulates on microstructural and tensile properties of elemental Mg', *Mater Sci Engin*, **A425**, 22–27.

Hassan SF, Gupta M (2008), 'Effect of submicron size Al2O3 particulates on microstructural and tensile properties of elemental Mg', *J Alloys Comp*, **457**, 244–250.

Kumar S, Ingole S, Dieringa H, Kainer KU (2003), 'Analysis of thermal cycling curves of short fibre reinforced Mg-MMCs', *Compos Sci Techn*, **63**, 1805–1814.

Lee CJ, Huang JC, Hsieh PJ (2006), 'Mg based nano-composites fabricated by friction stir processing', *Scr Mater*, **54**, 1415–1420.

Luo A (1995), 'Processing, microstructure and mechanical behaviour of cast magnesium metal matrix composites', *Met Mat Trans*, **26A**, 2445–2455.

Morisada Y, Fujii H, Nagaoka T, Fukusumi M (2006), 'MWCNTs/AZ31 surface composites fabricated by friction stir processing', *Mater Sci Engin*, **A419**, 344–348.

Purazrang K, Abachi P, Kainer KU (1994), 'Investigation of the mechanical behaviour of magnesium composites', *Comp*, **25**(4), 296–302.

Purazrang K, Kainer KU, Mordike BL (1991), 'Fracture toughness behaviour of a magnesium alloy metal-matrix composite produced by the infiltration technique', *Comp*, **22**(6), 456–462.

Salmang H, Scholze H (1983), *Keramik, Teil 2: Keramische Werkstoffe*, 6. Auflage, Springer-Verlag Berlin, Heidelberg, New York, Tokyo.

Saravanan RA, Surappa MK (2000), 'Fabrication and characterisation of pure magnesium-30 vol% SiCp particle composite', *Mat Sci Engin*, **A276**, 108–116.

Schröder J, Kainer KU, Mordike BL (1989), 'Particle reinforced magnesium alloys', *Proceedings of the 3rd European Conference on Composite Materials ECCM3*, 221–226.

Shimizu Y, Miki S, Soga T, Itoh I, Todoroki H, Hosono T, Sakaki K, Hayashi T, Kim YA, Endo M, Morimoto S, Koide A (2008), 'Multi-walled carbon nanotubes-reinforced magnesium alloy composites', *Scr Mat*, **58**, 267–270.

Tiwari S, Balasubramaniam R, Gupta M (2007), 'Corrosion behavior of SiC reinforced magnesium composites', *Corr Sci*, **49**, 711–725.

Trojanova Z, Gärtnerova V, Jäger A, Namesny A, Chalupova M, Palcek P, Lukac P (2009), 'Mechanical and fracture properties of an AZ91 magnesium alloy reinforced by Si and SiC particles', *Comp Sci Techn*, **69**, 2256–2264.

Tun KS, Gupta M (2007), 'Improving mechanical properties of magnesium using nano-yttria reinforcement and microwave assisted powder metallurgy method', *Compos Sci Techn*, **67**, 2657–2664.

Tun KS, Gupta M (2008), 'Effect of extrusion ratio on microstructure and mechanical properties of microwave-sintered magnesium and Mg/Y2O3 nanocomposite', *J Mater Sci*, **43,** 4503–4511.

Zhang Q, Chen DL (2004), 'A model for predicting the particle size dependence of the low cycle fatigue life in discontinuously reinforced MMCs', *Scripta Mater*, **51**, 863–867.

10

Applications: use of magnesium in medical applications

F. WITTE, Hannover Medical School, Germany

DOI: 10.1533/9780857097293.342

Abstract: This chapter introduces the use of magnesium and its alloys as temporary implant materials in medical applications. The chapter describes the fundamental concepts of biodegradation and toxicology of magnesium before discussing how magnesium implants can be investigated *in vivo* and *in vitro* in future developments of biodegradable metals.

Key words: biodegradable metal, concept of biodegradation, *in vivo* corrosion, cytotoxicity of magnesium alloys, medical application of biodegradable metals.

10.1 Introduction to biodegradable implants based on metals

Biodegradable implants provide a temporary mechanical support until an injured tissue is healed. Then, in an ideal case, the implant materials are intended to gradually degrade and finally completely disappear, to leave behind a fully recovered tissue.

These biodegradable implants have been produced based on polymeric and ceramic materials or (decellularized) organic materials. However, especially in applications which require high mechanical properties, the current biodegradable implants have significant limits, which make biodegradable metals based on elements such as magnesium very promising.

In the history of metals in surgical fixing of fractured bones and wound closure, most metals have been used as permanent implant materials and were selected on a trial-and-error basis. The basic idea of the temporary use of metal implants was reported by the Romans, who used metal clips for the adaptation of skin (Schuster, 1975). Various metals were used at the beginning of osteosynthesis, such as gold, silver, platinum, copper, lead and iron (Schuster, 1975). Since then, however, these metals have been rejected for surgical use for various reasons. Gold and platinum were desirable from a corrosion resistance standpoint, but were very expensive and suffered from poor mechanical properties, while lead was rejected due to its toxicity. Silver and iron were generally considered suitable biomaterials (Schuster, 1975). However, pure silver was mechanically insufficient for osteosynthesis, although its antibacterial effect was appreciated even at that time. There have been controversial discussions on the biocompatibility of iron since metallosis has been observed after iron implants. Metallosis is the local destruction of soft and hard tissue based on the mechanical–biological, electro-energetic, and chemo-toxic effects of metal implants (Schuster, 1975). Metallosis is also occasionally observed as an infiltration of periprosthetic soft tissues and bone by metallic debris resulting from wear of joint arthroplasties. Metallosis is often associated with significant osteolysis. Therefore, the identification of metallosis is an indicator for revision surgery (Heffernan *et al.*, 2008). These clinical observations have lead to the paradigm that metal implants should be generally corrosion resistant.

However, some early and also recent findings have shown that implants made of magnesium-based and iron-based materials may be suitable as temporary biomaterials which degrade *in vivo* by corrosion (Peuster *et al.*, 2001; Heublein *et al.*, 2003; Witte *et al.*, 2005; Witte, 2010). In fact, this is a paradigm-breaking approach to a new class of temporary implants – biodegradable metals. The basic concept of biodegradable metals is to compose the metals of elements which can be cleared from the body by physiological pathways and which do not exceed the toxicity limits during the corrosion process.

Research interest is rapidly growing in magnesium-based alloys for medical applications. However, in the words of the author of Ecclesiastes, 'there is nothing new under the sun' (Mantovani and Witte, 2010). The first time that magnesium was mentioned as a biodegradable implant material was in 1878, in a report about absorbable ligatures for the closure of bleeding vessels (Huse, 1878), while pure iron had been used for implants long before. The first studies on degradable metals in the musculoskeletal field were published at the beginning of the last century based on experiences with osteosynthesis implants made of magnesium alloys (Witte, 2010). At that time, no attention was paid to metal allergies because of more severe surgical complications at that time, such as infection or implant failure. Skin sensitizing reactions to metal implants were reported more often after the introduction of aseptic surgery and less corrosive osteosynthesis materials, such as stainless steel, and are clinically still apparent in 10–15% of all implanted metals (Witte *et al.*, 2008a). However, biodegradable magnesium alloys have shown no skin sensitizing potential in animal experiments (Witte *et al.*, 2008a). The advantage of biodegradable metals over existing biodegradable materials, such as polymers, ceramics or bioactive glasses, in load-bearing applications is higher tensile strength and a Young's modulus that is closer to that of bone (Table 10.1). These facts provide enough evidence to investigate selected metallic biomaterials as temporary implant materials.

Table 10.1 Mechanical properties of implant materials

Material	Density (g/cm^3)	Young's modulus (GPa)	Tensile strength (MPa)	Elongation (%)
Surgical steel (X2CrNiMo18164)	8.0	193	585	55
Surgical titanium (TiAl6V4)	4.43	100–110	930–1140	8–15
PEEK	1.28	3.6	92	50
Cortical bone	1.7–2.0	3–30	80–150	3–4
DL-PLA (DL-polylactide)	1.24	1.9	29	5.0
Magnesium alloy AZ9E, casted	1.81	45	240	3.0
Magnesium alloy Mg10Gd, casted	N/A	44	131	2.5
Magnesium alloy RS66, rapid-solidified, extruded	N/A	45	400	23.5

10.1.1 Brief historical introduction to biodegradable metals

Today, pure metals are used as biomaterials in some selected applications, such as silver coatings for antibacterial surface properties, copper for contraceptive coils, CP-Ti for dental applications, and Pt electrodes in, for example, cardiovascular applications. Recently, magnesium, iron and tungsten have been investigated as biomaterials for temporary implants (Peuster *et al.*, 2001; Heublein *et al.*, 2003; Peuster *et al.*, 2003; Witte *et al.*, 2005). Magnesium is the most extensively investigated element, especially for cardiovascular stent applications and musculoskeletal fixative devices. Tungsten has been investigated as a cardiovascular coil, to close persistent postnatal shunts in animal experiments (Peuster *et al.*, 2003). More promising, but still in the experimental stage, are iron-based cardiovascular devices, especially stents.

After magnesium was first mentioned as an absorbable implant material for the closure of bleeding vessels in 1878 (Huse, 1878), it was used in the early 20th century for fast absorbable wound closure and fixing fractured bones(Seelig, 1924; Witte, 2010). Payr, Chlumpsky and Lespinasse used magnesium for anastomosis of blood vessels and intestines. Payr used magnesium tubes for vessel and nerve suture, as well as for tiny arrows to treat hemangioma (Witte, 2010). Andrews and Seelig used magnesium for the ligature of blood vessels, and as a surgical suture material. Even though Seelig obtained large amounts of hydrogen cavities while using pure magnesium implants, he could not detect any negative systemic effects (Seelig, 1924). In musculoskeletal surgery, pure magnesium was first used by Lambotte in 1906 for fracture fixation plates (Lambotte, 1932). He observed a fast degradation of the magnesium plate within eight days when he combined the magnesium plate with steel screws. This report was probably the first indication that magnesium implants degrade *in vivo* by a corrosion process.

10.2 Fundamental concepts of biodegradation

The control and adaptation of the implant degradation rate is crucial, since the resorption capacity of the tissue is limited. Moreover, the local physiology of the implant environment determines the maximal degradation rate of a temporary implant.

There are many reasons that contribute to the corrosion of metals when implants are placed inside the human body. For magnesium corrosion *in vitro* and in technical applications, the most critical factor seems to be the local pH; however, in an *in vivo* environment there might be even more influencing factors, such as local blood flow (Witte *et al.*, 2008b). After surgery, the pH surrounding the implant is reduced to a value between 5.3 and 5.6, typically due to the trauma of surgery (Witte *et al.*, 2008b). This process may accelerate initial magnesium corrosion, while infectious microorganisms and crevices formed between components can reduce the local oxygen concentration. The main challenge with iron implants is to accelerate corrosion, while magnesium corrosion is generally too fast *in vivo*. An option to design appropriate magnesium alloys is to slow down the initial implant corrosion by appropriate alloying, microstructure design, processing, and/or an additional coating. However, it has to be kept in mind that the corrosion rate will be further reduced *in vivo* after implantation, due to adherent proteins and inorganic salts such as calcium phosphates, which stabilize the corrosion layer (Witte *et al.*, 2005; Eliezer and Witte, 2010b). Based on this theory, a magnesium implant with an initially reduced corrosion rate could finally lead to an arrested corrosion process *in vivo*. Thus, the right balance of a reduced corrosion rate and an assured complete corrosion *in vivo* would create a useful biodegradable magnesium implant – otherwise, parts of the implant may persist locally and will act as long-term biomaterials.

10.2.1 Corrosion *in vitro* and *in vivo*

The interdependence of the implant corrosion behavior and its human environment is scientifically of the highest interest in this growing field. However, accessibility of direct electrochemical measurements in humans ranges from difficult (in the mouth) to virtually impossible (e.g., for orthopedic devices) because of ethical, safety, legal and regulatory considerations (Bundy, 1995). Consequently, much effort has been devoted to identifying alternative environments which simulate the corrosion conditions *in vivo*. A possible alternative is laboratory testing performed under cell-culture conditions, but such tests are less related to real life conditions than tests on laboratory animals. The experimental parameters that simulate the corrosion *in vitro* are not fully known. Thus, the *in vitro* corrosion test system always needs to be adapted to the corresponding *in vivo* application.

The physiological temperature of 37°C is an important factor influencing *in vivo* electrochemical behavior (Eliezer and Witte, 2010b). Dissolved salts, particular chlorides, are probably the most influential for implant corrosion *in vivo* (Bundy, 1995). Also, gases dissolved in body fluids play an important role in implant alloy corrosion. Oxygen is one of the most important physiological gases; its partial pressure varies widely within the body, from about 2.67×10^2 to 1.33×10^4 Pa (Bundy, 1995). Thus, an implant surface can be in contact with anatomical environments of widely different PO_2, creating the possibility of developing different aeration cells (Bundy, 1995). Carbon dioxide is another gas that can be important for *in vivo* corrosion, because of its influence on the pH value (Bundy, 1995). While the pH is usually homeostatically regulated at about pH 7.4, the pH may fall to values below 4.5 at sites of inflammation, for a period of hours, or longer if acute inflammation processes convert into chronic inflammation (Bundy, 1995). These initial pH changes after surgery or during inflammation are especially critical for the corrosion resistance of magnesium implants.

The role played by proteins regarding corrosion *in vivo* is one of the most important aspects of the unique environment within the body, and can cause differences between the corrosion behavior in laboratory, chloride-containing environments, and *in vivo* (Bundy, 1995; Witte *et al.*, 2006b; Eliezer and Witte, 2010b). Especially for magnesium corrosion, the influence of albumin and other serum proteins on the formation of a corrosion-protective film has been investigated (Mueller *et al.*, 2009, 2010; Rettig and Virtanen, 2009; Eliezer and Witte, 2010a). In general, from the moment an implant is inserted into the body, it becomes covered with a layer of adsorbed proteins (Rudee and Price, 1985). The properties of this layer change with time, because (i) both thermodynamic and kinetic factors are involved in protein adsorption (Neumann *et al.*, 1980) and (ii) the adjacent cells are actively involved in protein synthesis (Clark and Williams, 1982). The effects of proteins on metal corrosion are complex and may increase or decrease the corrosion rate (Clark and Williams, 1982). The *in vivo* environment is characterized by dynamic, constantly changing, chemical and physiological processes, mechanical loading patterns, and bioelectric potentials (Bundy, 1995). There is certainly more than one standardized *in vivo* environment. However, in current biodegradable research no considerable attention has been devoted to the interaction between mechanical loading and corrosion phenomena yet. However, corrosion fatigue and stress corrosion cracking

(SCC) are actual force patterns *in vivo* that may increase the corrosion rate, while the mechanical integrity of the biomaterial is temporarily maintained (Bundy, 1995). Special attention has to be given to the implant design, which can alter the corrosion performance of alloys *in vivo*. Localized corrosion could occur, for example, between screws and plates for crevice corrosion, or with parts in relative motion for fretting corrosion, or both (Bundy, 1995; Eliaz, 2008).

10.2.2 Methods to measure corrosion *in vivo*

The elemental components of biodegradable magnesium alloys (Mg, Al, Li, Zn, rare earth elements) have been investigated by various analytical methods in histological sections, bone, tissue and body fluids (Witte *et al.*, 2008b). However, the application of these methods for trace and ultra-trace analysis in small sample volumes is typically hampered by the detection limits of the elemental concentration in physiological liquids and tissues. Current detection limits are given in serum (< 1 μg/L to about 1 mg/L), and in liver and bone (< 1 mg/kg up to about 500 mg/kg) (Witte *et al.*, 2008b). Further limitations are caused by time-consuming sample preparation (AES, OES, ICP–MS), the access to appropriate methods (NAA, synchrotron-based methods), the lack of sufficient lateral resolution for solid sample analysis (GD–OES) or challenging interferences during the measuring process (AAS, AES, ICP–MS, XRF) (Witte *et al.*, 2008b).

Magnesium corrosion can be determined *in vivo* using microtomography. In particular, synchrotron-based microtomography (SRμCT) is a non-destructive method with a high density and high spatial resolution (Witte *et al.*, 2006a). Furthermore, element-specific SRμCT can determine the spatial distribution of the alloying elements during *in vivo* corrosion (Witte *et al.*, 2006a). The remaining non-corroded metal volume, as well as the surface morphology, can be determined in three dimensions non-destructively on a micrometer scale using SRμCT. The reduction of the metallic implant volume can be converted into a corrosion rate by using a modification of the ASTM G31–72, 2004 Equation [10.1] for weight loss measurements:

$$\mathrm{CR} = \frac{W}{At\rho} \qquad [10.1]$$

where CR (mm/year) is the corrosion rate, W (g) is the weight loss of the metal or alloy, A (mm^2) is the initial surface area exposed to corrosion, ρ (g/mm^3) is the standard density of the metal or alloy, and t (365 days = 1 year) is the time of immersion. Herein, the weight loss will be substituted by the reduction in volume (ΔV) multiplied by the standard density (ρ) resulting in Equation [10.2]:

$$CR = \frac{\Delta V}{Atc} \qquad [10.2]$$

where ΔV (mm^3) is the reduction in implant volume that is equal to the remaining metal volume subtracted from the initial metal implant volume. This method provides a general corrosion rate of the implanted metal. A better analysis of the local corrosion can be obtained if the corrosion rates are calculated based on the pitting depth (Witte *et al.*, 2009).

10.3 Magnesium-based biodegradable metals

In general, magnesium and its alloys are light metals and provide a density of about 1.74 g/cm^3, which is 1.6 times less dense than aluminum, and 4.5 times less dense than steel (Witte *et al.*, 2008b). The fracture toughness of magnesium is higher than ceramic biomaterials such as hydroxyapatite, while the elastic modulus and compressive yield strength of magnesium are closer to those of natural bone than other commonly used metallic osteosynthesis materials (Table 10.1). Furthermore, magnesium has a low corrosion resistance, which is due its electrochemical standard potential of $E_o = -2.363$ V *vs* SHE, or alternatively $E_o = -2.159$ V *vs* SHE if magnesium oxides or magnesium hydroxides are formed. According to Pourbaix (Mueller *et al.*, 2009), these potentials may be shifted if the Mg ions are complexed in the corrosion layer. It is obvious that magnesium carbonates and phosphates are formed during Mg corrosion under cell-culture conditions and has been detected in a complex corrosion layer *in vivo*. Thus, complexed Mg is truly changing the local potential and promotes local pitting *in vivo*. However, detailed Mg corrosion processes remain currently uncovered *in vivo*. If exposed to air, the surface of magnesium is passivated by a growing thin grey layer of magnesium hydroxide, which reduces further by chemical reactions. Magnesium hydroxides are slightly soluble in water; however, severe corrosion occurs in saline media, as well as in the human body, where high chloride ion concentrations of about 150 mM are present. Magnesium hydroxide accumulates on the underlying magnesium matrix as a corrosion-protective layer in water, but when the chloride concentration in the corrosive environment rises above 30 mM (Shaw, 2003), magnesium hydroxide starts to convert into highly soluble magnesium chloride. Therefore, severe pitting corrosion can be observed on magnesium alloys *in vivo* (Witte *et al.*, 2005, 2006b; Xu *et al.*, 2007). Importantly, magnesium corrosion is relatively insensitive to various oxygen concentrations around implants in different anatomical locations (Witte *et al.*, 2008b). The overall corrosion reaction of magnesium in aqueous environments is given as:

$$Mg_{(s)} + 2H_2O_{(aq)} \rightleftharpoons Mg(OH)_{2(s)} + H_{2(g)} \quad [10.3]$$

This overall reaction may include the following partial reactions:

$$Mg_{(s)} \rightleftharpoons Mg^{2+}_{(aq)} + 2e^- \quad \text{(anodic reaction)} \quad [10.4]$$

$$2H_2O_{(aq)} + 2e^- \rightleftharpoons H_{2(g)} + 2OH^-_{(aq)} \quad \text{(cathodic reaction)} \quad [10.5]$$

$$Mg^{2+}_{(aq)} + 2OH^-_{(aq)} \rightleftharpoons Mg(OH)_{2(s)} \quad \text{(product formation)} \quad [10.6]$$

10.3.1 Physiological and toxicological aspects

During the corrosion process, a mass of magnesium ions are released from the magnesium implant, which can be eliminated from the body very rapidly via the blood serum and the kidneys (Witte *et al.*, 2008b). Magnesium can also be stored in muscle (39% of total Mg) or bone (60%), which are the natural storages of the 21–35 g of elemental magnesium of an average adult person who weighs about 70 kg (Witte *et al.*, 2008b). The level of magnesium in the extracellular fluid is kept constant at levels between 0.7 and 1.05 mM. While serum magnesium levels exceeding 1.05 mM can lead to muscular paralysis, hypotension and respiratory distress, cardiac arrest occurs only for severely high serum levels of 6–7 mM. However, magnesium is essential to human metabolism and is the fourth most abundant cation in the human body. Furthermore, magnesium is also a co-factor for many enzymes, and stabilizes the structures of DNA and RNA (Witte *et al.*, 2008b).

10.3.2 Evolution of hydrogen during *in vivo* corrosion

In magnesium and its alloys, impurities and cathode sites with a low hydrogen overpotential facilitate hydrogen evolution (Song and Atrens, 1999), thus causing substantial galvanic corrosion and potential local gas cavities *in vivo* (Witte *et al.*, 2005). However, there are contradicting reports on the occurrence of gas cavities after magnesium implantation. Gas cavities have been observed after subcutaneous implantation while intravasal application showed no local gas accumulation. An explanation for this observation may be based on the diffusion and solubility coefficient of hydrogen in biological tissues, which has been widely reviewed (Lango *et al.*, 1996). The solubility of hydrogen in tissues is influenced by the content of lipids, proteins and salinity, but in fat and oils, the solubility seems to be approximately independent of temperature in the physiological range (Piiper *et al.*, 1962; Lango *et al.*, 1996). Not only viscosity, but also different tissue components and structures like lipids, proteins and glycosaminoglycans influence the numeric value of the hydrogen diffusion coefficient (Vaupel, 1976; Lango *et al.*, 1996). Depending on experimental configuration, the diffusion coefficient may be underestimated in both stagnant and flowing media due to a boundary layer formation, which increases the effective diffusion distance (Lango *et al.*, 1996). This finding may be important for intravascular magnesium applications. Correlating the hydrogen diffusion coefficients from various biological media having fractional water contents from about 68% to 100%, it is found that the diffusion coefficient of hydrogen increases exponentially with the increasing water fraction of the tissue (Vaupel, 1976). The tissue water content increases from adipose tissue to skin, and from bone to muscles, in animals, and humans are similar for the same tissue regardless of the species (Witte *et al.*, 2008b). This can explain why different corrosion rates and gas cavities were observed for magnesium alloys in different anatomical implantation sites (Wen *et al.*, 2004; Witte *et al.*, 2005, 2007; Xu *et al.*, 2007). In an animal study with rats, it was shown that the adsorption of hydrogen gas from subcutaneous gas pockets was limited by the diffusion coefficient of hydrogen in the tissue; the overall hydrogen adsorption rate was determined as 0.954 mL per hour (Piiper *et al.*, 1962). Thus, the local blood flow and the water content of the tissue surrounding the implant are important parameters, which need to be considered in designing biodegradable magnesium alloys with an appropriate corrosion rate. Concomitantly, it can be assumed that local hydrogen cavities occur when more hydrogen is produced per time interval than can be dissolved in the surrounding tissue, or diffuse from the implant surface into the extracellular medium, which is renewed depending on the local blood flow. This means that magnesium alloys are corroding *in vivo* with an appropriate corrosion rate when no local gas cavities are observed during the implantation period in a specific anatomical site.

10.4 Recent research and future product development

Magnesium has been intensively investigated as a biodegradable material in cardiovascular application for several years. The German company *Biotronik* is producing an absorbable magnesium stent (AMS-1) that has been already used in several clinical trials covering peripheral and coronary indications (Bosiers *et al.*, 2005; Erbel *et al.*, 2007). However, despite the good clinical results, Biotronik is still improving its magnesium stent (AMS-2) by prolonged mechanical stability, improved stent design, modified alloy, and enhanced

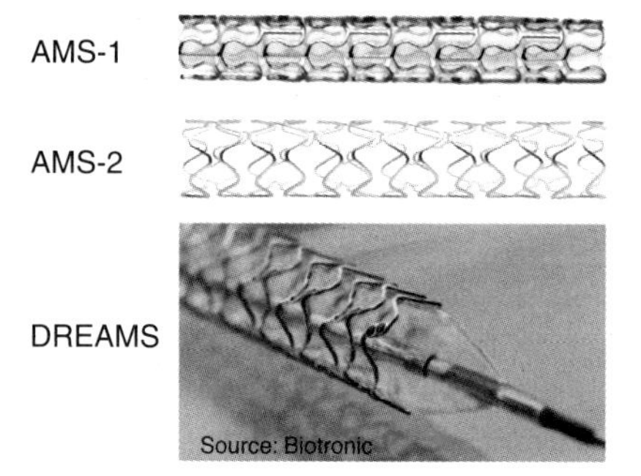

10.1 Following preliminary clinical trials, the absorbable metal stent (AMS-1) has been improved in regard to alloy composition, mechanical stability and design, resulting in a mechanically improved AMS-2 stent. Finally, a drug(DR)-eluting(E) absorbable(A) Metal(M) Scaffold(S) has been developed, which is now under investigation in clinical studies. With kind permission from Biotronik AG.

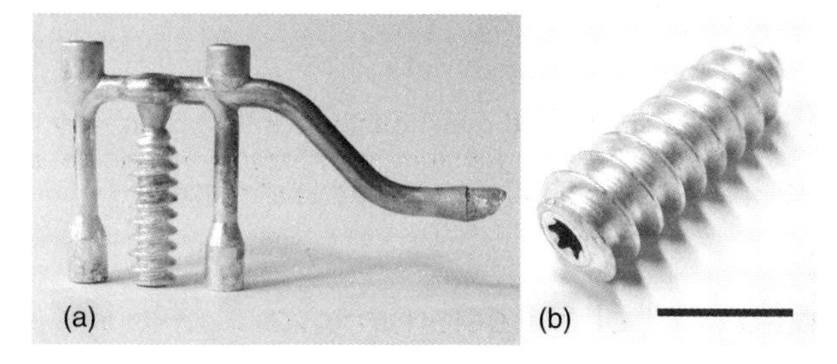

10.2 A high-pressure die-casting approach has been developed (a) to produce a prototype of an interference screw (b) (scale bar = 1 cm) for musculoskeletal applications. With kind permission from AAP Biomaterials GmbH.

surface passivity (Fig. 10.1). The latest development of Biotronik has been the successful reduction of neointima hyperplasia by creating a drug-eluting biodegradable magnesium stent (AMS-3, DREAMS). However, in this context, the biodegradable stent is envisaged to perform more as an assisting healing device and is called 'scaffold' instead of 'stent'. Following the first successful attempts to control the corrosion rate of magnesium implants *in vivo,* the industrial attention to biodegradable magnesium implants in musculoskeletal applications is rising. The first prototypes of magnesium implants for musculoskeletal applications have already been produced (Fig. 10.2). However, although the first biodegradable prototypes have been developed, and some basic knowledge has been summarized in reviews (Witte *et al.*, 2008b; Zeng *et al.*, 2008; Witte, 2010; Chen and Zhao, 2011; Virtanen, 2011; Xin *et al.*, 2011; Zhu *et al.*, 2011), there are still several open questions about the corrosion process *in vivo* and the local tissue-material interaction.

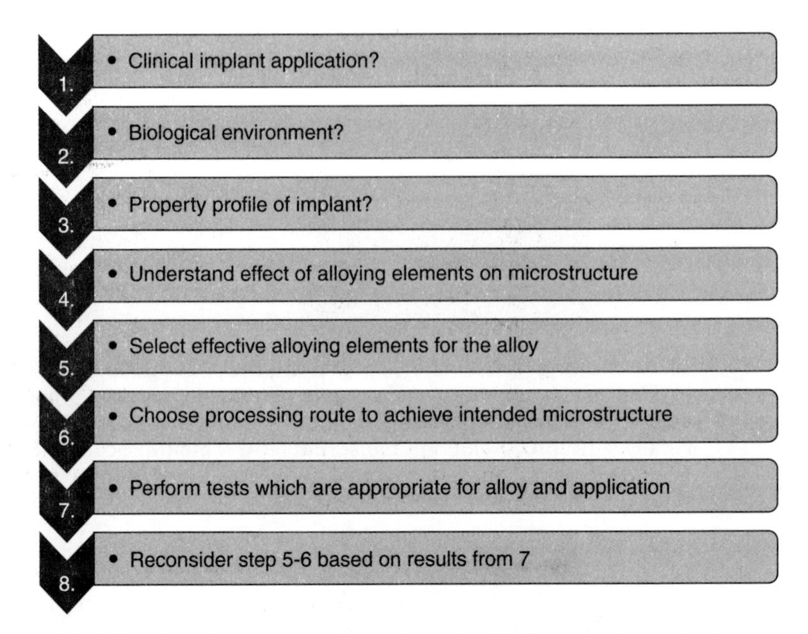

10.3 Following the steps and questions in this flow chart may help to select the appropriate magnesium alloy for the intended implant application. Modified diagram from Witte *et al.* (2008b).

10.4.1 How to control corrosion *in vivo*

Magnesium corrosion *in vivo* can be controlled by alloying, processing and coating. The first empirical approach in biodegradable stent development has been successfully leading to magnesium alloys containing rare earth elements. However, a more systematic approach is needed. For use in humans, it is recommended to use aluminum-free magnesium alloy systems, while aluminum-containing Mg alloys may be used for research purposes for example, to investigate the corrosion-protective effect of specific coatings by measuring the release of aluminum. As indicated in Fig. 10.3, it seems to be of major importance that an interdisciplinary team of researchers design magnesium alloys and their production processes according to the intended clinical application, they to review available data from literature, and they perform a critical analysis of their results and method used (Fig. 10.3). The suggested diagram in Fig. 10.3 may help to select and produce the best magnesium alloy based implant for a specific clinical application and ensures quality control.

10.5 Sources of further information and advice

Valuable sources of information for further reading are the special issues on biodegradable metals which have been published in Acta Biomaterialia

(Vol. 6, Issue 5, 2010) and in Materials Science and Engineering Part B (Vol. 176, Issue 20, 2011). Both issues result from the Symposium on Biodegradable Metals (www.biodegradablemetals.org), which was founded in 2009 in Berlin by Diego Mantovani and Frank Witte and is now an annual event. This symposium has a very active and interdisciplinary science community, which results in very intense and lively discussions at the meetings. The symposium is addressing academia, industry and regulatory agencies. As a result of this interactive work a workshop on biodegradable metals has been organized at US-FDA (http://erc.ncat.edu/biodegradablemetals/). Similar workshops with European regulatory agencies are currently planned.

10.6 References

Bosiers, M., Deloose, K., Verbist, Ü. and Peeters, P. 2005. First clinical application of absorbable metal stents in the treatment of critical limb ischemia: 12-month results. *Vascular Disease Management*, **2**, 86–91.

Bundy, K. 1995. In Vivo. *In:* Baboian, R. (ed.) *Corrosion Tests and Standards.* Fredericksburg, VA: ASTM.

Chen, H. and Zhao, X. X. 2011. Biodegradable magnesium-alloy stent: current situation in research. *Journal of Interventional Radiology*, **20**, 62–64.

Clark, G. C. and Williams, D. F. 1982. The effects of proteins on metallic corrosion. *Journal of Biomedical Materials Research,* **16**, 125–34.

Eliaz, N. 2008. Biomaterials and corrosion. *In:* Raj, U. K. M. A. B. (ed.) *Corrosion Science and Technology: Mechanism, Mitigation and Monitoring.* New Delhi: Narosa Publishing House.

Eliezer, A. and Witte, F. 2010a. Corrosion behavior of magnesium alloys in biomedical environments. *Journal of Advanced Materials Research*, **95**, 17–22.

Eliezer, A. and Witte, F. 2010b. The role of biological environments on magnesium alloys as biomaterials. *In:* Kainer, K. U., ed. *7th International Conference on Magnesium Alloys and Their Applications*, Dresden. Wiley-VCH, 822–827.

Erbel, R., di Mario, C., Bartunek, J., Bonnier, J., de Bruyne, B., Eberli, F. R., Erne, P., Haude, M., Heublein, B., Horrigan, M., Ilsley, C., Bose, D., Koolen, J., Luscher, T. F., Weissman, N. and Waksman, R. 2007. Temporary scaffolding of coronary arteries with bioabsorbable magnesium stents: a prospective, non-randomised multicentre trial. *Lancet*, **369**, 1869–75.

Heffernan, E. J., Hayes, M. M., Alkubaidan, F. O., Clarkson, P. W. and Munk, P. L. 2008. Aggressive angiomyxoma of the thigh. *Skeletal Radiology*, **37**, 673–678.

Heublein, B., Rohde, R., Kaese, V., Niemeyer, M., Hartung, W. and Haverich, A. 2003. Biocorrosion of magnesium alloys: a new principle in cardiovascular implant technology? *Heart*, **89**, 651–656.

Huse, E. C. 1878. A new ligature? *The Chicago Medical Journal and Exam*, 172, 2–3.

Lambotte, A. 1932. L'utilisation du magnésium comme matériel perdu dans l'ostéosynthèse. *Bull Mém Soc Nat Cir*, **28**, 1325–1334.

Lango, T., Morland, T. and Brubakk, A. O. 1996. Diffusion coefficients and solubility coefficients for gases in biological fluids and tissues: A review. *Undersea and Hyperbaric Medicine*, **23**, 247–272.

Mantovani, D. and Witte, F. 2010. Editorial to Special Issue on Biodegradable Metals. *Acta Biomaterialia*, **6**, 1679.

Mueller, W.-D., Mele, M. F. L. D., Nascimento, M. L. and Zeddies, M. 2009. Degradation of magnesium and its alloys: Dependence on the composition of the synthetic biological media. *Journal of Biomedical Materials Research Part A*, **90A**, 487–495.

Mueller, W. D., Lucia Nascimento, M. and Lorenzo de Mele, M. F. 2010. Critical discussion of the results from different corrosion studies of Mg and Mg alloys for biomaterial applications. *Acta Biomaterials*, **6**, 1749–55.

Neumann, A. W., Hum, O. S., Francis, D. W., Zingg, W. and van Oss, C. J. 1980. Kinetic and thermodynamic aspects of platelet adhesion from suspension to various substrates. *Journal of Biomedical Materials Research,* **14**, 499–509.

Peuster, M., Fink, C., Wohlsein, P., Bruegmann, M., Gunther, A., Kaese, V., Niemeyer, M., Haferkamp, H. and Schnakenburg, C. 2003. Degradation of tungsten coils implanted into the subclavian artery of New Zealand white rabbits is not associated with local or systemic toxicity. *Biomaterials*, **24**, 393–9.

Peuster, M., Wohlsein, P., Brugmann, M., Ehlerding, M., Seidler, K., Fink, C., Brauer, H., Fischer, A. and Hausdorf, G. 2001. A novel approach to temporary stenting: degradable cardiovascular stents produced from corrodible metal – results 6–18 months after implantation into New Zealand white rabbits. *Heart*, **86**, 563–569.

Piiper, J., Canfield, R. E. and Rahn, H. 1962. Absorption of various inert gases from subcutaneous gas pockets in rats. *Journal of Applied Physiology*, **17**, 268–274.

rettig, R. and Virtanen, S. 2009. Composition of corrosion layers on a magnesium rare-earth alloy in simulated body fluids. *Journal of Biomedical Materials Research Part A*, **88A**, 359–369.

Rudee, M. L. and Price, T. M. 1985. The initial stages of adsorption of plasma derived proteins on artificial surfaces in a controlled flow environment. *Journal of Biomedical Materials Research*, **19**, 57–66.

Schuster, J. 1975. *The Metallosis,* Stuttgart, Ferdinand Enke Verlag.

Seelig, M. G. 1924. A study of magnesium wire as an absorbable suture and ligature material. *AMA Archives of Surgery*, **8**, 669–80.

Shaw, B. A. 2003. Corrosion resistance of magnesium alloys. *In:* D, Stephen (ed.) *ASM handbook volume 13a: corrosion: fundamentals, testing and protection.* UK: ASM International.

Song, G. L. and Atrens, A. 1999. Corrosion mechanisms of magnesium alloys. *Advanced Engineering Materials*, **1**, 11–33.

Vaupel, P. 1976. Effect of percentual water-content in tissues and liquids on diffusion-coefficients of O2, CO2, N2, and H2. *Pflugers Archiv-European Journal of Physiology*, **361**, 201–204.

Virtanen, S. (2011) Biodegradable Mg and Mg alloys: Corrosion and biocompatibility. *Materials Science and Engineering B*, **176**, 1600–1608.

Wen, C. E., Yamada, Y., Shimojima, K., Chino, Y., Hosokawa, H. and Mabuchi, M. 2004. Compressibility of porous magnesium foam: dependency on porosity and pore size. *Materials Letters*, **58**, 357–360.

Witte, F. 2010. The history of biodegradable magnesium implants: A review. *Acta Biomaterialia*, **6**, 1680–1692.

Witte, F., Abeln, I., Switzer, E., Kaese, V., Meyer-Lindenberg, A. and Windhagen, H. 2008a. Evaluation of the skin sensitizing potential of biodegradable magnesium alloys. *Journal of Biomedical Materials Research Part A*, **86**A, 1041–1047.
Witte, F., Fischer, J., Nellesen, J. and Beckmann, F. Microtomography of magnesium implants in bone and their degradation. *Progress in Biomedical Optics and Imaging – Proceedings of SPIE*, 2006a San Diego, CA.
Witte, F., Fischer, J., Nellesen, J., Crostack, H. A., Kaese, V., Pisch, A., Beckmann, F. and Windhagen, H. 2006b. In vitro and in vivo corrosion measurements of magnesium alloys. *Biomaterials*, **27**, 1013–1018.
Witte, F., Fischer, J., Nellesen, J., Vogt, C., Vogt, J., Donath, T. and Beckmann, F. 2009. In vivo corrosion and corrosion protection of magnesium alloy LAE442. *Acta Biomaterials*, 6, 1792–9.
Witte, F., Hort, N., Vogt, C., Cohen, S., Kainer, K. U., Willumeit, R. and Feyerabend, F. 2008b. Degradable biomaterials based on magnesium corrosion. *Current Opinion in Solid State and Materials Science*, **12**, 63–72.
Witte, F., Kaese, V., Haferkamp, H., Switzer, E., Meyer-Lindenberg, A., Wirth, C. J. and Windhagen, H. 2005. In vivo corrosion of four magnesium alloys and the associated bone response. *Biomaterials*, **26**, 3557–3563.
Witte, F., Ulrich, H., Rudert, M. and Willbold, E. 2007. Biodegradable magnesium scaffolds: Part I: Appropriate inflammatory response. *Journal of Biomedical Materials Research – Part A*, **81**, 748–756.
Xin, Y., Hu, T. and Chu, P. K. (2011) In vitro studies of biomedical magnesium alloys in a simulated physiological environment: A review. *Acta Biomaterialia,* **7,** 1452–1459.
Xu, L., Yu, G., Zhang, E., Pan, F. and Yang, K. 2007. In vivo corrosion behavior of Mg-Mn-Zn alloy for bone implant application. *J Biomed Mater Res A*, **83**, 703–11.
Zeng, R., Dietzel, W., Witte, F., Hort, N. and Blawert, C. 2008. Progress and challenge for magnesium alloys as biomaterials. *Advanced Engineering Materials,* **10**, B3–B14.
Zhu, Y. Y., Wu, G. M. and Zhao, Q. 2010. Research progress of magnesium-based alloy in biomedical application. *Chinese Journal of Biomedical Engineering*, **29**, 932–938.

10.7 Appendix: list of abbreviations

AAS Atomic absorption spectroscopy
AES Atomic emission spectroscopy
GD-OES Glow-discharge optical emission spectroscopy
ICP-MS Mass spectrometry with inductively coupled plasma
NAA Neutron activation analysis
XRD X-ray diffraction
XRF X-ray fluorescence analysis

Index